MODERN ASPECTS OF ELECTROCHEMISTRY
No. 9

LIST OF CONTRIBUTORS

A. J. APPLEBY
Laboratoires de Marcoussis (C. G. E.)
Marcoussis, France

H. BLOOM
Chemistry Department
The University of Tasmania, Hobart
Tasmania, Australia

S. D. HAMANN
CSIRO Division of Applied Chemistry
Melbourne, Australia

P. KEBARLE
Chemistry Department
University of Alberta, Edmonton
Alberta, Canada

R. M. REEVES
Department of Chemistry
University of Bristol
Bristol, England

I. K. SNOOK
Chemistry Department
Royal Melbourne Institute of Technology, Melbourne
Victoria, Australia

A Continuation Order Plan is available for this series. A continuation order will bring delivery of each new volume immediately upon publication. Volumes are billed only upon actual shipment. For further information please contact the publisher.

MODERN ASPECTS OF ELECTROCHEMISTRY

No. 9

Edited by

B. E. CONWAY
Department of Chemistry
University of Ottawa
Ottawa, Canada

and

J. O'M. BOCKRIS
School of Physical Sciences
The Flinders University
Adelaide
South Australia

PLENUM PRESS • NEW YORK-LONDON

The Library of Congress cataloged the first volume of this title as follows:

Modern aspects of electrochemistry. no. [1]–
Washington, Butterworths, 1954–
 v. illus. 23 cm.

No. 1–2 issued as Modern aspects series of chemistry.
Editors: no. 1– J. Bockris (with B. E. Conway, no. 3–)
Imprint varies: no. 1, New York, Academic Press.—No. 2, London, Butterworths.

1. Electrochemistry—Collected works. I. Bockris, J. O'M., ed.
II. Conway, B. E., ed. (Series: Modern aspects series of chemistry)

QD552.M6   54—12732

Library of Congress [67r65c²5]

Library of Congress Catalog Card Number 54-12732
ISBN 0-306-37649-0

© 1974 Plenum Press, New York
A Division of Plenum Publishing Corporation
227 West 17th Street, New York, N.Y. 10011

United Kingdom edition published by Plenum Press, London
A Division of Plenum Publishing Company, Ltd.
4a Lower John Street, London W1R 3PD, England

All rights reserved

No part of this book may be reproduced, stored in a retrieval system, or transmitted, in any form or by any means, electronic, mechanical, photocopying, microfilming, recording, or otherwise, without written permission from the Publisher

Printed in the United States of America

Preface

As the subject of electrochemistry moves into the final quarter of the century, a number of developed areas can be assessed in depth while some new areas provide quantitatively and qualitatively novel data and results. The first chapter, by Kebarle, deals with an example of the latter type of field in which new information of the energetics and equilibria of reactions between ions and solvent molecules is studied in the gas phase and provides interesting basic information for treatments of ions in solution, i.e., ionic solvation.

Chapter 2, by Hamann, discusses the behavior of electrolyte solutions under high pressures, a matter of intrinsic interest in relation to ion–solvent interaction and the structural aspects of the properties of ionic solutions, especially in water. This topic is also of current interest with regard to the physical chemistry of the marine environment, especially at great depths.

In the article by Bloom and Snook (Chapter 3), models for treatments of molten salt systems are examined quantitatively in relation to the structure of molten ionic liquids and to the statistical mechanical approaches that can be meaningfully made to interpret their properties and electrochemical behavior.

For many years, treatments of the electrical double layer at charged electrode interfaces were developed with very little reference to the properties of the solvent and with surprisingly little recognition of its ubiquitous presence. This situation has changed in recent years and modern treatments of the double layer in which solvent properties and orientation have been considered are reviewed critically and in detail by Reeves in Chapter 4. Understanding of the role of solvent, especially water, in the properties of charged interfaces is now also of great importance in the field of biologically significant interfaces, e.g., at membranes, as well as at metal electrodes.

The final chapter, by Appleby, reviews some of the fundamental aspects of electrocatalysis with special reference to electronic aspects of chemisorption and charge transfer, and to examination of the factors involved in some key fuel-cell catalysis reactions, in particular, oxygen reduction.

Ottawa B. E. Conway
January 1974 J. O'M. Bockris

Contents

Chapter 1

GAS-PHASE ION EQUILIBRIA AND ION SOLVATION

P. Kebarle

I. Introduction	1
1. Gas-Phase Hydration of Ions in Relation to Hydration in Solution	1
2. Ion–Molecule Reactions in the Gas Phase	3
II. Principles of Gas–Phase Ion Equilibrium Methods	6
III. Gas–Phase Studies of Acids and Bases. Proton Transfer Equilibria	13
IV. Enthalpies and Free Energies of Formation of Ions in the Gas Phase and Total Energies of Solvation of Single Ions	18
V. Hydration of Spherically Symmetric Ions. The Positive Alkali and Negative Halide Ions	21
VI. The Hydrogen Ion and the Hydroxyl Ion Hydrates in the Gas Phase	32
VII. Hydrogen Bonding to Negative Ions	36
VIII. Ion Solvation by Protic and Aprotic Solvents	41
References	44

Chapter 2

ELECTROLYTE SOLUTIONS AT HIGH PRESSURE

S. D. Hamann

I. Introduction	47
II. Physical Properties of Water and Other Solvents at High Pressures	48
1. Physical Properties of Water at High Pressures	49
2. Physical Properties of Other Solvents at High Pressures	55
III. Electrical Conductivity of Electrolyte Solutions under Pressure	57
1. Experimental Methods	57
2. Results	65
IV. Ionization Equilibria under Pressure	76
1. Thermodynamics of Equilibria in Solution at High Pressures	78
2. Experimental Methods for Measuring Ionization Constants under Pressure	83
3. Discussion of Results	94
V. Properties of Electrolyte Solutions at High Shock Pressures	111
1. Elementary Theory of Shock Waves	112
2. Experimental Methods of Generating Strong Shock Waves	113
3. Measurements on Shock-Compressed Materials	115
4. Disadvantages and Advantages of Shock-Wave Methods	116
5. Electrical Conductivities of Weak Electrolytes in Shock Waves	117
6. Electrical Conductivities of Solutions of Strong Electrolytes in Shock Waves	124
7. Ionization Constant of Water at High Shock Pressures	126
Appendix	132
References	150

Chapter 3

MODELS FOR MOLTEN SALTS
H. Bloom and I. K. Snook

I. Introduction	159
1. Models	159
2. Radial Distribution Functions (RDF)	160
II. Operational Models	168
1. Hole Models	168
2. Liquid Free-Volume Model	173
3. Relationship between Free Volume from Different Models and the Hole Volume	176
4. The Adam and Gibbs Configurational-Entropy Theory	177
5. The Significant Structures Model	179
III. Models Involving Intermolecular Forces	182
1. Intermolecular Potentials in Molten Salts. Basic Theory	182
2. Statistical Mechanics of Molten Salts	186
References	235

Chapter 4

THE ELECTRICAL DOUBLE LAYER: THE CURRENT STATUS OF DATA AND MODELS, WITH PARTICULAR EMPHASIS ON THE SOLVENT
R. M. Reeves

I. Introduction	239
1. Basic Double-Layer Model	240
2. The Diffuse Layer	241
II. Some Considerations of the Properties of a Solvent in the Region Adjacent to a Surface	244
1. Introduction	244
2. General Properties of a Solvent Near Interfaces	244

    3. The Aqueous–Air Interface 246
    4. The Role of the Metal 248
    5. The Surface and the Work Function 250
III. Double-Layer Characteristics at Mercury........... 256
    1. Introduction................................ 256
    2. Classical Double-Layer Analysis................ 256
    3. Solvent Excesses 262
    4. Surface Excesses of Entropy and Volume 264
    5. Ionic Systems: General Characteristics 271
    6. The Fluoride Ion—Is Its Behavior Anomalous?... 273
    7. Capacitances over the Entire Concentration Range 277
    8. Maxima in the Capacitance–Potential Function... 279
    9. Anion Adsorption............................ 283
    10. Organic Systems 287
IV. Models of the Double Layer...................... 288
    1. Introduction................................ 288
    2. Earlier Theories of Adsorption 289
    3. Organic Systems: Classical Treatments 299
    4. Summary of Basis for Recent Developments 300
    5. Recent Developments......................... 301
    6. Recent Ionic Models......................... 324
    7. Intermediate Models......................... 335
    8. Organic Systems 338
    9. The Gallium–Solution Interface 347
V. Discussion and Conclusions 351
VI. Recent Advances Not Directly Applicable to Metal–
    Solution Interfaces............................ 360
References 362

Chapter 5

ELECTROCATALYSIS
A. J. Appleby

I. Introduction................................... 369
II. Electron Transfer at the Metal–Solution Interface ... 369
    1. General 369
    2. Thermal Theory 370

	3. Electrostatic Theory	378
	4. Improved Transition State Theories for Electrode Reactions	385
III.	Effect of Adsorption on the Rate of Reaction	391
IV.	Factors Other Than ΔH°_{ads} Affecting Reaction Rates	397
	1. Nuclear Transmission Coefficient	397
	2. Electron Transmission Coefficient	398
	3. Effects of the Diffuse Double Layer	398
	4. Effect of the Electronic Structure of the Electrode Material	398
	5. Effect of ΔH°_{ads} on the Entropy of Activation	399
V.	Experimental Rate Correlations	400
	1. General	400
	2. The Hydrogen Evolution Process	402
	3. Volcano Plots in the Hydrogen Evolution Reaction	406
	4. Heats of Adsorption and Frequency Factors in Hydrogen Evolution	415
	5. Electrocatalytic Studies in Other Systems	418
VI.	Electrocatalysis and the Oxygen Electrode	421
	1. Introduction	421
	2. The Oxygen Electrode on Platinum Oxide Surfaces	426
	3. The Oxygen Electrode on Other Oxidized Metals	430
	4. Electrocatalysis of the Oxygen Evolution Reaction	436
	5. Discussion of the Mechanism of the Oxygen Electrode on Oxidized Metals	439
VII.	The Kinetics and Mechanism of Oxygen Reduction on Phase-Oxide-Free Metals	443
	1. General	443
	2. Oxygen Reduction in Acid Solution	443
	3. Oxygen Reduction on Phase-Oxide-Free Palladium and Rhodium in Acid Solution	449
	4. Ruthenium, Iridium, and Osmium Electrodes	450
	5. Gold and Silver Electrodes	451
	6. Reaction Products—Effect of Impurities	452
	7. Electrocatalysis of the Oxygen Reduction on Phase-Oxide-Free Metals in Acid Solution	453
	8. Heats of Activation and Frequency Factors in the Oxygen Reduction Reaction	456
	9. Correlation between Heats of Activation and Estimated Heats of Adsorption	458

	10. The Compensation Effect	461
	11. Consequences of the Compensation Effect	466
VIII.	The Oxygen Electrode in Other Electrolytes	468
	1. Alkaline Solutions	468
	2. Oxygen Electrodes in Nonaqueous Media	470

References 471

INDEX 479

Gas–Phase Ion Equilibria and Ion Solvation

P. Kebarle

Chemistry Department, University of Alberta, Edmonton, Alberta, Canada

I. INTRODUCTION

1. Gas-Phase Hydration of Ions in Relation to Hydration in Solution

The nature of the interactions of the ion with the solvent is central to the study of ionic solutions and is thus of great importance in electrochemistry.

In examining the ion–solvent interactions, it is natural to consider first the properties of the isolated ion, i.e., radius (or geometry, if the ion is molecular) energetics, etc., then evaluate the interactions of the ion with the near-neighbor solvent molecules, and finally examine the interactions, structure modifications of solvent, etc., at larger distances. The approach from "isolated ion" to "ion in the solvent" is inherent in theoretical as well as experimental treatments. The central properties of interest of isolated monatomic ions are radius and charge in relation to enthalpy and free energy of formation. Generally for each of these quantities there is sufficient information, although there are difficulties in deciding what changes in radius should be made when the ion enters into the solution and thus experiences compressional interaction forces not present for the isolated ion in the gas phase. The situation regarding polyatomic ions, which may be inorganic like NH_4^+ and NO_3^- or organic like $C_6H_5NH_3^+$ and $C_6H_5COO^-$, is much less satisfactory. The enthalpies and free energies of formation of the gaseous species are often not known, while the geometries of the ions must be deduced from the geometries of the corresponding neutral molecules with the aid of theoretical assumptions. Sections III and IV of this chapter give a brief description of recent advances, particularly concerning the energetics, i.e., enthalpies and free energies of formation, for such species based on measurements of ionic equilibria in the gas phase.

While free energies of ionic solvation may be very approximately calculated on the basis of Born's continuum dielectric theory, any more sophisticated theory must take into account the finite size of the ion and that of the coordinating solvent molecules in the primary solvation shell. Also, how the solvent molecules are packed and oriented in this shell in relation to the solvent structure in the bulk of the surrounding solution is a matter of current interest.

Studies of ion–"solvent" association in the gas phase provide direct indications of the preferred coordination of ions by solvent molecules and the energy relations between solvate species in the gas phase containing different numbers of coordinated solvent molecules, i.e., different "solvation numbers" in the terminology of electrolyte solution theories.

One of the advantages of studies in the gas phase by means of mass spectrometry is that the stoichiometries of the various species that can be detected can be deduced exactly, while in solution they are usually ambiguous. The mass spectrometric method also permits measurements of gas phase ion equilibria and their temperature dependence. These data in turn lead to free energies, enthalpies, and entropies for ionic reactions.

The experimental characterization of stable ion–solvent complexes in the gas phase by mass spectrometry therefore provides a meaningful experimental basis for theoretical calculations of ion–solvent molecule interactions.

The strong interactions of ions with the nearest-neighbor molecules have been traditionally treated on the basis of classical ion-dipole and induced dipole interactions, e.g., in the work of Bernal and Fowler or Eley and Evans. More recently, quantum mechanical calculations of varying degrees of refinement have also been applied to systems including an ion and one or two molecules, e.g., the work of Clementi and Popkie.[1] As shown in the following sections, *ab initio* calculations often provide values in good agreement with the

Introduction

gas phase equilibrium data. Good theoretical results are particularly valuable since they also provide structural information.

The experimental ion equilibria studies represent the principal topic of the present article. The principle of the ion equilibria measurements is described in Section II, while the results on ion–solvent molecule complexes are given in Sections IV–VIII.

To workers in the field of ionic solutions, experimental studies of isolated ion–solvent molecule complexes may seem impossible at first glance. However, it is necessary to reflect only a little to realize that ion–solvent molecule "clusters" held by ion-dipole and induced dipole forces should exist in the gas phase and that therefore it should be possible to generate and study them.

Knowledge of the nature and stability of gas-phase solvate complexes also provides some basis for identification of significant species in solution in those cases where thermodynamic and other evidence may not be unambiguous, which is often the case: A special example of interest in this regard is the entity H_3O^+ (or $H_9O_4^+$) corresponding to the hydration of the proton in the bulk water solvent. Information on the stability and coordination of ions by solvent molecules in the primary layer is also a matter of great interest in other aspects of electrochemistry concerned with (a) double-layer structure and specific adsorption of ions at charged electrode interfaces (see Chapter 4) and (b) the act of desolvation or "reorganization" of the solvate layer around ions in solution (see Chapter 5) in the activation process of electron transfer, e.g., in metal electrodeposition or electrochemical redox reactions.

Of course, the further interactions which ion–solvate complexes (characterized in the gas phase) will undergo with the remainder of bulk solvent upon complete solvation are always important, but prior knowledge of favored configurations in the gas phase will provide a better basis for the understanding of the process of overall solvation of ions in the liquid solvent phase and the nature of the kinetically significant entities in ionic solutions.

2. Ion–Molecule Reactions in the Gas Phase

As mentioned above, the work described in the following sections of this chapter is based on measurements of ion–neutral molecule reaction equilibria in the gas phase. Before considering how such equilibria can be studied, it will be worthwhile to recall some of the early work on gaseous ions. Investigations of electrical discharges in rarefied gases started about 1750 and led, through the work of L. Hittdorf, G. Goldstein, W. Wien, and most notably J. J. Thomson, to recognition of the existence of electrons and the presence of positive and negative ions in the gas phase. Thomson, through his studies of gas discharges and the effect of electric and magnetic fields on the trajectories of the ions, was led to construct the first mass spectrograph. For various technical reasons the pressure in the ion source of Thomson's instrument was higher than that used in modern analytical instruments. At higher pressure the primary ions created by the ionizing medium (ionization is most often obtained by electron impact) may collide with neutral molecules and react with them. A possible reaction sequence following the primary ionization of a water molecule by a fast electron is shown in reactions (a) to $(n-1, n)$:

$$e + H_2O \rightarrow H_2O^+ + 2e \quad \text{primary ionization} \quad (a)$$

$$e + H_2O \rightarrow OH^+ + H + 2e \quad \text{primary ionization} \quad (b)$$

$$H_2O^+ + H_2O \rightarrow H_3O^+ + OH \quad \text{ion–molecule reaction} \quad (c)$$

$$OH^+ + H_2O \rightarrow H_3O^+ + O \quad \text{ion–molecule reaction} \quad (d)$$

$$H_3O^+ + H_2O \rightarrow H^+(H_2O)_2 \quad \text{forward clustering reaction} \quad (1, 2)$$

$$H^+(H_2O)_{n-1} + H_2O \rightarrow H^+(H_2O)_n \quad \text{clustering equilibrium } (n-1, n)$$

Reactions of the type shown above did occur in Thomson's instrument (at least up to the step c) and consequently the product ions of such reactions were observed by him.

Ion–molecule reactions obviously interfere with the analytical uses of the mass spectrometer. Thus, for example, the determination of isotope ratios would be greatly complicated if the ions were to participate in a variety of ion–molecule reactions. For this reason Aston, Dempster, Nier, and other pioneers of the mass spectro-

Introduction

metric technique made all efforts to reduce the pressure in the ion source to eliminate these undesirable ion–molecule reactions. The technical problems were soon solved and eventually instruments were developed which operate at pressures lower than 10^{-5} Torr, that is, under conditions when most ions can leave the ion source without colliding and reacting with other molecules.

For many years interest in mass spectrometry centered on the analytical applications, and the importance of investigating ion–molecule reactions was not appreciated. Therefore, in spite of the early discovery of such reactions, their systematic study was initiated only around 1952 by the work of Tal'roze and Lubimova,[1a] rapidly followed by Stevenson and Schissler,[2] Hamill and co-workers,[3] and Field et al.[4] Since then valuable information on the reactivity of gaseous ions has been provided by many laboratories.

Most of this work was done at low reaction chamber (ion source) pressures ($p < 10^{-3}$ Torr). Under such conditions the average ion suffers less than one collision with the neutral reactant gas. Therefore only very fast reactions which occur at every collision could be observed. Furthermore, the important class of reactions requiring thermal activation or deactivation could not be studied. This precluded the observation of ionic equilibria. Apparatus operating at sufficiently high ion source pressures ($p = 1$–20 Torr) where the study of ionic equilibria was possible was developed at the University of Alberta around 1963–64. The high-pressure mass spectrometer had initially been constructed for the study of ionic reactions in the radiation chemistry of gases. A chance observation[5] of $H^+(H_2O)_n$ as impurity ions in gases containing traces of moisture led to an examination of their equilibria.[6] A switch to the study of ion equilibria and ion–solvent molecule clustering reactions was made at that time in the belief that unique fundamental information could be obtained from such studies. The equilibrium concept and the thermodynamic data resulting from measurement of equilibria represent one of the oldest and most useful branches of physical chemistry. It is therefore not surprising that attempts predating those at the University of Alberta were made to measure ionic equilibria in the gas phase. For example, Chupka[7] accidentally observed the ions K^+ and $K^+(H_2O)$ in the mass spectrometric study of KCl and $(KCl)_n$ neutrals emitted from a heated oven coated with KCl. He studied the equilibrium $K^+ + H_2O = K^+(H_2O)$ and obtained an approximate equilibrium constant which was later verified by measurements in the present laboratory.[8] However, Chupka did not try to extend the ion equilibrium measurements. Probably there exist other such isolated observations in the early literature. The earlier experiments of the Alberta group are summarized in an article in the Advances of Chemistry series.[9] More recently work was summarized in a chapter of the book *Ions and Ion Pairs in Organic Reactions*[10] and another chapter in *Ion Molecule Reactions*.[11]

Independently of the ion equilibria studies in Alberta, which were mostly directed to ion–solvent molecule interactions, Conway (see Ref. 12) developed mass spectrometric apparatus and measured the equilibria $O_2^+ + O_2 = O_4^+$, the oxygen ions being of importance in the ionosphere and troposphere. Ion equilibria measurements from other laboratories were somewhat slow to follow, but in the last few years a number of workers have turned to such experiments (see next section) and it is clear that further rapid development of the field and a wealth of important measurements lie ahead.

II. PRINCIPLES OF GAS–PHASE ION EQUILIBRIUM METHODS

The work described in the subsequent sections is based on measurements of the equilibrium constants and their temperature dependence for the following types of reactions:

$$B_1H^+ + B_2 = B_1 + B_2H^+ \qquad (1)$$

$$A_1H + A_2^- = A_1^- + A_2H \qquad (2)$$

$$M^+(Sl)_{n-1} + Sl = M^+(Sl)_n \qquad (3)$$

Reactions (1) and (2) are proton transfer reactions which are the gas-phase analogs of Brönsted acid–base reactions. Reaction (1) involves neutral bases, while reaction (2) involves negatively charged bases. Results from the gas phase measurements of such reactions are discussed in sections 3 and 4. Reaction (3) is a process in which a positive (or negative) ion M^+ acquires stepwise ligand molecules Sl. The Sl stands for gas-phase solvent molecules like H_2O, NH_3, acetonitrile, etc. The equilibrium constants for the reactions (1)–(3) are given by the equations

$$K_1 = \frac{[B_1][B_2H^+]}{[B_2][B_1H^+]}, \quad K_2 = \frac{[A_1^-][A_2H]}{[A_2^-][A_1H]},$$

$$K_{n-1,n} = \frac{[M^+(Sl)_n]}{[M^+(Sl)_{n-1}][Sl]}$$

Once the equilibrium constants have been determined the free energy change is obtained from the equation $\Delta G° = -RT \ln K$, the enthalpy change is obtained from a Van't Hoff plot (i.e., ln K versus $1/T$), and the ΔS change from the relationship $\Delta G = \Delta H - T\Delta S$.

In order to determine the equilibrium constants, one needs to know the concentrations of the neutral reactants and the ratio of the concentrations of the two ionic reactants. The determination of the concentration of the neutral reactants does not present any special problem, since these concentrations are prepared by the experimenter. The ions are created by some form of ionizing radiation and the system must be so selected as to lead to the desired ionic reactants. In order to observe the equilibrium concentrations of the two reactant ions, one must create conditions for which the reactions coupling the equilibrium are faster than other reactions in which the ions might be engaged.

The reactions (1)–(3) can be easily conducted under conditions where their half-lives in the forward and reverse directions are in the 10–100 μsec range. This is achieved by working at neutral pressures in the Torr range. Pressures in the Torr range are also required for the thermalization of the reactant ions. Competitive reactions always present in the gas phase are ion discharge on the walls and recombination of the positive ions with electrons or negative ions. The discharge to the wall is reduced at gas pressures in the Torr range where diffusion to the wall is slowed down. The positive–negative charge recombination can be made unimportant by working at very low ionizing intensities, i.e., very low ion concentrations (less than 10^8 ions cm^{-3}). Under such conditions the half-lives of ions toward recombination become longer than milliseconds.

Finally, the sampling of the ions must be done in such a way that one either samples only ions which have reached equilibrium, i.e., ions which have stayed sufficiently long in the ion source, or one can pulse the ionizing beam and follow the establishment of equilibrium by observing the time dependence of the ion concentrations after the pulse. Evidently the conditions for measuring ionic equilibria cannot be met by conventional mass spectrometers constructed for analytical work and operating typically in the 10^{-6} Torr range.

The apparatus used for ionic equilibria measurement [10,11,13,14] is shown in Fig. 1. The 2000-V electrons, produced by the electron gun, enter the ion source (2) through a narrow slit. There some of the neutral gas provided by the gas supply lines (7) is ionized by electron impact. The created ions diffuse to the walls, reacting on their way with gas molecules and eventually reaching ion–molecule equilibrium (provided that the conditions are selected correctly). Some of the ions diffuse to the vicinity of the ion exit slit at the bottom of the ion source and escape into the evacuated region outside. Suitably shaped electrodes (3) capture the ions, which are then accelerated in the region (5) and subjected to conventional mass analysis (6) (magnetic deflection or quadrupole mass filter). The neutral gas escaping through the electron entrance and ion exit slits is evacuated by two high-speed pumping systems (8) and (9) which maintain a pressure differential of 10^5 between the ion source and the region outside and 10^2 between the region outside the ion source and the mass analysis system.

Pulsing the electron beam permits observation of the approach to equilibrium. Figure 2 shows a time profile of the negative ions escaping out of the ion exit slit after the short electron pulse. The concentration of escaping ions quickly reaches a maximum and then fades away over some milliseconds. The disappearance of the ions is due to diffusion to the walls of the ion source. The concentrations of neutral species is generally selected to lead to reactive change which is faster than ion loss to the walls. Such reactive changes are shown in Fig. 3a taken from a study[15] of the kinetics of the hydration of NO_2^- and NO_3^-. Oxygen gas at 2 Torr was used containing traces of water and ethyl nitrate. The electrons, slowed down by collisions with the oxygen gas, are captured by the ethyl nitrate and lead to the capture reaction (4):

$$e + C_2H_5ONO_2 = C_2H_5O + NO_2^- \qquad (4)$$

$$NO_2^- + H_2O + O_2 = NO_2^-(H_2O) + O_2 \qquad (5)$$

$$NO_2^-(H_2O) + H_2O + O_2 = NO_2^-(H_2O)_2 + O_2 \qquad (6)$$

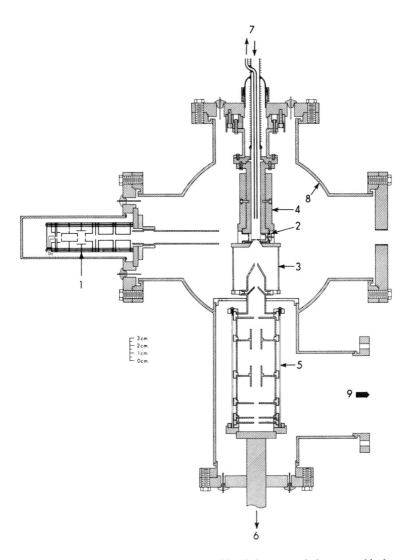

Figure 1. Apparatus: (1) electron gun assembly; (2) ion source in ion source block; (3) bottom lid of ion source and electrostatic shielding mesh; (4) outer mantle of ion source with heater wells and heaters; (5) ion acceleration tower; (6) to magnetic mass analysis; (7) slow flow circulation gas supply to ion source. (8) 8-in. pumping lead to 6-in. pump; (9) 4-in. differential pumping of accelerating and focusing region.

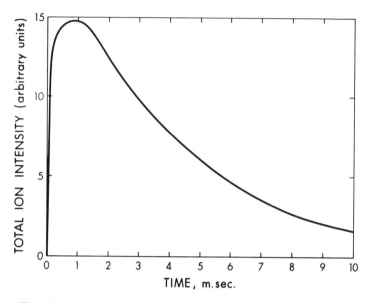

Figure 2. Time profile of negative ions of all masses escaping from ion exit slit after a 10-μsec electron pulse. Ion source pressure 4 Torr O_2; distance from electron beam to ion exit slit 4 mm.

The nitrate ion created in this manner reacts with water to form the monohydrate. Since (5) is exothermic, the excess energy must be removed by collisional stabilization with the third body O_2. The monohydrate reacts further to form the dihydrate (6). Equilibrium is established at long reaction times where the concentrations of the ions become stationary. The data in Fig. 3a have been normalized, i.e., in order to correct for the varying total ion concentration (see Fig. 2), the observed ion intensity of an ion with given mass was divided by the total ion intensity (ions of all masses) at time t. The resulting plot predicts the approximate kinetic behavior of a reaction system in which the ions are involved only in ion/molecule reactions. Results like those given in Fig. 4 allow approximate determinations of rate constants of the reactions involved.[13,15] The rate constants are of interest in fields where such reactions occur, i.e., ionosphere, troposphere (atmospheric electricity), and radiation chemistry.

The stationary concentrations (actually ion ratios) observed in plots like that shown in Fig. 3a are substituted in the equilibrium constant expressions (1) or (2) and the equilibrium constants are evaluated under the assumption that the ion ratios represent the equilibrium ion concentration ratios in the reaction system. Variation of the pressure of the neutral reactants should lead to equilibrium constants which are independent of pressure. A plot of such data is shown in Fig. 3b. The equilibrium constants determined at different temperatures are then used for Van't Hoff plots whose slope leads to the enthalpy changes of the reactions. A typical plot is shown in Fig. 6.

Recently three other types of apparatus have been used to study the kinetics of approach and the position of gas-phase ion equilibria. Howard et al.[16] and Bohme et al.[17] have applied the "flowing afterglow method" to the study of ion equilibria. The flowing afterglow method, developed to high perfection by Ferguson et al.,[18] consists of a gas flow reaction system (pressure

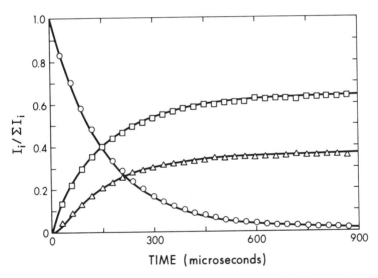

Figure 3a. Time dependence of NO_2^- and hydrates depicting reaction sequence: $NO_2^- \rightleftarrows NO_2^-(H_2O) \rightleftarrows NO_2^-(H_2O)_2$. (○) NO_2^-, (□)$NO_2^-(H_2O)$, (△) $NO_2^-(H_2O)_2$. The solid line connecting the experimental points represents a computer fit of the reaction system which leads to evaluation of the rate constants involved. $p(O_2) = 2.14$ Torr, $p(H_2O) = 27$ mTorr, $T = 300°K$. (From Ref. 15.)

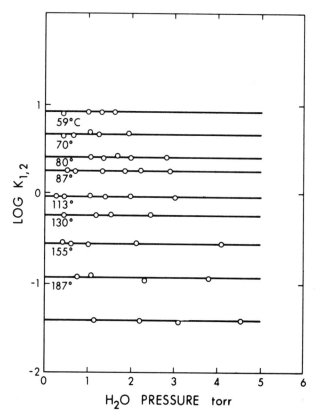

Figure 3b. Plots of $K_{1,2}$ for equilibrium $Cl^-H_2O + H_2O = Cl^-(H_2O)_2$ as a function of water pressure.

~1 Torr). The ions carried by an inert gas are made to react with gases added at different positions down stream. At the end of the reactor the gas is sampled by bleeding part of it through a small orifice into an evacuated mass analysis system.

Ion drift methods have also been applied to ion equilibria measurements.[20] These systems can also operate at pressures in the Torr range. The gas is stationary but the ions (of a given charge) are made to drift down the drift tube by the application of weak uniform electric fields. The ions may react during the drift time. The gas is sampled at the end of the drift tube through a leak leading to an evacuated mass analysis system.

All apparatus described above, including that used in the author's work, suffers from the drawback that the ions are not detected *in situ* but are sampled by bleeding into an evacuated mass analysis system. This sampling method may lead to errors due to discrimination against a given ion, collisional dissociation of the ions in the imperfectly evacuated mass analysis system, dissociation of weakly bonded ion–molecule complexes in the vacuum, etc. In carefully conducted experiments the above sources of error are believed to be small. However, it is fair to say that the experimental area is a difficult one and work with different apparatus avoiding the sampling problem would be of considerable value.

The only apparatus with which analysis of the ions is achieved *in situ* is the ion cyclotron mass spectrometer. In the ICR technique the resonance absorption of the ions orbiting in a magnetic field is used for mass determination.[21] Unfortunately, the resonance peaks become broadened at higher pressures so that the measurements at the present state of the art have been limited to below 10^{-3} Torr. The ion lifetimes can be extended by ion trapping techniques in the 100-msec range. The observation of ion equilibria is possible with such an apparatus. Pulsed electron ionization allows one to observe the establishment of the equilibrium. Since the pressures are very low, the number of collisions between reactive collisions is low (less than 100) even at the long trapped ion reaction time, so that some of the ions might not be truly thermalized. In spite of these difficulties, the ICR technique is extremely valuable since it avoids the sampling problems. Thus agreement between the ICR technique and the high-pressure techniques must almost certainly mean that both types of measurements are correct. Such agreement has been obtained for proton transfer equilibria desscribed in the next section.

III. GAS–PHASE STUDIES OF ACIDS AND BASES. PROTON TRANSFER EQUILIBRIA

Acid–base interactions probably represent the most important class of ionic reactions occurring in solution. The acids or bases are most often not spherically symmetric species so that conventional electrostatic treatments, like the Born equation or ion-dipole interaction calculations, are difficult to apply in a meaningful way. Information from gas-phase studies is therefore of particular importance. The gas-phase ion equilibrium method is capable of establishing an absolute gas-phase acidity or basicity scale which is independent of the solvent. Furthermore, gas-phase measurements can provide the enthalpies and free energies of formation of charged gas-phase acids BH^+ or corresponding bases A^-. From these results the single ion heats of total solvation can be calculated provided that the absolute enthalpy of formation of one reference ion is known and the conventional enthalpy of formation of BH^+ or A^- in solution is known. The same is true for the free energy and entropy changes. Examples of such applications will be given below and in Section 4.

The gas-phase method involving positive ions, i.e., the reaction

$$B_1H^+ + B_2 \overset{K_1}{=} B_1 + B_2H^+ \qquad (7)$$

consists in determining the equilibrium constant K_1 by measuring the concentrations of the ions B_1H^+ and B_2H^+ at equilibrium (see preceding section). An example of the information that can be obtained is given by the recent determinations of the relative basicities of a number of amines and other organic nitrogen-containing bases. The free energy changes obtained from the equation $\Delta G° = -RT \ln K$ for proton transfer to ammonia ($NH_3 = B_2$) are shown in Table 1. The results quoted were obtained in four different laboratories. Bowers et al.[22] and Taagepera and co-workers[23,24] used the pulsed electron ion cyclotron resonance technique (see preceding section) at pressures up to 10^{-3} Torr, while the author's experiments[25] were done with the high-pressure instrument (~ 5 Torr) described in the preceding section. The temperature dependence for some of the proton transfer equilibria measured with the high-pressure technique was determined[25] and is shown in Fig. 4. Evidently $\Delta G° (= \Delta H° - T \Delta S°)$ changes very little with temperature, i.e., $\Delta S°$ is very small. A small $\Delta S°$ is expected for proton transfer reactions. The results are not accurate enough to permit an evaluation of the small entropy change. Least squares treatment of the line with largest slope (line C in Fig. 4) gives $\Delta S° = 3.7$ e.u. The other lines indicate lower entropy changes.

Table 1
Relative Gas–Phase Basicities of Nitrogen Bases[a]

Base B	$\Delta G°_{(gas)}$[b]	$\Delta G°_{(aqua)}$[c]	PA[d]
Ammonia	0	0	207[e]
Methylamine	10.8,[25] 9.5[23]	1.9	217.8,[25] 216.5[23]
Dimethylamine	18.3,[25] 15.8[23]	2.1	225,[25] 222.8[23]
Trimethylamine	23.3,[25] 20.4[23]	0.75	230.3,[25] 227.4[23]
Ethylamine	12[23]	1.96	219[23]
i-Propylamine	14.4[23]	1.94	221.4[23]
t-Butylamine	16.5[23]	1.96	223.5[23]
4-NO$_2$-pyridine	1[24]	−10.8	208[24]
Pyridine	18[24]	−5.3[24]	225[24]
4-Methylpyridine	23[24]	−4.4[24]	230[24]
4-Methoxypyridine	26[24]	−3.6[24]	233[24]
Piperidine	23.9,[22] 24.5[26]	2.76	230.9,[22] 231.5[26]
Aniline	8.9[25]	−6.2	215.9[25]
P Methoxyaniline	15.1[26]	—	222.1
N Methylaniline	15.1[26]	−6.1	222.1
Cyclohexylamine	19.8[25]	2.2	226.8

[a] All values in kcal mole^{-1}.
[b] $\Delta G°$ for proton transfer to ammonia in gas phase.
[c] $\Delta G°$ for proton transfer to ammonia in aqueous solution obtained from $-RT \ln K_{NH_3}/K_B$, where K_B are base constants in aqueous solution (Ref. 29).
[d] Proton affinity of B is $\Delta H°$ for gas-phase reaction $BH^+ = B + H^+$.
[e] M. E. Haney and J. Franklin, *J. Chem. Phys.* **50**, 2028 (1969).

Obviously $\Delta G° \approx \Delta H°$ to a good approximation. The high-pressure data were obtained in the range 500–700°K. At lower temperatures the clustering reactions $BH^+ + B \rightleftharpoons (B)_2H^+$ reduced the concentrations of the monomers below the limits of measurement. The ICR measurements were done at room temperature. The low pressure greatly reduced the formation of clusters. As seen from Table 1, there is relatively good agreement between the determinations obtained with the two different methods.

The alkylamines grouped together in the table illustrate the increase of basicity with alkyl substitution. Thus, introducing CH_3 and moving from NH_3 to $(CH_3)_3N$, one finds a monotonic increase of basicity, each new methyl group having a strong but gradually decreasing effect. The basicity of primary amines increases regularly with the size of the alkyl substituent, i.e., from methyl to ethyl to isopropyl and t-butyl. The increase of basicity is easily understood on the basis of well-known electron-releasing ability of alkyl

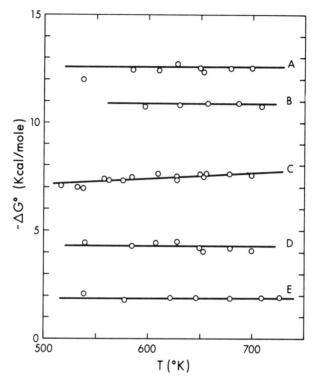

Figure 4. Temperature dependence of $\Delta G° = -2.3\,RT \log K$ for proton transfer equilibria: $M_1H^+ + M_2 = M_1 + M_2H^+$. M_1, M_2 are: (A) CH_3NH_2, $(CH_3)_3N$; (B) NH_3, CH_3NH_2; (C) CH_3NH_2, $(CH_3)_2NH$; (D) CH_3NH_2, $C_6H_5 \cdot NH \cdot CH_3$; (E) $C_6H_5NH_2$, CH_3NH_2. (From ref. 25.)

groups. The above conclusions follow earlier work done by Munson[27] and Brauman et al.[28] in which the order of basicities, but not the values of the equilibrium constants, had been established. It is important to note that the basicities of the alkylamines in aqueous solution do not follow any simple order (see second column in Table 1). Thus methylamine is a stronger base than ammonia, dimethylamine is slightly stronger than methylamine, and then the trend is reversed with trimethylamine becoming almost as weak a base as ammonia. This is the so called "amine anomaly." Evidently solvation effects mask the underlying electronic effects so clearly noticeable in the gas phase. Apart from the scatter of the proton transfer free energies in water, it is noticeable that they in all cases remain much smaller than those in the gas phase. Obviously the electronic stabilizing effect of the alkyl group in the gas phase is (at least partly) compensated by an adverse effect on the free energies of solvation. A question of interest is whether this effect is due to solvation changes of BH^+ or of the neutral species B. Arnett et al.[29] have made an analysis of the enthalpy, free energy, and entropy changes on basis of the Born–Haber type cycle shown below for the $\Delta G°$ changes:

$$NH_4^+ + B \rightarrow NH_3 + BH^+ \quad (\Delta G° \text{ gas-phase proton transfer})$$
$$-\Delta G_h°(NH_4^+) \uparrow \qquad \qquad \downarrow \Delta G_h°(BH^+)$$
$$-\Delta G_h°(B)_s \qquad \qquad \downarrow \Delta G_s°(NH_3)$$
$$NH_4^+ + B \rightarrow NH_3 + BH^+ \quad (\Delta G° \text{ proton transfer in water})$$

Using the proton transfer data for the gas phase and aqueous solution, one can calculate the single-ion free energy of hydration $\Delta G_h°(BH^+)$ relative to that of NH_4^+ if the free energies of solution of the gaseous neutral species are known. Similar calculations can be done for the other thermodynamic functions. The analysis[25] shows that the total free energies of hydration of the ions BH^+ become less exothermic with increasing alkyl substitution and size of the alkyl group and that this unfavorable change is almost of exactly equal magnitude to the gas-phase stabilization of the ions which it effectively cancels. The free energies of solvation of the molecular bases change very little with alkyl substitution. The lack of clear structural effects on the aqueous basicities is thus to be understood as resulting from the opposing effects of two large terms: The stabilization of the gaseous ion by the alkyl groups and the decrease of solvation of the stabilized ion due to the solvent-blocking effects of the alkyl groups. The gradually inverting order of aqueous base strength arises mainly from slight differences in the rate of change of the two large terms and the small effect of the solvation of the neutrals.

The basicities of the pyridines and anilines (Table 1) also reveal some interesting trends. In the traditional physical organic explanation of the lower basicity of aniline relative to ammonia electronic effects are invoked. It is stated that resonance structures contribute to the withdrawal of electronic charge from the nitrogen. The partial positive charge thus acquired is unfavorable to binding of the proton, i.e., it reduces the basicity. The gas-phase data show that aniline in the gas phase is more basic than ammonia. Thus the overall effect of the phenyl group is to stabilize and not destabilize BH^+. Evidently this electronic effect in the isolated species is obscured by solvation. One can, to a certain extent, avoid the overpowering effect of solvation by comparing molecules of similar size. Thus the gas-phase base strength of aniline is much lower than that of cyclohexylamine (see Table 1). The lower base strength of aniline relative to cyclohexylamine shows that phenyl is less electron donating than the cyclohexyl group. Well-known substituent effect in aniline and pyridine (see Table 1 and Ref. 24), deduced from observations in solution, are reproduced by the gas-phase measurements since one is comparing systems with similar carbon content and geometry.

Similar comparisons between the acidities of organic acids $HA = H^+ + A^-$ in the gas phase and in solution can be made. Gas-phase work due to Brauman and Blair[30] has shown that alkyl groups have a stabilizing effect on the negative ion A^- and thus increase the acidity in the gas phase. Extensions of this work by Bohme et al.[31] using the flowing afterglow technique have also led to the proposal[32] of an "absolute acidity scale" based on gas-phase acidity measurements.

IV. ENTHALPIES AND FREE ENERGIES OF FORMATION OF IONS IN THE GAS PHASE AND TOTAL ENERGIES OF SOLVATION OF SINGLE IONS

The gas-phase proton transfer equilibria discussed in the preceding section represent a method for determining enthalpies and free energies of formation of the ions B_2H^+ or A_2^-

$$B_1H^+ + B_2 = B_1 + B_2H^+$$
$$A_1^- + A_2H = A_1H + A_2^-$$

provided that the enthalpies or free energies of formation of the other species involved in the reaction are known. Thus the ion equilibrium method is becoming a very useful and efficient adjunct to the classical physical methods by which the energies of gaseous ions of interest to solution workers are determined. Apart from positive ions like Na^+, Ca^{2+}, etc., whose ionization potentials have long been established, the important positive ions appearing in solution are protonated bases BH^+. The classical physical methods like appearance potentials, etc., used for the determination of the heats of formation of $BH^+_{(g)}$ have been described recently by Haney and Franklin.[33] While the equilibrium measurements depend on reference values obtained by these methods, it is becoming clear that the equilibrium measurements will represent the efficient, versatile, and probably more accurate method for BH^+ measurements in the future.

The situation concerning the energies of formation of negative ions A^- might be similar. The classical physical method involves a determination of the electron affinity of A. Until recently, reliable electron affinities for even the most elementary negative ions were lacking (see review articles on electron affinities by Pritchard[34] and Berry[35]). Fortunately this situation changed with the advent of the photodetachment method.[36] Nevertheless, difficulties associated with the presence of excitation energy in A or A^- delayed application of the method to only a few systems.[36] Recent improvements of the laser photodetachment method[31] do bring great promise for electron affinity determinations. The developments are toward more accurate photoelectron energy analysis allowing vibrational and rotational states to be distinguished,[38] dye laser application with variable photon frequency near the photodetachment threshold,[39] and thermal energy A^- ions obtained from a high-pressure ion source.[40] Brauman[41] is also expecting to obtain electron affinities of many organic radicals by an application of the photodetachment technique in an ion cyclotron resonance mass spectrometer. However, it is likely that the ion equilibrium method will also play an important part for the determination of the energetics of gas-phase negative ions in extending the information to more systems and providing entropy differences.

Once the enthalpy or free energy of formation of a gaseous ion is known calculation of the corresponding single-ion energy of solvation is possible by the equation

$$\Delta H_h(A^-) = \Delta H_f(A^-, aq)_{abs} - \Delta H_f(A^-, g)$$

where

$$\Delta H_f(A^-, aq)_{abs} = \Delta H_f(A^-, aq)_{conv} - \Delta H_f(H^+, aq)_{abs}$$

The equations are written for the special case of hydration and enthalpy change. $\Delta H_f(A^-, aq)$ is the absolute heat of formation of the aquated ion A^-, while $\Delta H_f(A^-, aq)_{conv}$ is the conventional enthalpy of formation relative to $\Delta H_f(H^+, aq)_{conv} = 0$. For the absolute enthalpy of formation of the H^+ aquated ion $\Delta H_f(H^+, aq)_{abs} = 95.6$ kcal mole^{-1} can be used. This value is based[42a] on $\Delta H_f(H^+, g) = 365.6$ and $\Delta H_h(H^+) = -270$ kcal mole^{-1} Table 2 gives total enthalpies of hydration of a number of positive

Table 2

Absolute Single-Ion Enthalpies of Hydration
(in kcal mole^{-1})

Ion	$-\Delta H_h^\circ$	Ion	$-\Delta H_h^\circ$
H^+	270[a]	F^-	113[a]
Li^+	132[a]	Cl^-	81[a]
Na^+	106[a]	Br^-	78[a]
K^+	86[a]	I^-	64[a]
Rb^+	80[a]	OH^-	116[d]
Cs^+	72[a]	NO_2^-	73[d]
H_3O^+	112[b]	NO_3^-	68[d]
NH_4^+	84[c]	CN^-	~78[d]
$CH_3NH_3^+$	75[c]		
$(CH_3)_2NH_2^+$	70[c]		
$(CH_3)_3NH^+$	63[c]		
$C_2H_5NH_3^+$	75[c]		
$(C_2H_5)_2NH_2^+$	68[c]		
$(C_2H_5)_3NH^+$	61[c]		

[a]From Desnoyers and Jolicoeur.[42a]
[b]Based on separation due to Randles and $\Delta H = 168$ kcal mole^{-1} for gas-phase reaction $H_3O^+ = H^+ + H_2O$.[33]
[c]Based on separation due to Randles and data from Arnett[29] and $\Delta H = 207$ kcal mole^{-1} for gas-phase reaction $NH_4^+ = H^+ + NH_3$.[33]
[d]Based on separation due to Randles and data from Payzant et al.[44]

and negative ions. Many of the data for various ions were taken from the compilation of Desnoyers and Jolicoeur.[42a],* The new values are those of NH_4^+, the amines, NO_2^-, and NO_3^-. The separation of

*Also see Ref. 42b.

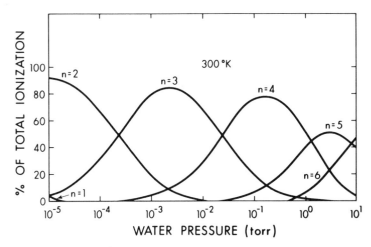

Figure 5. Relative equilibrium concentrations of sodium hydrates $Na(H_2O)_n$ in the gas phase at 300°K and variable water pressure.

the enthalpies of hydration for a given pair like NaCl used by Desnoyers and Jolicoeur is based on the experimental $\Delta G_h^\circ(K^+)$ obtained by Randles[43] from measurement of the EMF between an electrolyte solution and another surface separated from the solution by an inert gas. An independent support for the Randles result is given in the next section, which describes ion–water molecule interactions in the gas phase.

V. HYDRATION OF SPHERICALLY SYMMETRIC IONS. THE POSITIVE ALKALI AND NEGATIVE HALIDE IONS

The alkali and halide ions have played an important role in the studies of ionic hydration and solvation, and in theories of electrolyte solutions. Alkali halides form simple salts, soluble in water, and their ions have spherical symmetry and vary sufficiently in size to permit studies of the dependence of solvation parameters on the ionic diameter. For this reason the gas-phase studies of water clusters around the alkali[45,46] and halide[47] ions are of particular interest. The experimentally determined relative ion intensities of the sodium ion hydrates at various partial pressure of water at 300°K are shown in Fig. 5. It is assumed that equilibrium is established in the investigated systems and that the relative intensities are proportional to the relative stabilities of the clusters. The results indicate that three or even four different types of clusters may coexist at comparable concentrations. Furthermore, the number of ligands is not restricted to any fixed "coordination" number like four or six or eight; the higher the partial pressure of water, the larger the number of ligands and no cluster appears to be much more stable than any other.

These findings could have been predicted in principle on the basis of simple electrostatic and statistical thermodynamic calculations; nevertheless, they appear somewhat surprising since it is generally believed that certain structures with a discrete number of ligand molecules are the most stable. For example, it is commonly assumed that four, six, or eight (generally even numbers) of ligands form stable complexes but three, five, or seven never do. This attitude apparently arose from the experience gained in the studies of crystal structure. There three-dimensional extension imposes symmetry requirements which makes only certain structures allowed. Such restrictions do not apply to an isolated $M^+(H_2O)_n$ ion formed in the gas phase.

The symmetry requirements of the solid state are greatly relaxed in the liquid phase. Hence we might expect that the inner solvent sphere of an ion in a liquid could easily change the number of ligands by one or more units, similarly to the gaseous complex. In other words, the gas-phase data, by showing that the stabilities of hydrates differing by one molecule are comparable, support the view that for the alkali and halide ions the assignment of a fixed number of inner shell water molecules in aqueous solution is not realistic. The same conclusion probably holds for all weakly bonded complexes in which there is no significant d-orbital participation.

The equilibrium constants $K_{n-1,n}$ determined at different temperatures yield quantitative thermochemical data for the clustering reactions. Figure 6 shows typical Van't Hoff plots obtained for the various equilibrium constants of the $Na^+(H_2O)_n$ formation process. The slopes and intercepts of these plots lead to the relevant $\Delta H°_{n-1,n}$ and $\Delta S°_{n-1,n}$ values. The thermodynamic data obtained in this manner for other alkali hydrates and the halide hydrates are summarized in Tables 3 and 4.

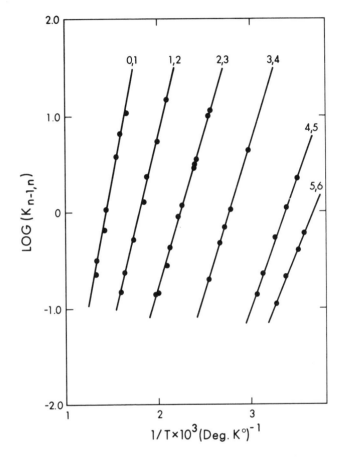

Figure 6. Van't Hoff plots for successive equilibrium constants $K_{n-1,n}$ of the reactions $Na^+(H_2O)_{n-1} + H_2O = Na(H_2O)_n$.

In Figs. 7 and 8 the variation of $\Delta H_{n,n-1}$ with n is shown for the alkali and halide ion hydration reactions. For each alkali ion $\Delta H_{n,n-1}$ decreases with increasing n, whereas for a constant n the enthalpy change is the largest for Li^+ and the smallest for Cs^+. Furthermore, the decrease of $\Delta H_{n,n-1}$ with n for a given ion is largest for Li^+, the ion with the smallest radius. These results could be expected. Similar trends are found for the halide ions.

Table 3

Enthalpy[a] and Entropy[b] Changes for Gas-Phase Reactions:
$$M^{\pm}(H_2O)_n = M^{\pm}(H_2O)_{n-1} + H_2O$$

$M^{\pm}$	$n, n-1$: 1,0	2,1	3,2	4,3	5,4	6,5	7,6	Ref.
H^+	168	32 (24.4)	20 (22)	17 (27)	15 (32)	13 (30)	12 (20)	14, 57
Li^+	34 (23)	26 (21)	21 (25)	16 (30)	14 (31)	12 (32)	—	46
Na^+	24 (21.5)	20 (22)	16 (22)	14 (25)	12 (28)	11 (26)	—	46
K^+	18 (21.6)	16 (24)	13 (23)	12 (25)	11 (25)	10 (26)	—	45, 46
Rb^+	16 (21)	14 (22)	12 (24)	11 (25)	10 (25)	—	—	46
Cs^+	14 (19.4)	12 (22)	11 (24)	10.6 (25)	—	—	—	46
NH_4^+	17.2 (20)	14.7 (22)	13.4 (25)	12.2 (27)	9.7 (22)	—	—	78
OH^-	25 (20.8)	16.4 (21)	15 (25)	14.2 (29)	14.1 (33)	—	—	44
F^-	23.3 (17.4)	16.6 (18.7)	13.7 (20.4)	13.5 (37)	13.2 (30.7)	—	—	47
Cl^-	13.1 (16.5)	12.7 (20.8)	11.7 (23.2)	11.1 (25.8)	—	—	—	47
Br^-	12.6 (18.9)	12.3 (23)	11.5 (25)	10.9 (27)	—	—	—	47
I^-	10.2 (16.3)	9.8 (19.0)	9.4 (21.3)	—	—	—	—	47
NO_2^-	14.3 (21)	12.9 (23.7)	10.4 (21.2)	—	—	—	—	44
NO_3^-	12.4 (19.1)	—	—	—	—	—	—	44
CN^-	13.8 (19.8)	—	—	—	—	—	—	44
O_2^-	18.4 (20)	17.2 (25)	15.4 (28)	—	—	—	—	47

[a] In kcal mole^{-1}.
[b] Entropy values in e.u. shown in parentheses below enthalpy value. Standard state, 1 atm.

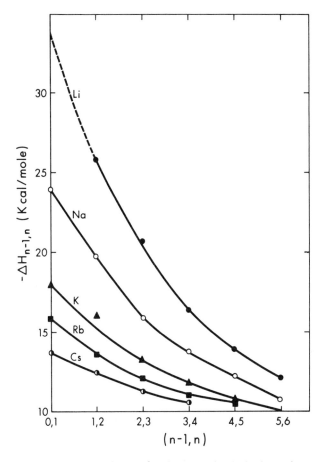

Figure 7. $\Delta H_{n-1,n}$ changes for the successive hydrations of the alkali ions: $M^+(H_2O)_n = M^+(H_2O)_{n-1} + H_2O$.

The alkali and halide ion results can be correlated with single-ion hydration enthalpies. In the previous section and Table 2 values for the single-ion enthalpies were quoted which had been listed by Desnoyers and Jolicoeur[42a] and were based on a measurement by Randles.[43] Prior to the Randles measurement, the single-ion enthalpies of Latimer et al.[48] were accepted. According to the Latimer separation, the ΔH_h for the negative ion was substantially

Table 4
Enthalpy Changes for Hydrogen-Ion and Hydroxyl-Ion Hydrate Dissociation in the Gas Phase[a] for Loss of One Molecule H_2O from $H_{2n+1}O_n^+$ and $H_{2n-1}O_n^-$

n	$(H_{2n+1}O_n)^+$		$(H_{2n-1}O_n)^-$	
	Experimental	Theoretical[b]	Experimental	Theoretical
1	168^{33}			
2	$36,^{57} 32,^{14} 32^{67}$	$32.3,^{64} 36.9,^{65} 43.5^{66}$	$25,^{44} 35^{67}$	$24,^{68} 41^{66}$
3	$20,^{57} 22,^{14} 22^{67}$	31^{66}	$18,^{44} 23^{67}$	30^{66}
4	$17,^{14,57} 17^{67}$	26^{66}	$15,^{44} 18^{67}$	23^{66}
5	$15,^{14,57} 15^{67}$	17.7^{66}	14^{44}	21^{66}
6	13^{57}			
7	12^{57}			

[a] All values in kcal mole^{-1}.
[b] Theoretical values correspond to electronic binding energy. In order to be compared with the experimental values, they should be corrected for zero-point energy. This would reduce them by some 3–5 kcal mole^{-1}.

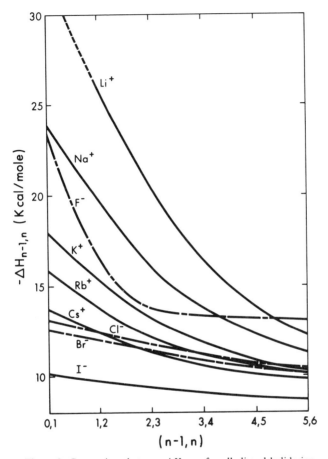

Figure 8. Comparison between $\Delta H_{n-1,n}$ for alkali and halide ion hydration. The initial hydration interactions of the isoelectronic positive ion are higher; however, for large n a crossover occurs indicating more favorable interactions for negative ions at large n.

larger than that for the positive isoelectronic ion. Thus $\Delta H_h(Br^-) = 81.4$, whereas $\Delta H_h(Rb^+) = 69.4$ kcal mole^{-1}. Two basically different explanations of Latimer's data were proposed. Some authors[49] assumed that the arrangement of water molecules around the negative ion is relatively more favorable, while Buckingham[50] suggested that water molecules possess an electrical quadrupole moment of significant magnitude. The orientation of the moment was assumed to be such that it resulted in a repulsion of positive ions, hence an attraction to negative ions. The present gas-phase results, which lead to higher energies of interaction for the positive ions [i.e., $\Delta H_{1,0}(K^+) > \Delta H_{1,0}(Cl^-)$] run counter to the orientation of the quadrupole moment assumed by Buckingham. Had there been a quadrupole orientation favoring stronger water attachment to negative ions, then its effect should have been most pronounced in the close-range interactions, i.e., at low n. Recent measurements of the water quadrupole moment by Verhoevan and Dymanus[51] are in agreement with the above conclusions. The quadrupole moment measured by these authors leads to attractive interaction with positive and not with negative ions as deduced by Buckingham. However, the magnitude of the quadrupole moment is small,[51] such that it becomes significant only for very small or multiply charged ions.

A correlation of the gas-phase data and the single-ion hydration energies of Latimer and Randles is shown in Fig. 9, which gives plots of $\Delta H_{n,0}(M^+) - \Delta H_{n,0}(X^-)$ for two isoelectronic pairs. The plots are based on the experimental $\Delta H_{n,n-1}$ whenever these were available (low n). For higher n, values extrapolated from Fig. 8 were used. Also indicated on the Figure are the differences of the corresponding total single-ion hydration energies based on Latimer's[48] and Randles'[42,43] data. It can be seen that the $\Delta H_{n,0}$ differences are closer to the Randles differences and seem to extrapolate for high n to those results. Since the $\Delta H_{n,0}$ differences should become equal to the ΔH_h differences for high n, the above results independently support Randles' single-ion hydration energies.

The Randles hydration energies for the positive and negative isoelectronic ions are almost equal (Table 2). Since the positive isoelectronic ion has a nuclear charge which is higher by two units, its radius should be smaller and lead to stronger interactions. As already mentioned, the $\Delta H_{n,n-1}$ values for the alkali ions are somewhat higher at low n. However, as demonstrated by Fig. 8, the $\Delta H_{n,n-1}$ values for the halide ions decrease more slowly with n and become more favorable at high n. The near equality of the total enthalpies of hydration thus seems to result from a cancellation of these two tendencies. One possible explanation for the slower fall off observed with negative ions would be the assumption that the

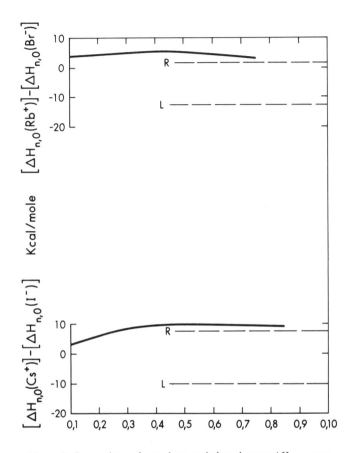

Figure 9. Comparison of gas-phase enthalpy changes $\Delta H_{n,0}$, corresponding to the process $M^+(H_2O)_n = M^+ + nH_2O$, with single-ion hydration enthalpies due to Latimer (L) and Randles (R). $\Delta H_{n,0}[Cs^+] - \Delta H_{n,0}[I^-]$ extrapolates for high n to the difference $\Delta H_h[Cs^+] - \Delta H_h[I^-]$ corresponding to Randles' single-ion hydration energies. A similar result is obtained for the Rb^+, Br^- pair.

hydrogen bonding of the water molecules to negative ions occurs with one H atom only and permits a closer packing of these molecules around the ion, i.e., the old idea expressed in Ref. 49.

The stepwise hydration energies $\Delta H_{n,n-1}$ of the alkali and halide ions (Table 3) can be compared with some calculated energies based on the electrostatic model or with MO calculations. Some electrostatic calculations done in the present laboratory accompanied the original publication of the experimental energies.[46,47] In these calculations the potential energy was expressed as a sum of terms resulting from the ion-dipole, ion-induced dipole, and van der Waals' attractive interactions and terms arising from the ion–water electron cloud repulsions and dipole–dipole repulsions. The absolute values of the calculated energies are greatly affected by the value assumed for the constant A appearing in the ion–water

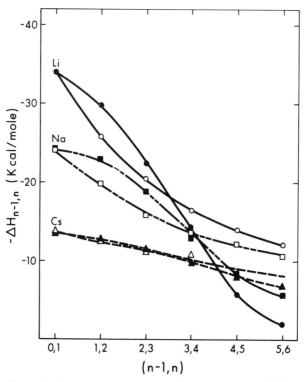

Figure 10. Comparison between experimental gas-phase $\Delta H_{n-1,n}$ values and calculated values on basis of electrostatic model. Electrostatic calculations give good fit for large ion, Cs^+, but predict different shape for small ion, Li^+. (Shaded points correspond to electrostatic calculations.)

repulsion term; A/R^{12}. The constant A could not be calculated from first principles; therefore its value was so adjusted as to give agreement with the experimental $\Delta H_{1,0}$. Such an A was then used for the calculation of the energies of the higher clusters of the same ion. The degree of agreement between the calculated and observed values is shown in Fig. 10. For the large Cs^+ the agreement is good. However, the calculated and experimental curves show distinctly different shapes for the two small ions Na^+ and Li^+. The initial rapid fall off of the experimental energies for Li^+, and to a lesser extent Na^+, was attributed to chemical bonding. More recently Spears[52] has made electrostatic calculations for the alkali and halide monohydrates and dihydrates. Spears used a two-parameter repulsive potential, $V(R) = Ae^{-bR}$. The values of A and b were obtained by adjusting to the experimental (Table 4) $\Delta H_{1,0}$ for K^+ and Br^-. The values for the other ions were obtained from these values and scattering repulsive potentials together with gas-phase spectral data for the alkali halides. The values obtained by Spears were in good agreement with the $\Delta H_{1,0}$ and $\Delta H_{2,1}$ of the alkali and the halide in the hydrates referred to in Table 4. It seems, therefore, that the softer repulsive potential obtained by Spears is better suited for the electrostatic calculations.

Diercksen and Kraemer[53,54] have made SCF MOLCGO calculations of the bonding energies in $Li^+(H_2O)$, $Na^+(H_2O)$, and $Li^+(H_2O)_2$. The values are $D_e(Li^+-OH_2) = 36$ kcal mole^{-1}, $D_e(Na-OH_2) = 25$ kcal mole^{-1}, and $D_e(H_2OLi^+-OH_2) = 31$ kcal mole^{-1}. The above values are for the electronic binding energy and should therefore be reduced by a few kcal mole^{-1} for the zero-point energy. Thus they are fairly close to the experimental values of Table 4, namely 34, 24, and 26 kcal, respectively.

Diercksen and Kraemer[55] have also calculated the binding energy of the F^- hydrate and obtained $D_e(F^--HOH) = 24.1$ kcal mole^{-1}. Corrected for zero-point energy, this value is also close to but somewhat lower than the experimental result (Table 4) 23.3 kcal mole^{-1}. The most stable structure found by the above authors for F^-HOH was the linear hydrogen-bonded structure, i.e. with the F, H, and O nuclei lying in a straight line. The bifurcated symmetrical structure in which both H atoms point symmetrically toward the negative ion was found to be of higher energy at equilibrium bonding distance, but became of equal energy at somewhat greater distances.

The electrostatic calculations of Arshadi et al.[47] and Spears[52] gave better bonding for the symmetric structure. However, it is felt that general hydrogen-bonding evidence favors the quantum mechanical result of Diercksen and Kraemer.

An application of the experimental values for the alkali and halide ions was made by Goldman and Bates.[56] These authors calculated total free energies and enthalpies of hydration by applying the Born equation not to the bare ions but to the hexahydrates. The hexahydrate energies were obtained from Table 4. The total enthalpies of hydration so calculated were found in good agreement with experimentally determined values. Goldman and Bates also used the experimental energies of Table 4 to obtain parameters for electrostatic calculations on a number of multiply charged spherical ions and calculated their total enthalpies and free energies of hydration.

VI. THE HYDROGEN ION AND THE HYDROXYL ION HYDRATES IN THE GAS PHASE

Studies of the gas-phase equilibria involving $H_{2n+1}O_n^+$ ion[14,57] and the $H_{2n-1}O_n^-$ negative ion[44,58] led to determination of the $\Delta H°_{n-1,n}$ and $\Delta S°_{n-1,n}$ values given in Tables 4 and 5. Considering first the proton hydrates, it is seen that the $\Delta H_{n,n-1}$, starting with 32 kcal mole^{-1} for the $H_3O^+-H_2O$ dissociation, decreases quite regularly with increase of n. The increase is initially rapid but slows down for larger n, for which the $\Delta H_{n,n-1}$ values have come down to 12 kcal mole^{-1} ($n = 7$) (see Fig. 11). The ideas concerning the proton hydrates, at the time when the first gas-phase equilibria data[57] were obtained, were quite different. The symmetric $H_3O^+(H_2O)_3$ structure proposed by Wicke et al.[59] from considerations of heat capacity results on aqueous solutions and supported by other measurements in aqueous solution[60-62] was considered to be exceptionally stable. Thus, one early theoretical treatment[63] of the isolated complex suggested that the three hydrogen bonds of the inner hydration shell contribute 45 kcal mole^{-1} each, while the first outer molecule in $H_3O^+(H_2O)_3-H_2O$ is held by only 9 kcal bonding energy.

That there are no such drastic changes in stabilities of the isolated complexes is also very directly demonstrated by the gas-phase equilibrium concentrations of the proton hydrates shown in

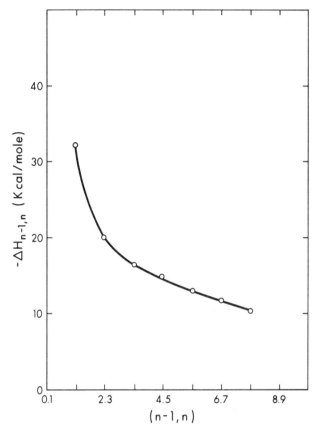

Figure 11. Plot of $\Delta H_{n-1,n}$ for reaction $H^+(H_2O)_n = H^+(H_2O)_{n-1} + H_2O$ in the gas phase. Regular decrease of enthalpy change indicates that no single structure has exceptional stability.

Fig. 12. The continuous distribution of the complexes with three, four, five, etc., water molecules observed as the water pressure is raised indicates a relatively gradual adjustment of structures rather than a sharp discontinuous drop after $n = 4$.

The values for the proton hydrate dissociation energies obtained by Friedman and co-workers[67] are also shown in Table 4. The experimental method used by Friedman consisted in determining the minimum kinetic energy required to collisionally

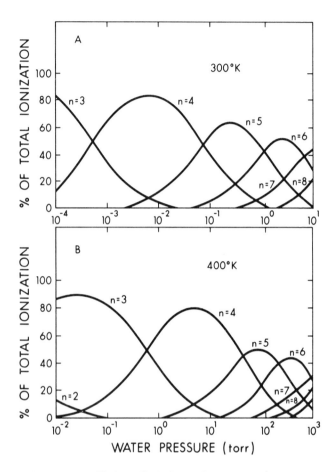

Figure 12. Equilibrium distribution of hydrates $H^+(H_2O)_n$ in the gas phase at different water pressures.

dissociate a beam containing a given clustered species. As can be noticed in Table 4, the results from these very different types of experiments are in close agreement with the measurements of the proton hydrate equilibria.

Unfortunately, the experimental energies do not contain any direct structural information. However, they are useful in establishing energy values with which theoretical calculations can be compared. Table 4 includes the results from a number of such calculations. Three theoretical values (all from *ab initio* MO calculations) are available for the $H_3O^+-H_2O$ energy. Kraemer and Diercksen's[64] leads to a value of 32.3 kcal mole^{-1}, while Kollman and Allen[65] obtain 36.9 and Newton and Ehrenson[66] 43.5 kcal mole^{-1}. For comparison with the experimental results, the above values should be reduced by some 5 kcal mole^{-1} zero-point correction.[66] This would mean that the Kraemer and Diercksen value is somewhat lower than the 32 kcal mole^{-1} experimental value, the Kollman result is very close, while the Newton and Ehrenson number is somewhat high. On the whole, the data may be considered very close considering the difficulties of the calculations.

An interesting result common to the calculations mentioned for $H_5O_2^+$ is the potential energy profile for the motion of the proton which is in line with the two O atoms. This potential for asymmetric motion of the proton is found to be very flat over a range of some 0.3 Å (O–O distance 2.39 Å).[64] The energy curve of Dierksen actually shows a double minimum corresponding to incipient asymmetric structures $H_2OH^+-OH_2$. However, the potential barrier between them is so small as to be almost negligible. This suggests a relatively easy continuous formation and splitting of systems $(H_2OHOH_2)^+$ which allow the above structures, and thus the positive charge, to translate relatively easily through the aqueous solution. It is interesting to note that crystals containing the $H_5O_2^+$ entity, and studied by neutron diffraction methods by Williams and Peterson,[69] also show that the proton lies in the middle of the two oxygens, indicating easy distortion in the asymmetric mode.

Theoretical studies for the higher proton hydrates were made only by Ehrenson and co-workers[66] (see Table 4). (These calculations use less extensive basic Gaussian type orbital sets and are probably less accurate.) Ehrenson's structural studies show that the most stable $H_9O_4^+$ structure is the symmetric Eigen structure $H_3O^+(H_2O)_3$. However, the other possible structures are not very much less stable and further addition of water leads to multiplicity of possible structures with similar energies. We think that these results also parallel the conclusions reached from the ion equilibrium studies.[57]

Considering the state of the proton in liquid water, the above experimental and theoretical studies might have the following meaning. Transition of the positive charge can be achieved by a multiplicity of structures. Assumption of a dominant inner shell hydrogen sphere of three molecules around an H_3O^+ entity is probably not necessary.

The ion equilibrium data for OH^- hydration are given in Table 4. Also shown are the results from Friedman's[67] collisional dissociation measurements and the theoretical calculations by Ehrenson and co-workers[66] and Kraemer and Dierksen.[68] Unfortunately, the agreement in these results is not as close as for the proton hydrates. Friedman's results for the OH^- hydrates are consistently higher than those from the equilibrium measurements. The Friedman data gave bonding energies for the OH^- hydrates which are very close to those for the H_3O^+ hydrates, while the equilibrium measurements[44] give somewhat lower values for the OH^- hydrates. It has been suggested by Friedman that the H_3O^+ and OH^- ions, being isoelectronic, should have very similar hydration energies. This argument is supported by the theoretical calculations of Ehrenson and co-workers[66] while the low value obtained from the equilibrium determinations for OH^-–H_2O is close to the calculated value obtained by Kraemer and Dierksen.[68] It is interesting to note that the total hydration enthalpy for OH^- appears to be 116 kcal mole^{-1}, while that for H_3O^+ is 112 kcal mole^{-1} (Table 2). However, these hydration enthalpies could be in error and it is conceivable that $\Delta H_h(H_3O^+) \approx \Delta H_h(OH^-)$. The low $\Delta H_{n,n-1}$ for OH^- obtained by the equilibrium experiments, if correct, taken together with the high $-\Delta H_h(OH^-)$, would mean that the hydration interactions of this ion for high n are more favorable than those for the H_3O^+ ion.

VII. HYDROGEN BONDING TO NEGATIVE IONS

The hydrogen bond dissociation energy $D(B^- - HR)$ is equal to the enthalpy change $\Delta H_{1,0}$ for the reaction

$$(1, 0) \quad (BHR)^-_{(g)} = B^-_{(g)} + HR_{(g)}$$

$K_{1,0}$ can be measured by the method discussed in Section II and Van't Hoff plots of $K_{1,0}$ provide $\Delta H°$ and $\Delta S°$ of the reactions. In a survey of hydrogen bonding to negative ions,[70] the equilibria of Cl^- with several hydrogen-bonding species, e.g., water, alcohols, and

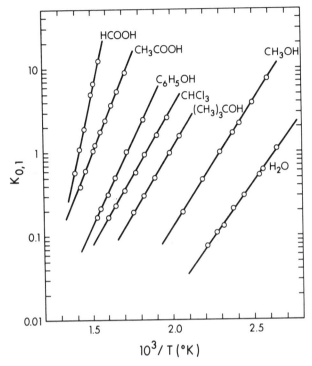

Figure 13. Van't Hoff plots of equilibrium constants for the gas-phase reaction $(ClHR)^- = Cl^- + HR$. Results show that the stability of the hydrogen-bonded complex $(ClHR)^-$ increases from H_2O to HCOOH, i.e., with gas-phase acidity of HR.

acids, were measured. The ion Cl^- was selected because it is easily produced in the gas phase and because it is a typical noble gas ion, neither too small like F^- nor too large like I^-. The Van't Hoff plots obtained are shown in Fig. 13. Examination of the figure immediately establishes the increasing stability of the Cl^- complexes in the order RH = HOH, CH_3OH, $(CH_3)_3COH$, Cl_3CH, C_6H_5OH, CH_3COOH, HCOOH. The $\Delta H^\circ_{1,0}$ and $\Delta S^\circ_{1,0}$ obtained from the plots are shown in Table 5.

We may consider the hydrogen-bonded complexes as species in which an incomplete proton transfer from the acid RH to the base B^- has occurred. This would suggest that the strength of the newly formed hydrogen bond B–H will depend on the easy

Table 5
Hydrogen Bonding to Negative Ions[a]

Reaction: $(ClHR)^- = Cl^- + HR$[b]			
HR	$\Delta H°$	$\Delta G°_{300}$	$D(H^+-R^-)$[b]
H_2O	13.1	8.2	390
CH_3OH	14.1	9.7	—
$(CH_3)_3COH$	14.2	11.1	—
Cl_3CH	15.2	10.8	356.7
C_6H_5OH	19.4	14.8	353
CH_3COOH	21.6	15.8	—
HCOOH	~37	25.4	—
Reaction: $(BHOH)^- = B^- + HOH$			
B^-	$\Delta H°$[c]	$\Delta G°$[c]	$D(B^--H^+)$[b]
OH^-	24	18	390
F^-	23	17.8	369
NO_2^-	14.3	8	339
Cl^-	13.1	8	333
Br^-	12.6	7	324
NO_3^-	12.4	6.4	323
I^-	10.2	5.3	314

[a] All values in kcal mole^{-1}, 300°K, standard state 1 atm.
[b] From Ref. 70.
[c] From Ref. 44.

availability of the proton and thus increase with the acidity of the proton donor RH. The order of H bond strengths should thus follow an order of increasing gas-phase acidity of RH. Obviously the strongest H-bonding species C_6H_5OH, CH_3COOH, and HCOOH are also the strongest acids in the group whose acidity increases in the order given. The acidity of HOH, CH_3OH, and $(CH_3)_3COH$ does not increase in this order in aqueous solution; however, the gas-phase acidity measurements discussed in Section III, and particularly the measurements of Brauman and Blair,[30] established that the gas-phase acidity of water and the alcohols follows the order shown above.

As a quantitative measure of the gas-phase acidity, one can select the free energy change for the gas-phase heterolytic dissociation since the dissociation constant K_a will be given for

Hydrogen Bonding to Negative Ions

HR = H$^+$ + R$^-$ by exp $-(\Delta G°/RT)$ ($= \Delta H° - T\Delta S°$). However, a good measure of (comparative) acidity can be obtained by considering the $\Delta H°$ term only.[30] The heterolytic bond dissociation energies for the RH compounds are also given in Table 5.

One may ask how the hydrogen bond energy changes in a series where the neutral acid HR is kept the same but the nature of B$^-$ is changed. Such a series in which HR is H$_2$O and B$^-$ is I$^-$, NO$_3^-$, Br$^-$, Cl$^-$, NO$_2^-$, OPh$^-$, CCl$_3^-$, F$^-$, and OH$^-$ is given in Table 5. The strength of hydrogen bonding increases in the order

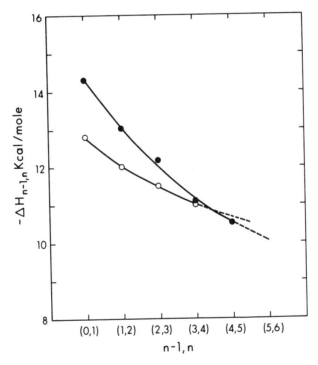

Figure 14. Enthalpy changes for reaction Cl$^-$(HR)$_n$ = Cl$^-$(HR)$_{n-1}$ + HR, where (●) HR = CH$_3$OH, (○) HR = HOH. Data show that although for small n the interaction with CH$_3$OH is stronger (corresponding to the higher gas-phase acidity of methanol), at high n the interactions with water become more favorable.

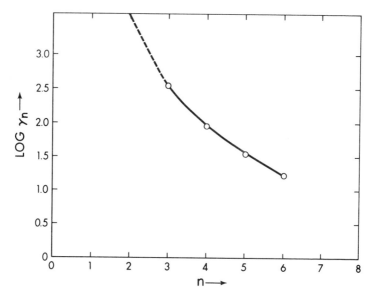

Figure 15. Plot of preference factor γ_n expressing preference for methanol M over water W in mixed clusters H^+L_n, where L is either M or W. Results show that a large preference for methanol in small clusters gradually decreases with growth of cluster. Extrapolation indicates preference for water for $n \approx 9$.

shown above. This is also the order of increasing heterolytic bond energy ($B^- - H^+$) or increasing basicity of the base B^-.

The gas-phase hydrogen bond energies of relatively simple acids HR and bases B^- thus reveal a very simple pattern, i.e., the bond strength increases with the acidity of the "proton donor" HR and increases with the basicity of the "proton acceptor" B^-. No such simple pattern has been found or can be expected in solutions where solvation effects would obscure the trends. In Section III it was found that electronic effects of substituents are clearly revealed in gas-phase acidities and basicities. Since H bonding to negative ions seems to follow gas-phase acidity and basicity orders, this would mean that substituent effects on H bonding in the gas phase may be expected also to follow a simple pattern.

The foregoing results showed that the interaction of Cl^- with one molecule of methanol (or any other alcohol) is stronger than that with water. It is further known that in the gas phase the reaction $H^+ + CH_3OH = CH_3OH_2^+$ is more exothermic than the corresponding reaction with water. Thus the gas-phase reaction

$$HCl + 2CH_3OH = CH_3OH_2^+ + Cl^-(HOCH_3)$$

is much more favorable than the corresponding reaction with two molecules of water. Yet in the liquid phase, HCl is much less dissociated in methanol than in water. The gas-phase clustering data do give an indication of this reversal of trends. Thus Fig. 14 gives measured $\Delta H_{n-1,n}$ data for Cl^- complexes with H_2O and methanol. These data show that, while at low n the $\Delta H_{n,n-1}$ for methanol are more exothermic, the trend becomes reversed at high n ($n \approx 5$). Figure 15 gives a plot of preference for methanol in positive ion mixed clusters containing the proton, water, and methanol.[72] The figure shows that the preference for methanol decreases rapidly with cluster size and indicates that water begins to be taken up preferentially at $n \approx 9$. Thus for both the negative and positive ion, solvation by water becomes more favorable at higher n. The behavior in the liquid phase thus seems to be predictable on the basis of a limited number of successive additions of solvent molecules.

VIII. ION SOLVATION BY PROTIC AND APROTIC SOLVENTS

Significant differences between the solvent effects of protic solvents (HOH, CH_3OH, etc.) and dipolar aprotic solvents like dimethyl formamide, dimethyl sulfoxide, acetonitrile, and others have been observed. Parker[73] has prepared summaries of such observations. Many of the effects can be explained by the assumption that aprotic solvents, in spite of their high dipole moment, solvate negative ions rather poorly. The energetics of negative and positive ion solvation by aprotic solvents have been examined by various conventional methods. Thus Choux and Benoit[74] have determined the enthalpies of transfer of salts from aqueous solution to aprotic solvents, while Coetzee and Campion[75] have studied relative activities of reference cations and anions in acetonitrile and water. These studies also indicate that aprotic solvents solvate negative ions less well than positive ions. Unfortunately, some of the extrathermodynamic assumptions that have to be made in order to separate the data to

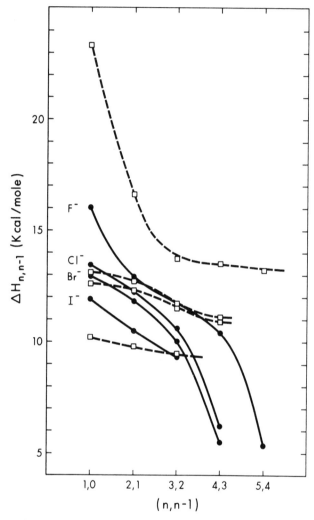

Figure 16. $\Delta H_{n,n-1}$ for halide ions solvated by acetonitrile (———) and water molecules (— —).

give individual contributions for the positive and negative ions are not free of objections.

A study of the gas-phase solvation of the halide[76] and the alkali ions[77] by acetonitrile provides some interesting information on solvation by this aprotic solvent. The enthalpy changes for the stepwise solvation reactions $X^-(CH_3CN)_n = X^-(CH_3CN)_{n-1} + CH_3CN$ obtained in that work[76] are shown in Fig. 16. Also included for comparison are the enthalpies for the reactions $X^-(H_2O)_n = X^-(H_2O)_{n-1} + H_2O$ discussed earlier (Section V and Ref. 47). The data in the figure show that the acetonitrile solvation energies follow the expected order with ionic size, i.e., they decrease from F^- to I^-. Comparing the results for acetonitrile with those for water, one notices that the acetonitrile $\Delta H_{n,n-1}$ values for F^- remain lower than those for water over the total observed range. For Cl^- and Br^- the $\Delta H_{1,0}$ for acetonitrile is higher but then a crossover occurs such that the $\Delta H_{2,1}$, $\Delta H_{3,2}$, etc., become relatively larger. The I^-–acetonitrile values are initially considerably larger, so that a crossover occurs later, i.e., at $\Delta H_{3,2}$. This reveals two trends. First, solvation by water is favored in all cases at large n. Second, the acetonitrile interactions become relatively stronger as one moves from small to large ions, i.e., from F^- to I^-.

The observation that acetonitrile becomes the poorer solvent at high n is not surprising. One could understand that it has a steric effect caused by the large size of the molecule and the specific location of the dipole $CH_3 \cdot \overset{+}{C} \equiv \overset{-}{N}$ which is such as to put the positive charge on the relatively inaccessible carbon atom in the middle of the molecule. It should be noted that a charge distribution in which the positive pole is relatively inaccessible is common to all dipolar aprotic solvents.

The trend toward better solvation by acetonitrile relative to that by water with increasing size of the ion has been deduced also in liquid solutions. Transfer enthalpies derived by Choux and Benoit[74] (water → aprotic solvent) are positive for Cl^- and Br^- but decrease with size of the ion, becoming negative for I^-.

A rationalization of the large drop of $\Delta H_{n,n-1}$ values (Fig. 16) around 4, 3 and 5, 4 has been given in the original publication.

The studies of the solvation of positive ions by acetonitrile have not been completed. The results so far obtained[77] give $\Delta H_{1,0}$ = 24 kcal mole^{-1} for the potassium ion. This is about 80% higher than the $\Delta H_{1,0}$ for the isoelectronic Cl^- ion. The acetonitrile value is also considerably larger than the $\Delta H_{1,0}$ = 18 kcal for potassium and water. We may understand these results as being a consequence of the previously mentioned location of the acetonitrile dipole. The negative pole located on the nitrogen is very accessible and the large acetonitrile dipole moment leads to the large $\Delta H_{1,0}$ for K^+ and acetonitrile. The fall off of the $\Delta H_{n,n-1}$ for that system may be expected to be initially slow but then, as n increases, the large size of the acetonitrile molecules must lead to $\Delta H_{n,n-1}$ values which are lower than those with water. However, there can be little doubt that the gas-phase data support the stronger solvation of positive ions by aprotic solvents expected from the results in liquid solutions.

REFERENCES

[1] E. Clementi and H. Popkie, *J. Chem. Phys.*, **57** (1972) 1077; **58** (1973) 1689.
[1a] V. L. Tal'roze and A. K. Lubimova, *Dokl. Akad. Nauk SSSR* **86** (1952) 909.
[2] D. P. Stevenson and D. O. Schissler, *J. Chem. Phys.* **23** (1955) 1353.
[3] G. G. Meisels, W. H. Hamill, and R. R. Williams, *J. Chem. Phys.* **25** (1956) 790.
[4] F. H. Field, J. L. Franklin, and F. W. Lampe, *J. Am. Chem. Soc.* **79** (1957) 2419.
[5] P. Kebarle and E. W. Godbole, *J. Chem. Phys.* **39** (1963) 1131.
[6] P. Kebarle and A. M. Hogg, *J. Chem. Phys.* **42** (1965) 798.
[7] W. A. Chupka, *J. Chem. Phys.* **30** (1959), 458.
[8] S. K. Searles and P. Kebarle, *Can. J. Chem.* **47** (1969) 2619
[9] P. Kebarle, "Mass spectrometric study of ion solvent molecule reactions in the gas phase," in *Mass Spectrometry in Inorganic Chemistry* (Advances in Chemistry Series, No. 72), Am. Chem. Soc., 1968.
[10] P. Kebarle, "Ions and ion solvent molecule interactions in the gas phase," in *Ions and Ion Pairs in Organic Reactions*, Vol. I, Ed. by M. Szwarc, Wiley, New York, 1972.
[11] P. Kebarle, "Higher-order reactions—ion clusters and ion solvation," in *Ion Molecule Reactions*, Ed. by J. L. Franklin, Plenum, New York, 1972.
[12] J. H. Yang and D. C. Conway, *J. Chem. Phys.* **40** (1964) 1729.
[13] D. A. Durden, P. Kebarle, and A. Good, *J. Chem. Phys.* **50** (1969) 805.
[14] A. J. Cunningham, J. D. Payzant, and P. Kebarle, *J. Am. Chem. Soc.* (to be published).
[15] J. D. Payzant, A. J. Cunningham, and P. Kebarle, *Can. J. Chem.* **50** (1972), 2230.
[16] C. J. Howard, H. W. Rundle, and F. Kaufman, *J. Chem. Phys.* **55** (1971) 4772.
[17] D. K. Bohme, R. S. Hemsworth, H. W. Rundle, and H. I. Schiff, paper presented at 20th Mass Spectrometry Conference, Dallas, 1972.
[18] E. E. Ferguson, F. C. Fehsenfeld, and A. L. Schmeltekopf, *Adv. Atomic Mol. Phys.* **5** (1969) 1.
[20] J. L. Pack and A. V. Phelps, *J. Chem. Phys.* **44** (1966) 1870.
[21] J. L. Beauchamp, L. R. Anders, and J. D. Baldeschwieler, *J. Am. Chem. Soc.* **89** (1967) 4569.
[22] M. T. Bowers, D. H. Aue, and H. M. Webb, *J. Am. Chem. Soc.* **93** (1971) 4314.
[23] W. H. Henderson, M. Taagepera, D. Oltz, R. T. McIver, J. L. Beauchamp, and R. W. Taft, *J. Am. Chem. Soc.* **94** (1972) 4728.
[24] M. Taagepera *et al.*, *J. Am. Chem. Soc.* **95** (1972) 1369.
[25] J. P. Briggs, R. Yamdagni, and P. Kebarle, *J. Am. Chem. Soc.* **94** (1972) 5128
[26] R. Yamdagni and P. Kebarle, *Can. J. Chem.* (to be published).
[27] M. J. B. Munson, *J. Am. Chem. Soc.* **87** (1965) 2332.

References

[28] J. I. Brauman, J. M. Riveros, and L. K. Blair, *J. Am. Chem. Soc.* **93** (1971) 3514.
[29] E. M. Arnett et al., *J. Am. Chem. Soc.* **94** (1972) 4724.
[30] J. I. Brauman and L. K. Blair, *J. Am. Chem. Soc.* **92** (1970) 5986; **50** (1968) 6561.
[31] D. K. Bohme and L. B. Young, *J. Am. Chem. Soc.* **92** (1970) 3301.
[32] D. K. Bohme, E. Lee–Ruff, and L. Brewster Young, *J. Am. Chem. Soc.* **93** (1971) 4608.
[33] M. A. Haney and J. L. Franklin, *J. Phys. Chem.* **73** (1969) 4328. S. L. Chong, R. A. Mayers, Jr., and J. L. Franklin, *J. Chem. Phys.* **56** (1972) 2427.
[34] H. D. Pritchard, *Chem. Rev.* **52** (1953) 529.
[35] R. S. Berry, *Chem. Rev.* **69** (1969) 533.
[36] L. M. Branscomb, D. S. Burch, S. J. Smith, and S. Gelman, *Phys. Rev.* **111** (1958) 504; L. M. Branscomb, *Phys. Rev.* **148** (1966) 11; B. Steiner, *J. Chem. Phys.* **49** (1968) 5097, 5151.
[37] D. Brehm, M. A. Gusinow, and L. J. Hall, *Phys. Rev. Letters* **19** (1967) 737.
[38] R. G. Celotta, *Phys. Rev.* (to be published).
[39] W. C. Lineberger and B. W. Woodward, *Phys. Rev. Letters* (1970) 424.
[40] E. Beatie, Joint Institute for Laboratory Astrophysics, University of Colorado, Boulder, Colorado, private communication.
[41] J. I. Brauman and K. C. Smyth, *J. Am. Chem. Soc.* **91** (1969) 7778; K. C. Smyth, R. T. McIver, and J. I. Brauman, *J. Chem. Phys.* **54** (1971) 2758.
[42a] J. E. Desnoyers and C. Jolicoeur, in *Modern Aspects of Electrochemistry*, No. 5, p. 1, Plenum, New York, 1969.
[42b] B. E. Conway and J. O'M. Bockris, *Modern Aspects of Electrochemistry*, No. 1, Chapter 3, Butterworths, London, 1954.
[43] J. E. B. Randles, *Trans. Faraday Soc.* **52** (1956) 1573.
[44] J. D. Payzant, R. Yamdagni, and P. Kebarle, *Can. J. Chem.* **49** (1971) 3309.
[45] S. K. Searles and P. Kebarle, *Can. J. Chem.* **47** (1969) 2619.
[46] I. Dzidic and P. Kebarle, *J. Phys. Chem.* **74** (1970) 1466.
[47] M. Arshadi, R. Yamdagni, and P. Kebarle, *J. Phys. Chem.* **74** (1970) 1475.
[48] W. M. Latimer, K. S. Pitzer, and C. M. Slanski, *J. Chem. Phys.* **7** (1939) 108.
[49] E. T. Verwey, *Rec. Trav. Chim.* **61** (1942) 127; D. R. Rosseinsky, *Chem. Rev.* **65** (1965) 467.
[50] A. D. Buckingham, *Disc. Faraday Soc.* **24** (1957) 151.
[51] J. Verhoevan and A. Dymanus, *J. Chem. Phys.* **52** (1970) 3222.
[52] K. Spears, *J. Chem. Phys.* (to be published).
[53] G. H. F. Diercksen and W. P. Kraemer, *Theoret. Chim. Acta (Berlin)* **23** (1972) 387.
[54] W. P. Kraemer and G. H. F. Diercksen, *Theoret. Chim. Acta (Berlin)* **23** (1972) 393.
[55] G. H. F. Diercksen and W. P. Kraemer, *Chem. Phys. Letters* **5** (1970) 570.
[56] S. Goldman and R. G. Bates, *J. Am. Chem. Soc.* **94** (1972) 1476.
[57] P. Kebarle, S. K. Searles, A. Zolla, J. Scarborough, and M. Arshadi, *J. Am. Chem. Soc.* **89** (1967) 6393.
[58] M. Arshadi and P. Kebarle, *J. Phys. Chem.* **74** (1970) 1483.
[59] E. Wicke, M. Eigen, and Th. Ackermann, *Z. Phys. Chem. (NF)* **1** (1954) 340.
[60] E. Glueckauf, *Trans. Faraday Soc.* **51** (1955) 1235.
[61] D. G. Tuck and R. M. Diamond, *J. Phys. Chem.* **65** (1961) 193.
[62] K. N. Bascombe and R. P. Bell, *Disc. Faraday Soc.* **24** (1957) 158.
[63] R. Grahn, *Ark. Fys.* **21** (1962) 13.
[64] W. P. Kraemer and G. H. F. Diercksen, *Chem. Phys. Letters* **5** (1970) 463.
[65] P. A. Kollman and L. C. Allen, *J. Am. Chem. Soc.* **92** (1970) 6101.
[66] M. De Paz, S. Ehrenson, and L. Friedman, *J. Chem. Phys.* **52** (1970) 3362; M. D. Newton and S. Ehrenson, *J. Am. Chem. Soc.* **93** (1971) 4971.

[67] M. De Paz, J. J. Leventhal, and L. Friedman, *J. Chem. Phys.* **51** (1969) 3748; M. De Paz, A. G. Guidoni, and L. Friedman, *J. Chem. Phys.* **52** (1970) 687.
[68] W. P. Kraemer and G. H. F. Diercksen, *Theoret. Chim. Acta (Berlin)* **23** (1972) 398.
[69] J. Williams and S. W. Peterson, *J. Am. Chem. Soc.* **91** (1969) 776; in *Spectroscopy in Inorganic Chemistry*, Vol. 2, p. 1, Academic Press, New York and London, 1971.
[70] R. Yamdagni and P. Kebarle, *J. Am. Chem. Soc.* **93** (1971) 7139.
[71] R. Yamdagni and P. Kebarle (unpublished work).
[72] P. Kebarle, R. N. Haynes, and J. G. Collins, *J. Am. Chem. Soc.* **89** (1967) 5753.
[73] A. J. Parker, *Quart. Rev. Chem. Soc.* **16** (1962) 163; A. J. Parker and R. Alexander, *J. Am. Chem. Soc.* **90** (1968) 3313.
[74] G. Choux and R. L. Benoit, *J. Am. Chem. Soc.* **91** (1969) 6221.
[75] J. F. Coetzee and J. J. Campion, *J. Am. Chem. Soc.* **89** (1969) 6221.
[76] R. Yamdagni and P. Kebarle, *J. Am. Chem. Soc.* **94** (1972) 2940.
[77] W. R. Davidson and P. Kebarle (unpublished work).
[78] J. D. Payzant, A. J. Cunningham, and P. Kebarle (to be published).

2

Electrolyte Solutions at High Pressure

S. D. Hamann

CSIRO Division of Applied Chemistry, Melbourne, Australia

I. INTRODUCTION*

All chemical reactions and rate processes are affected to some degree by hydrostatic pressure, just as they are by temperature, and in recent years there has been a remarkable growth of interest in the behavior of chemical and electrochemical systems at high pressures. In 1950 there were only a few laboratories in the world engaged in high-pressure chemical research, but now there are several hundred. The equipment needed for experiments to about 3 kbar is cheap and readily available, and pressure is taking its proper place with temperature as an easily accessible thermodynamic variable.

The impetus behind the recent work has stemmed on the one hand from the practical needs of modern technology and of the geological sciences for knowledge of the behavior of materials under extreme conditions, and on the other from a realization by chemists

---

*Note on pressure units: In this article the unit of pressure has been taken to be the bar, which is close to 1 atm and is an exact multiple of the accepted SI unit, the pascal (newton per square meter):

$$1 \text{ bar} = 10^5 \text{ Pa} = 10^5 \text{ N m}^{-2} \equiv 0.98692 \text{ atm}$$
$$\equiv 1.01972 \text{ kg(wt) cm}^{-2}$$

In addition, two multiples of the bar are used, the kilobar and megabar:

$$1 \text{ kbar} = 10^3 \text{ bars} = 10^8 \text{ Pa} = 0.1 \text{ GPa}$$
$$\equiv 986.92 \text{ atm}$$
$$1 \text{ Mbar} = 10^6 \text{ bars} = 10^{11} \text{ Pa} = 0.1 \text{ TPa}$$
$$\equiv 986{,}920 \text{ atm}$$

that pressure effects can give new types of fundamental information. In electrochemistry, for example, compression provides the only practical way of isothermally changing the dielectric constant of a solvent without changing its chemical composition.

The purpose of this article is to review the effects of pressure on some of the properties of electrolyte solutions. It will deal particularly with changes which occur in the conductivities of electrolyte solutions, in the ionization equilibria of weak acids and bases, and in the formation of ion pairs and ion complexes. It will not cover the influence of pressure on electrode processes, which has been reviewed by Hills[1] and Hills and Ovenden,[2] nor will it be concerned with solutions of electrolytes in supercritical water at high temperatures but at relatively low densities ($\rho < 1.0$ g cm^{-3})— there have been good reviews of that subject by Franck[3] and Marshall.[4] Most of the results to be discussed have been obtained by conventional hydrostatic methods, but the final section is devoted to a description of new shock wave techniques which extend the range of attainable pressures and permit the combination of very high pressures with very high temperatures.

II. PHYSICAL PROPERTIES OF WATER AND OTHER SOLVENTS AT HIGH PRESSURES

To understand properly and to predict the effects of pressure on electrolyte solutions, it is essential to know how compression alters the physical properties of solvents, in particular those of water.

For a thermodynamically stable liquid, an increase of pressure at constant temperature invariably raises the density ρ and this change is usually, but not always, accompanied by increases in the dielectric constant (relative permittivity) ε, the refractive index n, and the viscosity η of the liquid. At present there is no sound way of calculating those changes theoretically for highly polar liquids like water, and it is necessary to rely upon and often to extrapolate experimental data measured over limited ranges of pressure and temperature: for many of the common solvents there are no data at all. However, there are indications that in the future a reliable and soundly based theory of the properties of water at normal and high pressures may emerge from statistical calculations of the type that have been performed by Barker and Watts[5] using the Monte Carlo method, and by Rahman and Stillinger,[6] using the molecular dynamics method. These calculations have the potential of predicting most of the important properties of water at any pressure and temperature provided its molecules are still predominantly un-ionized, and, in addition, they give interesting information about the three-dimensional "structure" of water. It should be possible to extend them to other polar liquids and to dilute solutions of molecules and ions.

1. Physical Properties of Water at High Pressures

The following is a brief critical survey of the published data for the more important physical properties of compressed water.

(i) Density and Specific Volume

Grindley and Lind[7] have recently measured the $P-V-T$ properties of liquid water to 8 kbar between 25 and 150°C with an accuracy of 0.01 % in the volume. Their results agree very well with precise data which Kell and Whalley[8] had obtained to a lower maximum pressure of 1 kbar, and are completely consistent with compressibility data derived independently from measurements of the speed of sound in compressed water. A few of the results are listed in Table 1.

There are other, less accurate, data available beyond the pressure and temperature limits of Grindley and Lind's measurements and these have been reviewed and summarized by Kennedy and Holser[9] in the form of tables of the specific volume of water between -10 and $+1000°C$ to a maximum static pressure of 50 kbar. Similar reviews and tables have been published by Jůza,[10] Sharp,[11] and Köster and Franck.[12]

(ii) Viscosity

For many years there were somewhat conflicting data for the viscosity η of water at high pressures, but the discrepancies have now been removed by the careful measurements of Bett and Cappi,[14] Wonham,[15] Agaev and Yusibova,[16] and Stanley and Batten,[17] whose results all agree well with each other. Bett and Cappi's work has covered the widest range of pressures and temperatures, from 0 to 10 kbar and from 0 to 100°C. Some of their data are listed in

Table 1
Some Physical Properties of Water at High Pressures at 25°C

Pressure, kbar	0.001	0.5	1	2	3	4	5	6	7	8	9	10
Density,[a] g cm^{-3}	0.9970	1.0184	1.0380	1.0722	1.1014	1.1270	1.1501	1.1710	1.1902	1.2079	1.2244	1.2398
Relative density[a]	1.0000	1.0215	1.0411	1.0754	1.1048	1.1304	1.1535	1.1745	1.1938	1.2115	1.2281	1.2435
Specific volume,[a] cm^3 g^{-1}	1.0030	0.9819	0.9634	0.9327	0.9079	0.8873	0.8695	0.8540	0.8402	0.8279	0.8167	0.8066
Relative specific volume	1.0000	0.9790	0.9605	0.9299	0.9052	0.8846	0.8669	0.8514	0.8377	0.8254	0.8142	0.8042
Viscosity,[14] cP	0.8937	0.8865	0.8900	0.9250	0.9827	1.0552	1.1402	1.2385	1.3506	1.4782	1.6208	1.7754
Relative viscosity,[14]	1.0000	0.9919	0.9959	1.0350	1.0996	1.1807	1.2758	1.3858	1.5112	1.6540	1.8138	1.9866
Dielectric constant[b]	78.36	80.15	81.73	84.88	87.91	—	—	95.63	—	100.21	—	104.53
Relative dielectric constant[b]	1.0000	1.0228	1.0430	1.0832	1.1219	—	—	1.2204	—	1.2788	—	1.3340
Refractive index at 588 nm[31]	1.3329	1.3399	1.3461	—	—	—	—	—	—	—	—	—
Relative refractive index[31]	1.0000	1.0053	1.0099	—	—	—	—	—	—	—	—	—
Relative coefficient of self-diffusion[34]	1.00	0.99	0.97	0.93	0.86	0.78	0.70	0.59	0.51	0.46	0.41	—

[a]The data to 8 kbar are taken from Grindley and Lind's results,[7] and those for 9 and 10 kbar from Bridgman's.[13]
[b]The data to 2 kbar are taken from the results of Owen et al.[19] and Dunn and Stokes,[20] and those above 2 kbar from Lees.[29]

Table 1. It will be seen that the viscosity at 25°C initially *decreases* slightly with increasing pressure, which is abnormal behavior for a liquid, although it is normal for polar gases at moderate pressures.[18] But it increases at higher pressures and above 40°C it increases at all pressures.

(iii) Dielectric Constant

The static dielectric constant ε of water is perhaps its most important property from the standpoint of electrolyte chemistry—it is essential to know it in order, say, to apply the Debye–Hückel theory (see Section III) or investigate ion association.

There are two sets of accurate data for the dielectric constant of water at moderate pressures. Owen et al.[19] measured ε between 0 and 80°C at pressures up to 1 kbar, and Dunn and Stokes[20] measured it between 10 and 65°C to 2 kbar. These results agree well with each other but differ considerably from the earlier results of Kyropoulos[21] and Scaife,[22] which showed an increase of ε with pressure that was about 15% larger than that given by the more recent work. The values reported by Schadow and Steiner[23] show pressure dependencies that are larger at 20°C and smaller at 45°C than those of Owen et al. and Dunn and Stokes.

It should be mentioned that Owen and Brinkley[24] derived the following empirical relationship between ε and P from the data of Kyropoulos:

$$\frac{\varepsilon_1}{\varepsilon_P} = 1 - 0.4060 \log_{10}\left(\frac{B+P}{B+1}\right)$$

where ε_1 and B depend on the temperature. This formula has been widely used in theoretical treatments of pressure effects on electrolyte solutions,[25–28] but it is now clear that it was based on unreliable data and should be replaced by formulas which properly fit the later results. Dunn and Stokes[20] found that their values could be fitted closely to a relation of the form

$$\varepsilon_P = \varepsilon_1 + [BP/(1+CP)]$$

where ε_1, B, and C depend on the temperature. Owen et al.[19] fitted their values to an expression involving powers of the pressure and temperature. Neither formula should be used for extrapolation beyond the upper limits of the experiments. The theoretical Clausius–

Mossotti relationship

$$\left(\frac{\varepsilon-1}{\varepsilon+2}\frac{1}{\rho}\right)_T = \text{const}$$

fails badly for compressed water.

The only results for pressures above 3 kbar are those of Scaife,[22] mentioned above, and some unpublished ones obtained by Lees.[29] Where they overlap, Lees' results agree well with those of Owen et al.[19]

Recently Franck and Heger (see Ref. 30) have extended the measurements of ε for water to 550°C and 5 kbar, but their results were limited to rather low densities.

(iv) Refractive Index

Workers at the U.S. National Bureau of Standards[31] have measured the refractive index n of water at a number of wavelengths and at pressures up to 1.5 kbar. The results are illustrated in rows 9 and 10 of Table 1. Despite a statement to the contrary,[32] they deviate considerably from the Lorentz–Lorenz relationship

$$\left(\frac{n^2-1}{n^2+2}\frac{1}{\rho}\right)_T = \text{const}$$

However, they can be fitted quite well by an empirical formula

$$n(\rho, t) = 1.334 + 0.334(\rho - 1) - 1.90 \times 10^{-5}\rho t$$

which Zel'dovich et al.[33] found gives a good representation also of some values of n measured at very high shock wave pressures. Here ρ denotes the density in g cm^{-3} and t the temperature in °C.

(v) Diffusivity

Benedek and Purcell[34] measured the self-diffusion coefficient D of water to 10 kbar at 25°C, using a nuclear magnetic resonance spin-echo method, and obtained the results shown in row 11 of Table 1. Their observation that D decreases with increasing pressure at 25°C (despite the small initial decrease in the viscosity of water— see row 6) has been confirmed by Hertz and Rädle,[35] who found, however, that at 0°C, D initially increases and has a maximum at about 1.5 kbar.

(vi) Vibrational Spectrum

Walrafen and others have studied the O–D stretching band of dilute solutions of HDO in H_2O at moderate temperatures and at pressures up to 7.2 kbar. The band has a maximum at about 2507 cm^{-1} and a weak shoulder at about 2650 cm^{-1} which Walrafen[36] has attributed to the presence of "non-hydrogen-bonded" water molecules. He found that although the shoulder became less pronounced when the pressure was raised, the whole peak remained asymmetric, and he calculated from a computer analysis that the "component percentages" are essentially unchanged by pressure. However, the spectra are open to other interpretations[37] than the oversimplified one of supposing that water consists of distinct states in chemical equilibrium. Franck and Roth[38] carried out some extremely interesting measurements of the infrared O–D stretching absorption from 30 to 400°C and from 50 to 4000 bars and concluded that there was no indication of non-hydrogen-bonded OD groups or of well-defined clusters of water molecules at densities above 0.1 g cm^{-3}, and that their results supported the idea of a continuous distribution of neighboring O–O distances. In other words, the results were consistent with Pople's single-state model for water.[39] On the general question of the influence of pressure on the "structure" of water, it would be very enlightening to apply the molecular dynamics method of Rahman and Stillinger[6] to gain an insight into the molecular state of compressed water.

(vii) Phase Transitions

Water is abnormal in that it expands when it freezes to ice I at normal pressure, so that the freezing temperature is lowered by an increase of pressure. But that trend is reversed at 2 kbar, when ice I transforms to a different crystal modification, ice III, which is denser than the liquid phase. Tammann and, later, Bridgman[40] discovered five other dense forms of ice at high pressures, and Whalley et al.[41] have recently found two additional ones. These phases exist under the conditions of pressure and temperature shown in Fig. 1.

The structures of the high-pressure ices are extremely interesting.[42] They are all 4-coordinated and fully hydrogen bonded, but the O–O–O bond angles cover a very wide range from 85 to 135°,

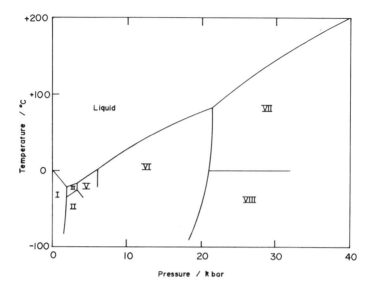

Figure 1. The phase diagram of liquid and solid water. Ice IV, which exists in a metastable state within the region of ice V, is not shown, nor is the low-temperature phase ice IX.

showing both how determined water is to maintain a hydrogen-bonded structure and how adaptable the structure is to distortion from the ideal and most stable tetrahedral arrangement. The structures of the densest forms, ices VI, VII, and VIII (which has a density of 1.63 g cm^{-3}), each consist of two interpenetrating but fully hydrogen-bonded lattices. One framework has cavities into which the molecules of the other fit. The frameworks are interpenetrating but not interconnecting.

The freezing of water at high pressures imposes a practical limit on the pressures that can be reached in liquid aqueous systems. For example, it is impossible to compress a dilute aqueous solution beyond about 10 kbar at 25° without causing the water to freeze to ice VI. For that reason electrochemical measurements above 10 kbar have been made either at higher temperatures[43,44] or in nonaqueous solvents[45] (see Section III) which either do not reach their thermodynamic freezing pressures or else "superpress" beyond them in the fluid state.

2. Physical Properties of Other Solvents at High Pressures

Most organic solvents are more compressible than water and their other physical properties are correspondingly more affected by an increase of pressure. The following figures illustrate the differences in compressibilities of a few common solvents[46–48]; they are the percentage contractions that occur when the pure liquids are compressed from 1 bar to 1000 bars at 30°C:

water	3.9	tetrahydrofuran	6.4
methanol	7.9	acetic acid	6.3
ethanol	7.2	pyridine	5.1
1-propanol	6.7	piperidine	5.6
2-propanol	7.1	ethylene glycol	2.9
acetone	7.9	propylene glycol	3.6

At higher pressures most organic liquids contract by 20–30% of their initial volumes at 12 kbar, and the few that can be compressed to

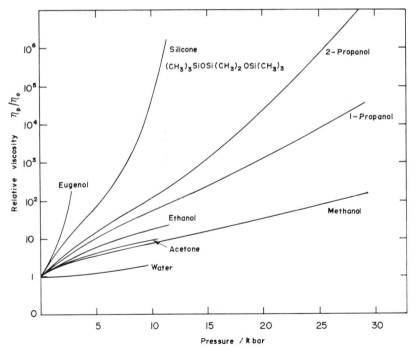

Figure 2. The viscosities of some compressed liquids at room temperature.

50 kbar without freezing contract by 36–39% at that pressure.[49] The glycols and glycerol are exceptional in being less compressible than water.

Concerning mixed solvents, it is a remarkable fact that the addition of a small amount of any organic liquid to water initially reduces its compressibility[46,47]: The mixture is less compressible than pure water even when the second component is very much more compressible than water (e.g., methanol). With increasing concentration of the second component the compressibility passes through a minimum at a volume fraction between 10 and 50% and then rises until it reaches the compressibility of the organic liquid. In other words, the excess compressibilities of aqueous solutions are negative and it follows that the excess volumes (volumes of mixing), which are also negative, become less so at high pressures.[46] In this sense, compression tends to make the mixtures more nearly ideal.

The viscosities of organic liquids increase almost exponentially with increasing pressure. This is illustrated in Fig. 2 by the results of some of Bridgman's measurements[50] to 12 and 30 kbar. The change is often very large and it varies considerably from one liquid to another. In a homologous series it seems to depend on the structural complexity of the molecules. This is shown by the curves in Fig. 2 for the series of hydroxy compounds: H_2O, CH_3OH, C_2H_5OH, 1-C_3H_7OH, 2-C_3H_7OH, $(C_{10}H_{11}O)OH$ (eugenol).

The dielectric constants ε of most organic solvents are lower than that of water, but they are increased more, proportionally, by pressure. A quantity which is important in electrostrictive effects in liquid systems (see Section IV) is the derivative $-\partial \varepsilon^{-1}/\partial P = (\partial \varepsilon/\partial P)/\varepsilon^2$, which has the values listed in Table 2 for some liquids at normal temperature and pressure.[19,23,24,51,52]

Table 2

Effect of Pressure on the Dielectric Constants of Liquids

	Water	Methanol	Ethanol	Acetone	Benzene	Ethyl ether
$\varepsilon(25°C)$	78.36	32.7	24.5	20.7	2.27	4.33
$10^3 \times \partial \varepsilon/\partial P$, bar^{-1}	3.69	3.5	2.3	3.0	0.13	1.1
$10^5 \times \partial(\ln \varepsilon)/\partial P$, bar^{-1}	4.71	10.6	9.2	14.3	5.6	25
$10^7 \times (\partial \varepsilon/\partial P)/\varepsilon^2$, bar^{-1}	6.01	32.4	37.6	69.1	250	590

III. ELECTRICAL CONDUCTIVITY OF ELECTROLYTE SOLUTIONS UNDER PRESSURE

The electrical conductivity of aqueous solutions was one of the first properties of fluids to be investigated under pressure. In 1827 Colladon and Sturm[53] reported that the resistances of solutions of some strong salts were not measurably altered by a pressure of about 30 bars. Later Herwig[54] also failed to detect any change at 18 bars, but in 1885 Fink[55] established that there is a small but definite increase of about 0.005 % per bar in the conductance of NaCl solutions between 1 and 500 bars and slightly larger increases for HCl and $ZnSO_4$. In 1894 Fanjung[56] found that the conductivities of solutions of weak carboxylic acids were increased much more by compression than those of strong salts, and from this he concluded that the strengths of the acids were increased by pressure in a way that Ostwald[57] had already predicted from density measurements at atmospheric pressure.

The early experiments were followed by fairly extensive work, which has been reviewed at different times by Tammann,[58] Brander,[59] Cohen and Schut,[60] Hamann,[61] Strauss,[62] Horne,[63] and Brummer and Gancy.[64] Broadly, it has had two aims: First, the study of the pressure dependence of conductivities, in itself, as a means of probing the mechanism of ion transport in solutions, and, second, the application of conductance methods in utilitarian ways to study other changes at high pressures—for example, ionization equilibria, micelle formation, reaction rates, and solubilities are all conveniently measured by calibrated conductivity methods. Some of these applications will be discussed in Section IV.

1. Experimental Methods

(i) Apparatus

The designs of high-pressure conductivity apparatus are varied and ingenious and there are many of them described in the references listed in the last paragraph. It will be sufficient here to discuss a few forms which happen to have been used by the writer but which exemplify most of the principles found in other designs.

The arrangement[65] shown in Figs. 3 and 4 is a convenient one for use to 3 kbar. The conductivity cell (A) is immersed in insulating

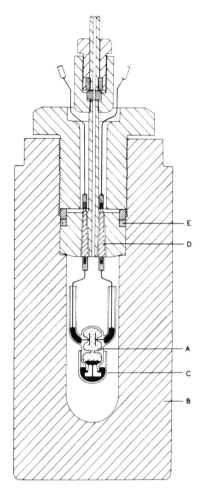

Figure 3. An apparatus for conductance measurements to 3000 bars.

oil inside a high-tensile steel pressure vessel (B) in a thermostat tank; the oil transmits pressure from a separate commercial hand pump and is sealed in the vessel by a Bridgman "unsupported area" packing (E)[48] of rubber. The conductivity cell has conventional square foil electrodes of platinized platinum mounted in a glass container and separated by a glass iris from mercury (C) which isolates the electrolyte solution from the oil. Electrical leads are introduced through glass-insulated steel cones (D), of the kind invented by Welbergen.[66] This arrangement and refinements of it[67] are capable of giving accurate results for most electrolyte solutions at moderate pressures, but they do have some failings and limitations. The mercury used to separate the electrolyte from the pressure fluid is not entirely inert and in some circumstances it can cause troublesome contamination. Also, the solubility, or more pertinently the rate of solution, of glass in water and in alkaline solutions is greatly enhanced at high pressures and can cause serious errors in dilute

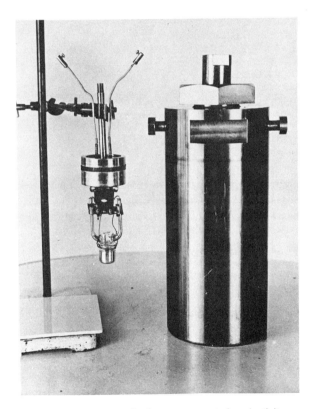

Figure 4. Apparatus for the measurement of conductivity to 3 kbar.

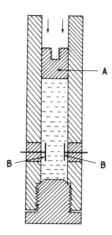

Figure 5. A Teflon conductivity cell for use to 12 kbar.

solutions. A pressure limitation is imposed by the difference in compressibilities of glass and platinum, which makes it impossible to use the cells much above about 3 kbar without cracks occurring at the glass/metal seals.

These disadvantages are eliminated in the cell[43,61] shown in Fig. 5, which is based on a design by Jamieson.[68] The body of the cell is made of Teflon (polytetrafluoroethylene), which remains sufficiently plastic to at least 12 kbar, to allow it to deform around the platinum seal without breaking. It has the additional advantage over glass of being insoluble in water and resistant to alkali at normal temperatures. The use of mercury is avoided by using a sliding Teflon plug (A) to transmit the pressure from the oil to the electrolyte, in the manner of a hypodermic syringe. However, the cell has the disadvantage that its cell constant is much more dependent on pressure than that of a glass cell, both because of the greater compressibility of Teflon and the fact that it undergoes phase transitions involving discontinuous contractions at high pressures.[69–71] At 25°C there is a phase change at 5.5 kbar with a 0.7% linear contraction, and another with a contraction of 0.5% at 250 bars, which in the writer's experience is much more troublesome because the Teflon does not always revert to its original form on

Electrical Conductivity of Electrolyte Solutions Under Pressure

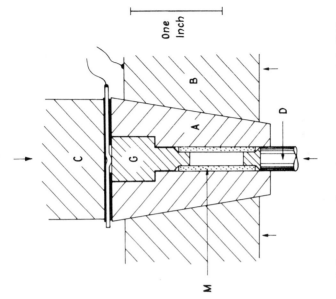

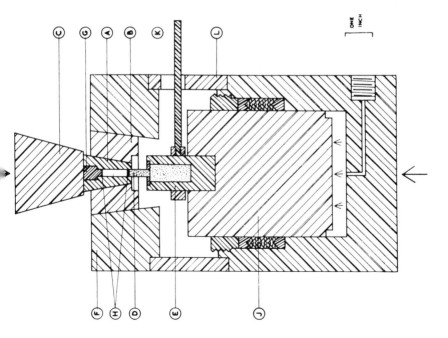

Figure 6. An apparatus for conductance measurements to 40 kbar.

decompression to 1 bar. The problem of the relatively large and imperfectly known change of cell constant can be solved by adopting Ellis's device[72] of mounting the electrodes on a separate glass spacer so that their separation is independent of the shrinkage of the Teflon body.

For work at pressures higher than 12 kbar, but of a much lower accuracy, the writer[73] has used the arrangement shown in Figs. 6 and 7. Here the solution is held in a polyethylene tube (M), sealed at each end by gold-plated steel plugs which form the electrodes of the cell. The tube is contained in a conical steel pressure vessel (A), supported by two conical rings (B, F). Internal pressure is generated by compressing the tube by the ram (J) and the tungsten carbide piston (D), which first pushes the plugs into the tube until they form line seals at the mouths and then compresses the polyethylene longitudinally, at the same time tightening the seals against leakage. As the internal pressure is raised, an external supporting pressure is

Figure 7. Apparatus for the measurement of conductivity to 40 kbar.

Electrical Conductivity of Electrolyte Solutions Under Pressure

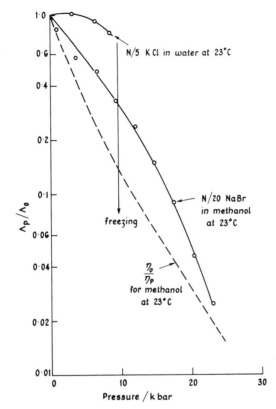

Figure 8. The relative molar conductivities of some electrolyte solutions at high pressures. Λ_P denotes the molar conductivity of the solute under pressure and Λ_0 the conductivity at atmospheric pressure; η_P and η_0 are the corresponding viscosities.

developed by forcing the conical pressure vessel (A) into its supporting rings using a separate 200-ton hydraulic press to act on the block (C): The angles of the cones are such that an internal pressure of, say, 40 kbar is supported by 20 kbar at the outside of (A) and by 10 kbar at the outside of (B). The change of cell constant under pressure can be calculated from the displacement of the piston in the pressure vessel, measured by the rod (K), and the estimated deformation of the polyethylene tube. The bottom electrode of the cell is in electrical contact with the piston and the top electrode is reached through the insulated plug (G). Figure 8 shows the results of some measurements made with this apparatus of the relative molar conductivities of potassium chloride in water to 8 kbar (water freezes to ice VI at 10 kbar at 23°C) and of sodium bromide in methanol to 25 kbar.

Measurements under conditions of combined high pressures and high temperatures are particularly difficult because of the very corrosive nature of hot, highly ionized, electrolyte solutions. Franck and his collaborators[74] have designed and used a variety of cells for those conditions and in one extreme instance[75] managed to make a few conductivity measurements on water to 100 kbar at 1000°C. The problem of corrosion can be by-passed by using fast techniques such as the shock wave method described in Section V.

(ii) Sources of Error

Before considering the results, it is desirable to review some of the sources of error that are peculiar to high-pressure conductivity measurements.

Temperature control. The act of compressing a liquid causes adiabatic heating which is typically about 10°C per kilobar. Since ionic conductivities are very sensitive to temperature changes and because heat transference is unavoidably slow in the massive pieces of apparatus that are used in high-pressure work, it is usually necessary to wait at least an hour for temperature equilibrium to be reached after a compression or a decompression step, which makes the experiments extremely tedious.

Solvent conductance. The conductivity of *pure* water increases by a factor of about 16 between 1 bar and 10 kbar, that is, to a value of about $9 \times 10^{-7} \Omega^{-1} cm^{-1}$ at 25°C, which is still small enough to be neglected in most measurements on aqueous solutions. However, the solutions are very seldom prepared in pure water; they are usually made in "conductivity water" with a conductivity of the order of $10^{-6} \Omega^{-1} cm^{-1}$, which is roughly trebled at 3 kbar.[27,43,59,72] The conductivity and its change are sensitive to minute traces of impurities and the best that can be done is to measure them under the conditions that apply in the rest of the experiments. The writer does not share the view of Brummer and Gancy[64] that because the solvent corrections are difficult to measure and to apply it is best to ignore them.

Cell constant. It has already been mentioned that the cell and its electrodes change their dimensions with increasing pressure and this alters the cell constant. The effect is particularly important and troublesome for Teflon cells [see Section III.1(i)] unless they have internal electrode spacers made of some other material.[72]

Contamination and leakage. Hamann and Strauss[43] drew attention to the danger of contamination by material from the pressure fluid (usually an oil) if this is in direct contact with the electrolyte solution, as it was in Zisman's experiments.[77] But their own measurements showed that similar, if less serious, contamination can arise when mercury is used to separate the two, and Gancy and Brummer[67] have found that dissolved oxygen seems to increase the dissolution of mercury. They also discovered a source of error in leakage of electrolyte through the platinum/glass seals, from where it crept over the outer surface of the cell and provided a parallel electrical path. It is unlikely that this problem would arise in cells with side arms of the type shown in Fig. 3.

Compressibility of solutions. Because the solutions contract under pressure, by as much as 10% at 3 kbar, the concentration of electrolyte expressed in moles dm^{-3} increases and a corrected value must be used in calculating the molar conductivity Λ. To a fair approximation it can be supposed that the compression of a dilute solution is the same as that of the solvent, but even then the reliability of the published compressions for organic solvents may be no better than 1%. For water it is certainly better than that, although results that in the past were corrected using Bridgman's data should be recorrected using Grindley and Lind's results (Table 1).

In view of these real or potential difficulties in high-pressure work it is not surprising that the results of different workers show discrepancies of the order of 1%. One paper[78], which drew attention to the discrepancies, itself reported results for the conductivity of aqueous potassium chloride solutions which were wrong by about 5% at 1 kbar.[63]

2. Results

(i) Conductivities at Infinite Dilution

From the standpoint of theories of ion transport in solution it is desirable to know the limiting molar conductivities Λ^∞ of salts at infinite dilution, or better, the individual limiting conductivities

λ_+^∞ and λ_-^∞ of the ions, in conditions where ion–ion interactions do not interfere with their motion.

To obtain Λ^∞, it is necessary to measure Λ over a range of concentrations c of a salt and extrapolate the results to infinite dilution using a relationship such as Kohlrausch's law:

$$\Lambda = \Lambda^\infty - Sc^{1/2} \tag{1}$$

where S is a function of pressure and temperature but not of concentration (in dilute solutions). The most extensive measurements of this kind at high pressures have been concerned with solutions of KCl in water at 25°C. The results are summarized in Table 3. There

Table 3
The Limiting Molar Conductivity $\Lambda^\infty (\Omega^{-1}\ cm^2\ mole^{-1})$ of KCl in Compressed Water at 25°C[a]

Pressure, bars	1	500	1000	1500	2000	3000	4000	5000
Buchanan and Hamann[65] (1953)[b]	149.9	—	152.7	—	149.9	146.6	—	—
Ellis[72] (1959)	149.9	151.5	152.0	—	149.4	144.7	—	—
Fisher[81] (1962)	149.9	152.3	152.6	151.4	149.4	—	—	—
Ovenden[80] (1965)	149.9	151.7	152.5	151.8	150.1	—	—	—
Gancy and Brummer[79] (1971)	149.9	151.8	152.2	151.4	149.7	—	—	—
Nakahara et al.[27] (1972)	149.9	152.6	153.4	152.9	151.7	146.9	140.5	133.1
λ_+^∞	73.5	73.6	73.2	72.5	71.4	—	—	—
λ_-^∞	76.4	78.2	79.0	78.9	78.3	—	—	—

[a] The values in this table are all based on the value $\Lambda_{KCl}^\infty = 149.88$ at 1 bar and 25°C, measured by Owen and Zeldes.[82]
[b] Recalculated from the *molal* conductivities listed by Buchanan and Hamann.[65]

is good agreement between the results of different workers, except for the recent data of Nakahara et al.,[27] who worked with unusually dilute solutions and considered that this reduced the errors of extrapolating Λ to infinite dilution; of course, it also increased the magnitude of the solvent correction, to about 7% in the most dilute solution at 3 kbar.

The separation of Λ^∞ into the individual ionic contributions λ_+^∞ and λ_-^∞ requires knowledge of the transport numbers t_+^∞ and t_-^∞ which have been measured for KCl solutions under pressure, by a modified moving boundary method, by Wall and Berkowitz,[83] Kay et al.,[84] and Matsubara et al.[85] The values at 25°C are given in Table 4. Combined with Gancy and Brummer's values[79] of Λ^∞, these yield the individual conductivities shown in the last two rows of Table 3.

Table 4

Pressure, bars	1	500	1000	1500	2000
t_+^∞	0.490	0.485	0.481	0.479	0.477
t_-^∞	0.510	0.515	0.519	0.521	0.523

Gancy and Brummer[79] have recently made careful measurements of the limiting conductivity ratios $\Lambda^\infty(P\ \text{bar})/\Lambda^\infty(1\ \text{bar})$ for ten 1:1 salts in water from 1 to 2300 bars and from 3 to 55°C, and their results can be combined with high-pressure transport numbers[83] to derive the limiting conductivities of most of the ions concerned. Fisher,[81] Fisher and Davis,[86a,b] Hamann et al.,[26] Inada et al.,[87] and others have reported values of Λ^∞ at high pressures for polyvalent salts of the 1:2, 2:2, 1:3, 2:3, and 3:3 types.

The behavior of Λ^∞ is remarkably uniform for most salts, even polyvalent ones. At 25°C its value initially increases slightly with increasing pressure by a few percent to 1 kbar, and then decreases almost linearly above 2 kbar until it reaches about 70% of its original value at the freezing pressure of water, 10 kbar. In the high-pressure range its pressure dependence is about the same as that of the fluidity $1/\eta$ of water (compare Tables 3 and 1, and see Fig. 7.2 of Ref. 61). At temperatures above about 50°C, Λ^∞ decreases at all pressures. These trends have been interpreted by some authors as implying drastic changes in the "structure" of water, and elaborate and detailed explanations of the results have been given in terms of structure making and structure breaking by the ions and by pressure and by temperature. In the writer's view, these approaches are essentially sterile in that they are incapable of making useful predictions, but a satisfying treatment may ultimately come from sound statistical calculations of the type that Rahman and Stillinger[6] have made for pure water. Similar comments apply to many of the treatments of the pressure and temperature dependence of the λ_i^∞ in terms of Arrhenius types of equations involving activation volumes $\Delta V^{\ddagger}$ and activation enthalpies $\Delta H^{\ddagger}$. It is a simple fact that the conductivities of electrolyte solutions do not, even roughly, obey the Arrhenius formula; the "activation enthalpy" of Λ^{∞} for KCl in water at 1 bar decreases by 40% between 0 and 50°C. Also, the transition-state models that have been used in justification of the Arrhenius relationships are much too crude to be realistic.

One interesting suggestion advanced by Horne[63,88–90] is that compression tends to "dehydrate" ions and that this change is responsible for the initial increase of Λ^{∞} with pressure. Above 2–3 kbar no further dehydration occurs and the behavior of Λ^{∞} then parallels the fluidity $1/\eta$ of water, as would be expected of simple spheres moving in a uniform medium according to the Stokes (or Walden) relationship

$$\Lambda^{\infty}\eta = \text{const} \qquad (2)$$

At first Horne[88] suggested that pressure stripped the ions of their outer hydration sheaths, leaving the innermost shells of tightly bound water intact, but later[89] he went further and implied that all local order was destroyed around the ions between 1 and 5 kbar. A fact which is consistent with Horne's general idea is that the smaller and more strongly solvated ions like Li^+ and F^- are the ones whose conductivities are the most increased initially by pressure.[79,91] However, the increase in conductivity is not nearly large enough to correspond with complete "dehydration." Between atmospheric pressure and 2 kbar, Λ^{∞}_{LiF} increases from 94.1 to 99.4 Ω^{-1} cm^2 mole^{-1} (unpublished measurements by the writer; see also the data for KF, LiCl, and KCl in Ref. 79), whereas Λ^{∞}_{RbBr} decreases from 155.9 to 151.8 Ω^{-1} cm^2 mole^{-1}, showing that although the mobilities of the Li^+ and F^- ions increase, they are still very much less than those of the larger and less strongly hydrated ions Rb^+ and Br^-. Nakahara et al.[27] have derived "hydration numbers" from their conductivity data for KCl solutions and concluded that they change very little with pressure; they certainly do not approach zero at 5 kbar.

Horne was aware of the contradiction between his supposed pressure dehydration and the accepted principle of electrostriction of solvent around ions (see Section IV), which requires that hydration in the sense of water–ion interaction be increased by compression. He circumvented the problem by postulating[89,90] a multilayer model comprising electrostricted water plus a cluster-enriched zone, or "iceberg" water, plus ordinary water, which is a flexible enough model to accommodate almost any observation. To the writer, there is no doubt that the *equilibrium* thermodynamic free energy of solvation becomes steadily more negative with increasing pressure (see Fig. 16) and hydration defined in that way always increases with compression. The confusion arises from trying to describe very complex *dynamic* effects simply in terms of some ill-defined and all-embracing "hydration."

The limiting conductivities of strong acids and bases in water are much more strongly affected by pressure than those of salts (see Fig. 9), and there is no doubt that the difference arises from the influence of pressure on the abnormal mobilities of the H^+ and OH^- ions.[43,61] This can be seen from the values[92] in Table 5 for $\lambda_{H^+}^\infty$ and

Table 5

Pressure, bars	1	1000	2000
$\lambda_{H^+}^\infty, \Omega^{-1} \text{cm}^2 \text{mole}^{-1}$	350	368	378
$\lambda_{H^+}^\infty - \lambda_{K^+}^\infty, \Omega^{-1} \text{cm}^2 \text{mole}^{-1}$	200	215	228

for the excess conductivity $\lambda_{H^+}^\infty - \lambda_{K^+}^\infty$ at 25°C. Hamann and Strauss[43,61] advanced an explanation of the effect, but it was later proved wrong by the work of Conway *et al.*[93] and Eigen and de Maeyer,[94] which established that the slow step in proton transference is the reorientation of a water molecule to receive a proton from an adjacent H_3O^+ entity. Conway[95] has pointed out that since compression accelerates this process, it might also be found to reduce the dielectric relaxation time of water, which has not so far been measured under pressure. It may be mentioned here that abnormal proton and hydroxyl conduction persists to very high pressures and high temperatures and the shock wave results presented in Table 15 show that $\lambda_{H^+} - \lambda_{K^+}$ is about $1200 \, \Omega^{-1} \, \text{cm}^2 \, \text{mole}^{-1}$ at 114 kbar and 670°C.

It was mentioned in Section II that the viscosities of nonaqueous solvents increase rather rapidly with increasing pressure (see Fig. 2), and the conductivities of dissolved strong electrolytes decrease correspondingly in rough accordance with the Stokes relationship (2). This behavior is to be seen in the early measurements of

Schmidt,[96] in the results shown in Fig. 8 for sodium bromide in methanol at a finite concentration, and in the results of other measurements in methanol,[45,97–99] ethanol,[100] acetone,[101–103] nitrobenzene,[99,101,104] various other organic liquids,[98,103,105–107] and their mixtures with water.[108–109,113]

(ii) Conductivities at Finite Concentrations

Apart from its influence on the limiting conductivities λ_i^∞ of ions, pressure can have a twofold effect on the conductivities λ_i at finite concentrations, first by modifying the influence of ion atmospheres on the mobilities of the ions, and second by changing the

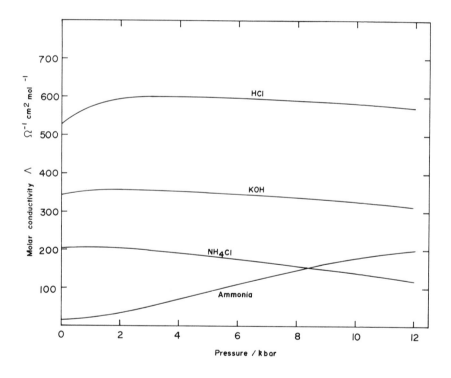

Figure 9. The molar conductivities of some electrolytes in water at high pressures at 45°C. The molality of each is about 0.01 mole kg^{-1}.

degree of ionization of incompletely dissociated weak electrolytes and ion pairs. The second effect is illustrated in Fig. 9, which compares the influence of pressure[43] on the conductivities of solutions of three strong electrolytes and one weak electrolyte, ammonia (see also Table 8). The effect of pressure on the excess conductivities of H^+ and OH^- ions, discussed in the last subsection, is also apparent in this diagram from the differences between the behavior of HCl, KOH, and NH_4Cl. Pressure-induced changes of ionization will be

Table 6
The Debye–Hückel–Onsager Parameters in CGS and SI Units[a]

Parameter	CGS units	SI units
A	$[2\pi Ne^6/1000(\ln 10)^2 k^3]^{1/2}(\varepsilon T)^{-3/2}$	$[Ne^6/32\pi^2(\ln 10)^2 k^3 \varepsilon_0^3]^{1/2}(\varepsilon T)^{-3/2}$
	$= 1.8248$	$= 5.7706$
	$\times 10^6$ mole$^{-1/2}$ dm$^{3/2}$ °K$^{3/2}$	$\times 10^4$ mole$^{-1/2}$ m$^{3/2}$ °K$^{3/2}$
	$\times (\varepsilon T)^{-3/2}$	$\times (\varepsilon T)^{-3/2}$
B	$[8\pi Ne^2/1000k]^{1/2}(\varepsilon T)^{-1/2}$	$[2Ne^2/k\varepsilon_0]^{1/2}(\varepsilon T)^{-1/2}$
	$= 5.0291$	$= 1.5904$
	$\times 10^9$ mole$^{-1/2}$ dm$^{3/2}$ °K$^{1/2}$ cm$^{-1} \times (\varepsilon T)^{-1/2}$	$\times 10^{10}$ mole$^{-1/2}$ m$^{1/2}$ °K$^{1/2}$ $\times (\varepsilon T)^{-1/2}$
B_1	$\left[\dfrac{8\pi Ne^6}{9000k^3}\right]^{1/2} \dfrac{q}{1+q^{1/2}}(\varepsilon T)^{-3/2}$	$\left[\dfrac{Ne^6}{72\pi^2 k^3 \varepsilon_0^3}\right]^{1/2} \dfrac{q}{1+q^{1/2}}(\varepsilon T)^{-3/2}$
	$= 2.8012$	$= 8.8582$
	$\times 10^6$ mole$^{-1/2}$ dm$^{3/2}$ °K$^{3/2}$	$\times 10^4$ mole^{-1} m$^{3/2}$ °K$^{3/2}$
	$\times (\varepsilon T)^{-3/2} q/(1+q^{1/2})$	$\times (\varepsilon T)^{-3/2} q/(1+q^{1/2})$
B_2	$[2e^2 F^4/9000\pi k N]^{1/2}(\eta^2 \varepsilon T)^{-1/2}$	$[e^2 F^4/18\pi^2 k N \varepsilon_0]^{1/2}(\eta^2 \varepsilon T)^{-1/2}$
	$= 41.243$ Ω^{-1} cm^2 mole^{-1}	$= 1.3042 \times 10^{-5}$ Ω^{-1} kg
	(mole dm$^{-3})^{-1/2}$ °K$^{1/2}$P	mole$^{-3/2}$ m$^{5/2}$ sec^{-1} °K$^{1/2}$
	$\times (\eta^2 \varepsilon T)^{-1/2}$	$\times (\eta^2 \varepsilon T)^{-1/2}$

[a] In these formulas, F = Faraday constant, e = elementary charge, k = Boltzmann constant, N = Avogadro constant, ε_0 = permittivity of vacuum, ε = relative permittivity (dielectric constant), η = viscosity, $q = \tfrac{1}{2}$ for symmetric electrolytes (otherwise, see Ref. 110, p. 137), T = absolute temperature.

Table 7
Debye–Hückel–Onsager Parameters[a] for Symmetric Electrolytes in Water at 25°C, Based on the Values in Tables 1, 3, and 6

Pressure, bars	ε	$10^2 \eta$, P	A	$10^{-7} B$	B_1	B_2	Λ^∞_{KCl}	$S = B_1 \Lambda^\infty_{KCl} + B_2$	$\Lambda^\infty_{KCl} - \Lambda_{KCl}$ for $I = 0.001$ at 1 bar
1	78.36	0.8937	0.5110	3.290	0.2297	60.38	149.85	95.21	3.01
500	80.15	0.8865	0.4940	3.253	0.2221	60.19	151.79	94.31	3.01
1000	81.73	0.8900	0.4797	3.222	0.2157	59.37	152.22	92.60	2.99
1500	83.34	0.9043	0.4659	3.190	0.2095	57.87	151.42	89.98	2.93
2000	84.88	0.9250	0.4533	3.161	0.2038	56.05	149.67	86.93	2.85

[a] The units of A and B_1 are $(\mathrm{dm^3\ mole^{-1}})^{1/2}$, of B are $(\mathrm{dm^3\ mole^{-1}})^{1/2}\ \mathrm{cm^{-1}}$, of B_2 and S are $\Omega^{-1}\ \mathrm{cm^2\ mole^{-1}\ (dm^3\ mole^{-1})^{1/2}}$, of Λ_{KCl} are $\Omega^{-1}\ \mathrm{cm^2\ mole^{-1}}$, and of I are $\mathrm{mole\ dm^{-3}}$.

considered in Section IV and the discussion here will be restricted to the first of the above effects.

In dilute solutions the influence of the ion atmosphere on the motion of a particular ion is adequately described by the Debye–Hückel–Onsager theory[110] in terms of an electrophoretic effect arising from the tendency of the electric field to move the ion atmosphere and its associated solvent in the opposite direction to the ion and its associated solvent, and a relaxation effect arising from the finite time required for the ion atmosphere to adjust to the motion of the ion. Both effects depend on the electrostatic forces between oppositely charged ions, which in turn vary inversely as the dielectric constant ε of the solvent and so are reduced by raising the pressure. Quantitatively, they are described by the formula

$$\Lambda = \Lambda^\infty - (B_1 \Lambda^\infty + B_2) I^{1/2} \tag{3}$$

where I denotes the ionic strength. The first term in brackets arises from the relaxation effect and the second from the electrophoretic effect. The quantities B_1 and B_2 are defined in Table 6, and Table 7 lists their numerical values for compressed water at 25°C, calculated from the dielectric constants and viscosities listed in Table 1.

Equation (3) has the same form as Kohlrausch's empirical relation (1). For 1:1 electrolytes the ionic strength I is simply the concentration of salt in moles dm^{-3} and the coefficient S equals $B_1 \Lambda^\infty + B_2$. The second last column in Table 7 lists this coefficient for the particular case of potassium chloride at 25°C, and the last column lists the theoretically derived difference between Λ_{KCl} and Λ_{KCl}^∞ for a 0.001 m solution (for which $I = c \approx 0.001\rho$, where ρ is the density of water in g cm^{-3}).

At atmospheric pressure, deviations from (3) become serious at ionic strengths greater than about 0.005 mole dm^{-3} and it is then necessary to use more refined theoretical treatments such as that of Fuoss and Onsager,[111] or semiempirical relationships of the type that Robinson and Stokes[110] and Davies et al.[112] have found useful. The equation of Robinson and Stokes is

$$\Lambda = \Lambda^\infty - [(B_1 \Lambda^\infty + B_2) I^{1/2}/(1 + aBI^{1/2})] \tag{4}$$

where a is an ion size parameter of about 4 Å for most 1:1 salts and B is the parameter in the Debye–Hückel expression for activity coefficients:

$$\log f_\pm = -AI^{1/2}/(1 + aBI^{1/2}) \quad (5)$$

The quantities A and B are defined in Table 6 and their values for water at high pressures are given in Table 7.

Experimentally, Hamann and Strauss[45,61] observed the expected decrease of S at high pressures for aqueous solutions and found also that the decrease is much more pronounced for solutions in methanol than in water. This is because compression causes a greater proportional increase in its viscosity (see Fig. 2) which reduces the electrophoretic effect by slowing down the movement of the ions. For sodium bromide in methanol at 25°C the measured values of S at 1 and 3000 bars were 273 and 53, respectively (in the units of Table 7), while the values calculated from (3) were 245 and 96. The agreement is probably as good as can be expected from the experimental uncertainties and from possible errors in the high-pressure values of η and ε for methanol.

Ellis[72] misinterpreted the theoretical effect of pressure on S as meaning that the ratio $\Lambda(P\ \text{bar})/\Lambda(1\ \text{bar})$ for a salt at finite concentration should be less than $\Lambda^\infty(P\ \text{bar})/\Lambda^\infty(1\ \text{bar})$ instead of greater, although he and Clark[113] later observed a trend in the right direction for solutions of salts and of hydrochloric acid in a 50% ethanol/water (w/w) mixture at pressures up to 3 kbar. More recently this trend has been confirmed for aqueous solutions by Gancy and Brummer,[64,79] who found that equation (3) is adequate for a number of salts to 2 kbar at low concentrations but that (4) is better at higher concentrations. Nakahara et al.[27] have also found that their experimental values of S for dilute solutions of potassium chloride agree closely with ones calculated from (4).

(iii) Conductivities at High Concentrations

In very concentrated solutions the trends initially predicted by the Debye–Hückel–Onsager relationship reverse and the ratios $\Lambda(P\ \text{bar})/(1\ \text{bar})$ become less than the limiting ratios. The explanation of this behavior is simply that the concentrated solutions are relatively viscous and compression produces a greater proportional increase in their viscosities than it does in pure water. Figure 10 illustrates the conductivity results of Molenat[114] for compressed concentrated solutions of NaCl in water at 25°C. Similar, but more

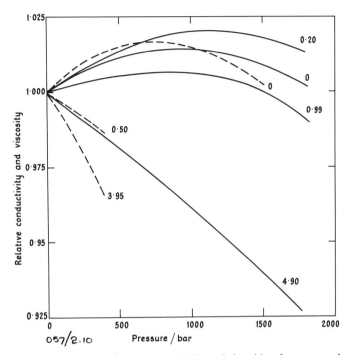

Figure 10. The relative molar conductivities and viscosities of concentrated solutions of NaCl in water at 25°C. The solid curves are plots of Λ_P/Λ_0 and the dashed curves are plots of η_0/η_P (these symbols are defined in the legend for Fig. 8). The number on each curve is the concentration of NaCl in mole dm^{-3} at 1 bar.

qualitative, results for concentrated solutions have been published by Körber[115] and Adams and Hall[116] in the form of tables of the relative resistances of their cells under pressure, uncorrected for the increase of concentration with pressure. Figure 10 also shows how the relative fluidities[117] $1/\eta$ of solutions of NaCl vary with pressure and concentration. The influence of high concentrations of solute is in agreement with Tammann's hypothesis[118] that a dissolved salt alters the properties of water in the same way as an additional high external pressure.

Although molten salts are outside the scope of this article, there have been a number of studies of their conductivities under pressure and good reviews of both the experimental and theoretical work are contained in recent papers by Cleaver et al.[119]

IV. IONIZATION EQUILIBRIA UNDER PRESSURE

One of the most pronounced and general of the chemical effects of pressure is the increase which occurs in the degree of ionic dissociation of weak electrolytes and ion pairs when the pressure is raised. It is related thermodynamically to the decrease in total volume that accompanies ionization reactions and in fact it was first predicted and discovered as a result of volumetric measurements.

In 1878 Ostwald[57] measured the volume changes that occur when acids are neutralized with bases in aqueous solution. He found that when both acid and base were strong, as in the neutralization of hydrochloric acid by potassium hydroxide

$$H^+ + Cl^- + K^+ + OH^- \rightarrow H_2O + K^+ + Cl^- \qquad (6)$$

$$\equiv H^+ + OH^- \rightarrow H_2O \qquad (7)$$

the reaction was accompanied by an almost constant expansion of about $+19.5 \text{ cm}^3 \text{ mole}^{-1}$, independent of the particular acid and base. On the other hand, if either the acid or base were weak, and sufficiently concentrated to be almost completely un-ionized, the change of volume was less and depended on the weak electrolyte. For instance, the changes for the neutralizations of acetic acid with potassium hydroxide and of ammonia with hydrochloric acid

$$CH_3CO_2H + K^+ + OH^- \rightarrow CH_3CO_2^- + K^+ + H_2O \qquad (8)$$

$$NH_3 + H^+ + Cl^- \rightarrow NH_4^+ + Cl^- \qquad (9)$$

were $+9.5$ and $-6.6 \text{ cm}^3 \text{ mole}^{-1}$, respectively. It follows by subtracting (7) from (8) and (9) that the ionization reactions

$$CH_3CO_2H \rightarrow CH_3CO_2^- + H^+ \qquad (10)$$

$$NH_3 + H_2O \rightarrow NH_4^+ + OH^- \qquad (11)$$

$$H_2O \rightarrow H^+ + OH^- \qquad (12)$$

must involve volume changes of -10.0, -26.1, and -19.5 cm^3 mole^{-1}, respectively, at the ionic strength of about unity used in the neutralization experiments (see Section IV.1 for a discussion of the influence of ionic strength on these values).

In 1884 Le Chatelier[120] published his general principle of chemical equilibria in the form: "The increase of condensation of the whole of a system maintained at constant temperature leads to modifications [of chemical equilibrium] which tend to reduce the condensation of the system." This statement becomes clearer when it is noted that Le Chatelier used the term "condensation" to mean either the pressure or the number of molecules per unit volume. In 1887 Braun[121] expressed the principle more explicitly in relation to the solubility of salts in liquids: "Substances which dissolve in their nearly saturated solutions with a contraction are more strongly dissolved by increasing the pressure." And in the same year Planck[122] published an exact quantitative relationship between the volume change accompanying a reaction and the influence of pressure in its equilibrium constant [equation (34)].

The significance of Ostwald's volume measurements was then obvious: the ionization of a weak electrolyte involves a contraction and so should be enhanced by raising the pressure. In 1894, at Ostwald's suggestion, Fanjung[56] made some direct measurements of the influence of pressure on the ionization of weak carboxylic acids AH in water. He determined the electrical conductivities of solutions of the acids to 270 bars at 13.5–18.6°C and found that they increased much more rapidly with increasing pressure than did the conductivities of solutions of strong electrolytes. He combined the data for strong and weak electrolytes to derive approximate ionization quotients K_c

$$K_c = (c_{A^-} c_{H^+})/c_{AH} \qquad (13)$$

and found that these increased by an average of about 12% between 1 and 270 bars. The change for acetic acid corresponded, on the basis of Planck's relationship,[122] to a contraction of $-10.6 \text{ cm}^3 \text{ mole}^{-1}$ for the ionization reaction (10). This was remarkably close to Ostwald's measured value of $-10.0 \text{ cm}^3 \text{ mole}^{-1}$.

Fanjung's work was followed closely by an important paper in which Drude and Nernst[123] explained the contraction and the pressure effect in terms of electrostriction of solvent around the ions that are formed by the dissociation reaction. They developed a quantitative electrostatic theory which is identical with Born's[124] later and better known treatment of the hydration of ions in solution. The experimental work was extended by Tammann and his collaborators[58,59,115] to a maximum pressure of 3 kbar and to a fairly wide range of acids and bases. Although this work has been largely superseded by more accurate measurements in the past twenty years, it gave some interesting insights into ion–solvent interactions and established the usefulness of "Tammann's hypothesis," which says that an ionized solute alters the behavior of water in much the same way as an additional external pressure.

The remainder of this section will be devoted, first, to a consideration of the thermodynamic relationship between the volume change for a reaction in solution and the influence of pressure on its equilibrium; second, to a brief account of some of the experimental methods used to measure the pressure effects; and third, to a review of the results and a discussion in terms of electrostatic theory and molecular models.

1. Thermodynamics of Equilibria in Solution at High Pressures

The thermodynamic condition for chemical equilibrium in a general reaction

$$a\text{A} + b\text{B} + \cdots \rightleftharpoons l\text{L} + m\text{M} + \cdots \qquad (14)$$

is that

$$a\mu_\text{A} + b\mu_\text{B} + \cdots = l\mu_\text{L} + m\mu_\text{M} + \cdots \qquad (15)$$

where $a, b, \ldots$ denote the stoichiometric numbers of molecules of the type A, B, ... taking part in the reaction, and the $\mu_\text{A}, \mu_\text{B}, \ldots$ are the molar chemical potentials of the species. Equation (15) is a complete expression of the equilibrium condition, but it is not a practical one in that form. For it to be useful it is necessary to relate the chemical potentials to experimentally determinable concentration scales such as the mole fraction scale or the molality scale.

The mole fraction x_J of a component J in the mixture is defined as

$$x_\text{J} = n_\text{J}/(n_\text{A} + n_\text{B} + \cdots) \qquad (16)$$

where the $n_\text{A}, n_\text{B}, \ldots$ represent the number of moles of each species in the mixture. For ideal systems obeying Raoult's law, μ_J is then

Ionization Equilibria under Pressure

given by

$$\mu_J = \mu_J^\circ + RT \ln x_J \quad (17)$$

where μ_J° is the molar chemical potential of the pure component J. For nonideal systems it is convenient to define an activity coefficient f_J such that

$$\mu_J = \mu_J^\circ + RT \ln f_J x_J \quad (18)$$

$$\lim_{x_J \to 1} f_J = 1 \quad (19)$$

Equation (15) may then be rewritten

$$l\mu_L^\circ + m\mu_M^\circ + \cdots - a\mu_A^\circ - b\mu_B^\circ - \cdots = -RT \ln K_x \quad (20)$$

where K_x is an equilibrium "constant" (which is actually a function of temperature and pressure) defined by

$$K_x = \frac{x_L^1 x_M^m \cdots}{x_A^a x_B^b \cdots} \frac{f_L^1 f_M^m \cdots}{f_A^a f_B^b \cdots} \quad (21)$$

From the relationship

$$\left(\frac{\partial \mu_J^\circ}{\partial P}\right)_T = V_J^\circ \quad (22)$$

where V_J° is the molar volume of J in its pure state, it follows that

$$-\partial(RT \ln K_x)/\partial P = lV_L^\circ + mV_M^\circ + \cdots - aV_A^\circ - bV_B^\circ - \cdots \quad (23)$$

$$= \Delta V^\circ \quad (24)$$

where ΔV° is the excess of the molar volumes of the products over those of the reactants, all in their pure states.

The molality scale of concentrations is a more convenient one than the mole fraction scale for dilute solutions, where one of the components, say S, is in large excess and can arbitrarily be called the solvent. The solvent S may or may not take part in the reaction. The molality m_J of a component J is defined by

$$m_J = 1000 n_J / M_S n_S \quad (25)$$

where M_S is the molecular weight of the solvent. The molality is related to the mole fraction by

$$m_J = 1000 x_J / M_S x_S \quad (26)$$

In the limit of very dilute solutions x_S is close to unity and

$$m_J \approx 1000 x_J / M_S \tag{27}$$

For such solutions it is convenient and conventional to define activity coefficients γ_J for *solute* species J by

$$\mu_J = \mu_J^\infty + RT \ln \gamma_J m_J \tag{28}$$

$$\lim_{\Sigma_J m_J \to 0} \gamma_J = 1 \tag{29}$$

The chemical potential term μ_J^∞ is different from μ_J° in (17) and (18); it is defined by

$$\mu_J^\infty = \lim_{\Sigma_J m_J \to 0} (\mu_J - RT \ln m_J) \tag{30}$$

and its derivative with respect to pressure,

$$\partial \mu_J^\infty / \partial P = V_J^\infty \tag{31}$$

is equal to the partial molar volume V_J^∞ of J at infinite dilution in the pure solvent.

Equation (27) can be combined with (15) to give relations analogous to (20) and (21):

$$l\mu_L^\infty + m\mu_M^\infty + \cdots - a\mu_A^\infty - b\mu_B^\infty - \cdots = -RT \ln K_m \tag{32}$$

$$K_m = \frac{m_L^l m_M^m \cdots \gamma_L^l \gamma_M^m \cdots}{m_A^a m_B^b \cdots \gamma_A^a \gamma_B^b \cdots} \tag{33}$$

and it follows from (31) that

$$-\partial(RT \ln K_m)/\partial P = lV_L^\infty + mV_M^\infty + \cdots - aV_A^\infty - bV_B^\infty - \cdots \tag{34}$$

$$= \Delta V^\infty \tag{35}$$

where ΔV^∞ denotes the excess of the partial molar volumes of the products over those of the reactants, all at infinite dilution in the solvent. Equation (34) is essentially the relationship which Planck[122] derived in 1887.

Experimentally it is not always practicable to measure the activity coefficients $\gamma_A, \gamma_B, \ldots$ or to adopt the alternative procedure of extrapolating measurements made over a range of concentrations of the solutes to zero concentration, where the γ are unity and K_m

Ionization Equilibria under Pressure

reduces to the first term K_m' in (33):

$$K_m' = m_L{}^l m_M{}^m \cdots / m_A{}^a m_B{}^b \cdots \tag{36}$$

It is sometimes only possible to determine K_m' at finite concentrations and at fairly high ionic strengths, and in these cases it is necessary to know what errors are introduced by neglecting the pressure dependence of the activity coefficient term $\Pi(\gamma)$:

$$\Pi(\gamma) = \gamma_L{}^l \gamma_M{}^m \cdots / \gamma_A{}^a \gamma_B{}^b \cdots \tag{37}$$

$$K_m = K_m' \Pi(\gamma) \tag{38}$$

The pressure derivative of $\Pi(\gamma)$ can be derived from relationships of the following kind[125,110]

$$\partial (RT \ln \gamma_J)/\partial P = V_J - V_J^\infty \tag{39}$$

where V_J denotes the partial molar volume of J in the actual solution of finite concentrations of the solute species. Combining these relations with (33) and (35) gives

$$-\partial(RT \ln K_m')/\partial P = \Delta V \tag{40}$$

where ΔV now denotes the excess of the partial molar volumes of the products over those of the reactants, all at the actual concentrations of the solution.

Formulas (35) and (40) show that whereas the pressure dependence of K_m is governed by the partial molar volumes of the solute species in an infinite volume of the solvent, that of K_m' is governed by the corresponding partial molar volumes in the particular equilibrium mixture. The difference between ΔV^∞ and ΔV is usually small in comparison with ΔV^∞, but it is measurable. Taking the simple case of the ionization in water of a weak 1:1 electrolyte such as acetic acid, HAc, it is possible to estimate the difference theoretically in the following way.

The equilibrium constant is

$$K_m = \frac{m_{Ac^-} m_{H^+}}{m_{HAc}} \frac{\gamma_{Ac^-} \gamma_{H^+}}{\gamma_{HAc}} \tag{41}$$

where it is usual to assume that γ_{HAc} is unity. If it is further assumed that γ_{Ac^-} and γ_{H^+} each have the limiting Debye value $\gamma_\pm$ given by

$$\log_{10} \gamma_\pm = -AI^{1/2} = -1.8248 \times 10^6 I^{1/2}/(\varepsilon T)^{3/2} \tag{42}$$

where T denotes the absolute temperature in °K, ε is the dielectric constant, and I is the ionic strength in moles dm^{-3}, then the derivative of the activity coefficient term in (41) can be found by differentiating (42) with respect to pressure, allowing for the changes of A (Table 7) and of I with pressure. In the particular case where the ionic strength arises predominantly from the presence of an added strong salt, as in an emf cell, its pressure dependence is close to that of the density of water. Introducing experimental values of the derivatives $\partial \varepsilon/\partial P$ and $\partial \rho/\partial P$ for water at 25°C and atmospheric pressure[8,19,20] then gives the result

$$\partial(RT \ln \gamma_\pm^2)/\partial P = 2.79 I^{1/2} \qquad (43)$$

$$= \Delta V - \Delta V^\infty \qquad (44)$$

where I is in moles dm^{-3} and ΔV is in cm^3 mole^{-1}. This formula yields the following estimates of $\Delta V - \Delta V^\infty$:

I, mole dm^{-3}:	0.0001	0.001	0.01	0.1	0.5
$\Delta V - \Delta V^\infty$, cm^3 mole^{-1}:	0.028	0.089	0.28	(0.89)	(1.98)

The values in parentheses are uncertain because the limiting Debye equation is inaccurate at high concentrations. For comparison, the total volume change ΔV^∞ is -11.50 cm^3 mole^{-1} for acetic acid; the values of $-\Delta V^\infty$ for other 1:1 acids and bases are usually in the range 10–30 cm^3 mole^{-1} and the accuracy of measurement is often about ± 0.5 cm^3 mole^{-1}. The positive sign of the coefficient in (43) means that ΔV is less negative than ΔV^∞ and so the influence of pressure on K_m' decreases as the ionic strength increases.

Finally, a word of caution should be added about the use of the *molarity* scale of concentrations in high-pressure work and, for that matter, in high-temperature work. It is illogical and misleading to use units which themselves vary with pressure and temperature. In particular, it is wrong to suppose that relationships of the form (35) and (40) apply when K is expressed in molar units (moles dm^{-3} or moles liter^{-1}). Instead, as Guggenheim[126] and the writer[61,127] have pointed out, if the equilibrium constant is defined in terms of these units

$$K_c = \frac{c_L^l c_M^m \cdots}{c_A^a c_B^b \cdots} \frac{y_L^l y_M^m \cdots}{y_A^a y_B^b \cdots} \qquad (45)$$

where the c are concentrations in molarities and the y are molar scale activity coefficients, then the pressure derivative of $\ln K_c$ is

$$-\partial(RT \ln K_c)/\partial P = \Delta V^\infty + (a + b + \cdots - l - m - \cdots)RT\kappa \quad (46)$$

where κ is the compressibility coefficient of the solvent, defined as

$$\kappa = -(1/V)\,\partial V/\partial P \quad (47)$$

For the ionization of a 1:1 electrolyte the second term in (46) is just $-RT\kappa$, which is $-1.1\,\text{cm}^3\,\text{mole}^{-1}$ for water and $-3.2\,\text{cm}^3\,\text{mole}^{-1}$ for methanol at 25°C and atmospheric pressure. It is ironical that a number of workers have specially measured the compressibilities of their particular solvent systems just to be able to calculate molar concentrations rather than molal ones. They have then calculated incorrect values of ΔV^∞ from a supposed relationship

$$-\partial(RT \ln K_c)/\partial P = \Delta V^\infty \quad (48)$$

Had they used the correct form (46), they would have obtained the right values, but in doing so would have exactly cancelled out their unnecessary compressibility "corrections."

2. Experimental Methods for Measuring Ionization Constants under Pressure

The usual methods of determining ionization constants by conductivity, electromotive force (emf), and spectroscopic measurements can all be applied at high pressures. The conductivity method is the oldest and is still the most convenient and widely used. The emf method is more difficult to use under pressure, although in some ways it is less laborious; its main value has lain in its application to very weak electrolytes. The spectroscopic method is less general than the other two, but has proved useful for some sparingly soluble acids and bases.

(i) Conductivity Method

The techniques of conductivity measurements under pressure have already been described in Section III. It was shown there that the molar conductivities Λ of strong electrolytes in water are not greatly affected by pressure, whereas those of weak electrolytes increase very markedly with increasing pressure. This contrast is shown by the values of Λ for ammonium chloride and ammonia plotted in Fig. 9 and listed in Table 8.

Table 8
Molar Conductances of Strong and Weak Electrolytes in Water at 45°C[a]

Pressure, kbar	0.001	2	4	6	8	10	12
Λ_{NH_4Cl} at $m = 0.0099$ mole kg^{-1}	201	203	191	172	155	138	121
Λ_{NH_3} at $m = 0.0109$ mole kg^{-1}	15.1	36.0	74.1	111	149	178	196
Degree of ionization α of NH_3	0.043	0.10	0.21	0.33	0.45	0.55	0.65
$10^5 K_m$, mole kg^{-1}	2.06	12.4	54	155	359	660	1018
$-\Delta V^\infty$, cm^3 mole^{-1}	28.0	21.2	15.9	12.0	9.4	7.0	5.2
$\Delta G(P$ bar$) - \Delta G(1$ bar$)$, kJ mole^{-1}	0	−4.75	−8.64	−11.4	−13.6	−15.3	−16.4

[a] The units of Λ are Ω^{-1} cm^2 mole^{-1}. The values listed in this table differ slightly from those originally published.[43,61] A correction has been applied for changes in the cell constant of the Teflon conductivity cell caused by high-pressure phase transitions.[69–71]

To estimate the degree of ionization α of a weak electrolyte, say acetic acid HAc, it is necessary to compare the value of its conductivity with the value it would have if it were fully ionized. As a first approximation, the latter can be taken to be the limiting conductivity of the acid at infinite dilution Λ_{HAc}^{∞}. It is impracticable to derive this value by extrapolation, but it can be estimated by applying Kohlrausch's law of independent migration of ions to the limiting conductivities of the strong electrolytes: sodium acetate, sodium chloride, and hydrochloric acid:

$$\Lambda_{HAc}^{\infty} = \lambda_{H^+}^{\infty} + \lambda_{Ac^-}^{\infty} = \Lambda_{NaAc}^{\infty} + \Lambda_{HCl}^{\infty} - \Lambda_{NaCl}^{\infty} \qquad (49)$$

each measured at the same pressure P. The method therefore consists in measuring separately the conductances of solutions of HAc and the related strong electrolytes, each over a range of concentrations and pressures. The values for the strong electrolytes are extrapolated to infinite dilution with the help of a relationship such as (1), and the first approximation to α for a particular molal concentration m of acid is obtained from Ostwald's relationship $\alpha \approx \Lambda/\Lambda_{HAc}^{\infty}$. This value is not exact because the acid solution contains a finite concentration $2\alpha m$ of ions, and Λ_{HAc}^{∞} should be replaced by Λ_{HAc}^{e}, the equivalent conductivity of a hypothetically fully ionized solution of HAc at a concentration αm. That correction can be made by applying equation (49) not to the values of the Λ at infinite dilution, but to those measured at a concentration αm. This gives an improved approximation to α and a new value for αm; the process is repeated until α becomes constant. The molal ionization quotient K_m' [equation (36)] can then be derived from

$$K_m' = \alpha^2 m/(1-\alpha) \qquad (50)$$

and the molal ionization constant K_m [equation (33)] obtained by extrapolating the values of K_m' to infinite dilution, where $\Pi(\gamma)$ is unity.

The above procedure was developed by Davies[128], Sherrill and Noyes,[129] and MacInnes and Shedlovsky[130] and was first applied in high-pressure work by Buchanan and Hamann[65] and Hamann and Strauss.[43,45] In their earlier work Tammann and his collaborators[58,59,115] used only the first approximation to α and did not attempt to extrapolate K_m' to infinite dilution.

The experimental effort can be reduced by using semitheoretical relationships for the extrapolations and interpolations needed to obtain Λ^∞, Λ^e, and K_m. For instance, Robinson and Stokes[110] have shown that if the conductivity Λ^e is assumed to obey equation (4) and the activity coefficients are assumed to obey the Debye–Hückel relationship (5), with $a = 4$ Å, then reliable estimates of K_m can be derived from measurements at only one or two concentrations of acid provided these are less than about 0.01 mole kg^{-1}. To apply the method at high pressures, it is only necessary to adjust the parameters A, B, B_1, and B_2 for the changes that occur in the dielectric constant and viscosity of the solvent when it is compressed (see the values in Table 7 for water).

Apart from weak acids and bases, the conductance method can be used to study the dissociation of ion pairs at high pressures. In

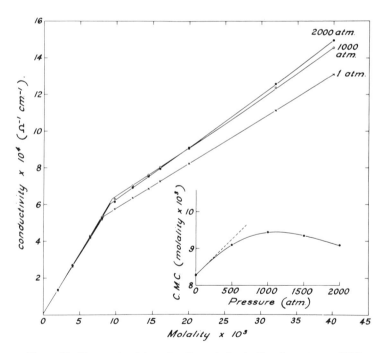

Figure 11. The conductivity of sodium dodecylsulfate in water at 25°C. Inset: the variation of the critical micelle concentration with pressure.

1909 Körber[115] observed that the conductance of a 0.2 N solution of zinc sulfate in water increased by about 40% between 1 and 3000 bars, whereas the conductances of solutions of most other salts increased by only about 10%. Later Tammann and Rohmann[131] interpreted this difference as arising from an increase in the dissociation of $ZnSO_4$, consistent with a dissociation constant of 0.005–0.06 mole kg^{-1} between concentrations of 0.0005 and 0.5 mole kg^{-1}, and with a volume change $\Delta V = -8.2$ cm^3 $mole^{-1}$, suggested by the density data of Hallwachs[132]. More recently Fisher,[81] Hamann et al.,[26] and others (see the appendix, Tables 20 and 23) have carried out quantitative studies at high pressures, applying modern theories of the dependence of conductivities and activity coefficients on ionic strength. The method involves the application of Kohlrausch's law to obtain a value of Λ^∞ and the correction of Λ^∞ to Λ^e by a suitable conductivity equation.

Conductivity measurements also provide a convenient means of studying the influence of pressure on the association of ions into micelles. The formation of micelles of a salt begins fairly abruptly at a "critical micelle concentration" (CMC) of the solute and is accompanied by a more or less sharp decrease in the rate of change of conductivity of the solution with increasing concentration. This is illustrated in Fig. 11, which shows also how the CMC for sodium dodecylsulfate varies with pressure[133] in water at 25°C.

(ii) EMF Method

Because of its experimental simplicity and its accuracy, the electromotive force method is the one normally used nowadays to measure ionization constants at atmospheric pressure. The most accurate results are obtained using hydrogen electrodes to measure the concentrations of hydrogen ions in buffer solutions, in cells without liquid junctions, of the type

$$H_2, Pt|AH, NaA, NaCl|AgCl, Ag \qquad (51)$$

where AH is a weak acid and NaA is its sodium salt.

Unfortunately, hydrogen electrodes are awkward to use at high pressures and their potentials change enormously with pressure simply because of the very large compressibility of gaseous hydrogen. For example, Hainsworth et al.[134] found that the emf of the cell

$$H_2, Pt\,(P\text{ bar})|HCl\,(0.1\text{ N})|Hg_2Cl_2, Hg \qquad (52)$$

increased by 100 mV between 1 and 1000 bars and that virtually the whole of the change could be explained by the increase in fugacity of compressed hydrogen, which tended to mask any more subtle changes in solution. The situation is complicated by the considerable solubility of compressed hydrogen in water and the high partial pressure of water vapor in the saturated gas phase: it is a very difficult theoretical problem to correct for the total change of free energy that occurs when hydrogen is compressed over an aqueous solution.

Hills and Kinnibrugh[135] avoided some of these difficulties by eliminating the gas phase and simply compressing a solution saturated with hydrogen at atmospheric pressure, assuming that the solution contained a constant concentration of dissolved hydrogen. They found that the emf of the above cell then changed by only 8 μV bar^{-1}, corresponding to a net volume change of 7.6 cm^3 mole^{-1} at atmospheric pressure. From the known molar volumes and partial molar volumes of Hg, Hg$_2$Cl$_2$, and HCl they concluded[135] that the partial molar volume of molecular hydrogen in water must be $+8$ cm^3 mole^{-1}. In fact, however, that value should be doubled for molecular hydrogen H$_2$, but even then it is considerably less than the directly measured value[136,137] $V_{H_2} = +25.2$ cm^3 mole^{-1} in water at 25°C, so that the use of this cell must still be considered suspect. More recently Heusler and Gaiser[138] have used a gas-free saturated hydrogen electrode in conjunction with a silver/silver chloride electrode, instead of a calomel electrode, to 2.5 kbar. The absolute emf of the cell was wrong by 2 mV, but the pressure response corresponded to a value $V_{H_2} = 24.5 \pm 0.5$ cm^3 mole^{-1}, in good agreement with the directly measured value.

The problems associated with the use of hydrogen electrodes have been partly by-passed by Distèche's[139a–139e] discovery that it is possible to use a glass electrode under pressure, provided the pressure is allowed to act equally on both sides of the membrane. He found that the electrodes retain their correct hydrogen ion response to at least 1 kbar, and he and Dubuisson[139b] used a glass electrode to measure the pH of sea water at great depths. He later showed in laboratory experiments[139c–e] that cells with glass electrodes give values for the ionization constants of weak acids and bases that agree well with those from high-pressure conductivity measurements. They have the great advantage over the conductance

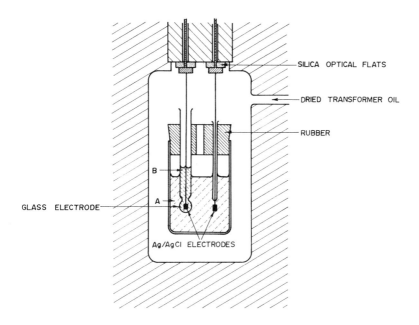

Figure 12. An emf arrangement for measuring the degree of ionization of water under pressure.

method of giving reliable values for the ionization constants of very weak acids and bases, such as boric acid and water itself. Figure 12 shows an experimental arrangement which the writer[140] has used to measure the value of the ionization constant K_w of water to 2 kbar. The cell was constructed as follows:

$$\text{Ag, AgCl} | \text{HCl (0.1 N)} | \text{glass} | \text{KCl (0.1 N)} + \text{KOH} | \text{AgCl, Ag} \qquad (53)$$
$$\phantom{\text{Ag, AgCl} | \;\;} B \phantom{\text{HCl (0.1 N)} | \text{glass} | \text{KCl (0.1 N)} + \text{KO}} A$$

where the molality of KOH was varied between 0.001 and 0.02. The emf of this cell is given by

$$\mathscr{E} = -\frac{RT}{F} \ln K_w + \frac{RT}{F} \ln m_{H^+(A)} m_{OH^-(B)}$$
$$+ \frac{RT}{F} \ln \left(\frac{\gamma_{H^+(A)} \gamma_{Cl^-(A)} \gamma_{OH^-(B)}}{\gamma_{Cl^-(B)}} \right) \qquad (54)$$

where F denotes the Faraday constant and the subscripts A and B refer to solutions outside and inside the glass electrode, respectively. The ionic strengths of the two solutions were almost the same and it was therefore justifiable to assume that all the activity coefficients were of about the same magnitude. To that approximation,

$$\mathscr{E} = -(RT/F)\ln K_w' + (RT/F)\ln m_{H^+(A)}m_{OH^-(B)} \quad (55)$$

where K_w' is the ionization product $m_{H^+}m_{OH^-}$ [cf. equation (36)]. The experimental results at 25°C and an ionic strength of 0.1 mole dm^{-3} (at atmospheric pressure) are given in Table 9.

Table 9

Pressure P, bars	1	500	1000	1500	2000
$K_w'(P\text{ bar})/K_w'$ (1 bar)	1.00	1.48	2.12	2.91	3.79
$K_w(P\text{ bar})/K_w$ (1 bar)	1.00	1.50	2.17 (2.20 ± 0.02)	3.00	4.14 (4.19 ± 0.25)

The values listed in the third row have been corrected by activity coefficients given by the Debye–Hückel equation (5), using the high-pressure parameters A and B in Table 7. The figures in parentheses were estimated indirectly by Kearns[141] from compressibility data for aqueous solutions. Whitfield[142] and Linov and Kryukov[143] have recently repeated the high-pressure measurements and obtained values of K_w' that agree very well with the above; Linov and Kryukov's data extend to 7.85 kbar, where $K_w(P\text{ bar})/K_w(1\text{ bar}) = 51.3$.

(iii) Spectroscopic Method

If the parent molecule AH and the ion A$^-$ of a weak acid have sufficiently different optical absorption spectra, the ionization constant can be estimated spectroscopically in the following way. The optical density D for unit path length at a particular wavelength is

$$D = (1-\alpha)\varepsilon_{AH}c + \alpha c\varepsilon_A \quad (56)$$

where c is the total concentration of the electrolyte, α is its degree of dissociation, and ε_{AH} and ε_{A^-} are the molar extinction coefficients of its two forms at the same wavelength as D. It follows that

$$\alpha = (D - \varepsilon_{AH}c)/(\varepsilon_{A^-} - \varepsilon_{AH})c \tag{57}$$

If the measurements are made in buffered solutions for which the hydrogen ion activity $m_{H^+}\gamma_{H^+}$ is known, then the ionization constant is given by

$$K_m = [\alpha\gamma_{A^-}/(1 - \alpha)]m_{H^+}\gamma_{H^+} \tag{58}$$

where m denotes the total molality of the acid. The method is particularly useful for sparingly soluble electrolytes since the ε are often of the order of 10^4 dm^3 mole^{-1}, so that c need only be of the order of 10^{-4} mole dm^{-3} to give optical densities near unity.

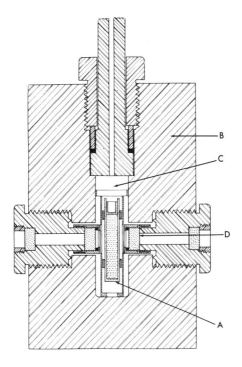

Figure 13. A high-pressure optical cell for spectroscopic determination of ionization constants.

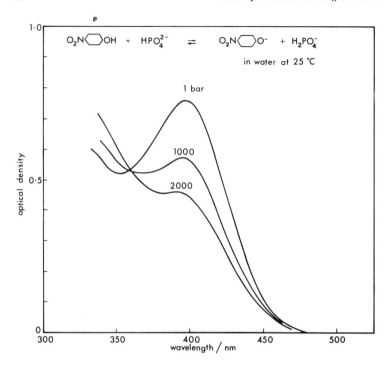

Figure 14. Visible and ultraviolet spectra of 4-nitro-phenol in phosphate buffer in water at 25°C. The number on each curve indicates the pressure in bars.

Hamann and Linton[144] have found that the method can be applied quite readily under pressure if the solution is contained in a temperature-controlled high-pressure optical cell of the type shown in Fig. 13. Here, the square silica cell (A) containing the aqueous solution is mounted in a pressure vessel (B) and surrounded by an optically clear and immiscible pressure fluid (C), such as a light paraffin oil. It is viewed through optically flat silica windows (D). The solution may either be in direct contact with the pressure fluid or, if mutual mixing presents a problem, separated from it by a sliding plug.

Some typical results are shown in Fig. 14 for the ionization of 4-nitro-phenol in an aqueous phosphate buffer solution. An increase of pressure lowers the absorption at 396 nm, due to the anions

Ionization Equilibria under Pressure 93

of the phenol, and raises the absorption at 295 nm, due to the parent molecules. In other words, it shifts the equilibrium shown at the top of Fig. 14 to the left, which means that compression increases the acidity of the phosphate ion $H_2PO_4^-$ more than it does that of the phenol. The experiments themselves are not particularly difficult but interpretation of the results is complicated by the fact that if a buffer is used, it is necessary to know how pressure affects the pH of the buffer solution, that is, how it affects the term $m_{H^+}\gamma_{H^+}$ in (58): it is not sufficient just to determine α. In the case of a few relatively strong acids such as picric acid ($K_m = 0.38$ mole kg^{-1}) it is possible to fix the pH so that it is nearly independent of pressure by using hydrochloric acid as the "buffer," but more generally it is necessary to use acetate, phosphate, or carbonate buffers. Fortunately the pressure dependences of the ionization constants of acetic acid and the ions $H_2PO_4^-$ and HCO_3^- have been measured by conductivity and emf methods, and the necessary corrections for acetate buffer can be applied in the form

$$m_{H^+}\gamma_{H^+} = K_m(\text{HAc})m_{\text{HAc}}/m_{\text{Ac}^-}\gamma_{\text{Ac}^-} \quad (59)$$

If this value is substituted in (58), it follows that since m_{HAc} and m_{Ac^-} are virtually independent of pressure, the pressure derivative of K_m is

$$-\frac{\partial(RT \ln K_m)}{\partial P} = -\frac{\partial[RT \ln K_m(\text{HAc})]}{\partial P} - \frac{\partial\{RT \ln[\alpha/(1-\alpha)]\}}{\partial P}$$

$$-\frac{\partial[RT \ln(\gamma_A-/\gamma_{\text{Ac}^-})]}{\partial P} \quad (60)$$

and since $\gamma_A-/\gamma_{\text{Ac}^-}$ is probably close to unity and independent of pressure, (60) can be written

$$\Delta V^\infty = \Delta V^\infty(\text{HAc}) - \partial\{RT \ln[\alpha/(1-\alpha)]\}/\partial P \quad (61)$$

so that the experimentally determined relationship between α and P leads directly to knowledge of the behavior of K_m for the acid AH being investigated, in terms of the behavior of $K_m(\text{HAc})$ of the acetic acid buffer.

The situation is not quite so straightforward for phosphate and bicarbonate buffers, since their ionization leads to the formation of doubly charged ions and the term γ_A/γ_{Ac^-} in (60) is replaced by a factor $\gamma_A \cdot \gamma_{HCO_3^-}/\gamma_{CO_3^{2-}}$, which is not necessarily close to unity (since γ_i is determined by the square of the ionic charge) or independent of pressure. In these circumstances it is necessary to use the Debye–Hückel theory to estimate the pressure dependence of the activity coefficient factor.

The spectroscopic method has been applied also in high-pressure measurements of the dissociation of ion pairs[28] and the formation and dissociation of ion complexes of various kinds.[145–147]

3. Discussion of Results

The appendix to this article contains a critical compilation of the results of measurements of ionization equilibria at high pressures. It lists values of the volume changes for ionization reactions derived both from the high-pressure data using equation (35) and from density and volumetric measurements made at atmospheric pressure. It covers the literature published before October 1972.

It will be seen from the tables that over 100 ionization reactions have been investigated under pressure and that values of ΔV^∞ at atmospheric pressure are known for about another 100 from density and dilatometric measurements. The results will be discussed here in the following order:

(i) Ionization of neutral molecules, for example,

$$CH_3CO_2H \rightleftarrows CH_3CO_2^- + H^+$$

$$NH_3 + H_2O \rightleftarrows NH_4^+ + OH^-$$

$$H_2O \rightleftarrows H^+ + OH^-$$

(ii) Ionization of charged molecules, for example, ions of dibasic and polybasic acids and of amino acids:

$$HCO_3^- \rightleftarrows CO_3^{2-} + H^+$$

$$H_3\overset{+}{N}CH_2CO_2^- \rightleftarrows H_2NCH_2CO_2^- + H^+$$

(iii) Ionization of molecules in excited states, for example,

$$\text{2-naphthol}^* \rightleftarrows \text{2-naphthoxide}^{*-} + H^+$$

(iv) Dissociation of associated ions, for example, micelles and ion pairs:

$$Mg^{2+}SO_4^{2-} \rightleftarrows Mg^{2+} + SO_4^{2-}$$

(i) Ionization of Neutral Molecules

The data in appendix Tables 17–19 and 22 show that *in every case* the ionization of an electrically neutral acid or base in solution involves a contraction and is enhanced by an increase of pressure. A few typical results are shown in Fig. 15, where the logarithm of the ratio $K_m(P \text{ bar})/K_m(1 \text{ bar})$ is plotted against the pressure. The plots are distinctly curved, showing that $|\Delta V^\infty|$ decreases with increasing pressure (see also Table 8).

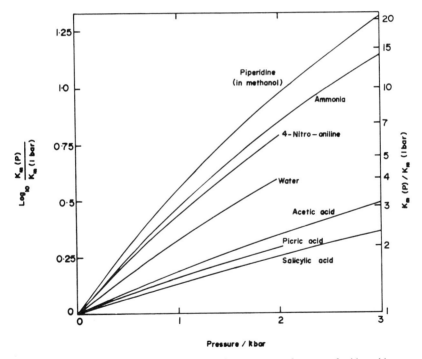

Figure 15. The effect of pressure on the ionization constants of a range of acids and bases in water at 25°C. The data are taken from appendix Tables 17, 18, and 22.

The explanation of the universal contraction lies in the strong interactions that exist between charged ions and the surrounding solvent. In an ionization reaction a neutral molecule is replaced by two ions whose electrical fields strongly attract the dipolar solvent molecules and compress them locally to a higher density than they had around the parent molecule. A thermodynamically equivalent way of describing the situation is to say that pressure, by raising the dielectric constant of the solvent (Section II), stabilizes the free ions with respect to the neutral molecule. To put this idea on a quantitative basis, Drude and Nernst[123] considered a simple model in which the solvent was treated as a continuum of dielectric constant ε and the ions were assumed to occupy spherical cavities of radius r. They concluded that the volume change ΔV_{el} (electrostriction) associated with the development of a charge ze on an ion is

$$\Delta V_{\text{el}} = (NV\kappa z^2 e^2/2r\varepsilon^2)\, \partial\varepsilon/\partial V \tag{62}$$

where N denotes Avogadro's constant, V is the volume of the solvent, and $\kappa = -(\partial V/\partial P)/V$ is its compressibility. The assumptions and approximations that underlie (61) are precisely those which Born[124] made in deriving his formula for the free energy of solvation of ions,

$$\Delta G_{\text{el}} = -(Nz^2 e^2/2r)[1 - (1/\varepsilon)] \tag{63}$$

and Krichevskii[151] rediscovered (62) in an equivalent form

$$\Delta V_{\text{el}} = -(Nz^2 e^2/2r\varepsilon^2)\partial\varepsilon/\partial P \tag{64}$$

by differentiating (63) with respect to pressure, assuming that only ε is pressure dependent. Later, the writer[65,61] introduced an additional term to allow for the change of r with pressure, that is, to allow for the compressibility of the ions:

$$\Delta V_{\text{el}} = -\frac{Nz^2 e^2}{2r\varepsilon^2}\frac{\partial\varepsilon}{\partial P} + \frac{Nz^2 e^2}{2r^2}\left(1 - \frac{1}{\varepsilon}\right)\frac{\partial r}{\partial P} \tag{65}$$

Other refinements to the original model have been discussed by Whalley,[148] Benson and Copeland,[149] and Desnoyers et al.[150]

When formula (65) is applied[61] to singly charged ions ($z = 1$, e = electronic charge) in water at atmospheric pressure and 25°C it yields a value $\Delta V_{\text{el}} \approx -10 \text{ cm}^3 \text{ mole}^{-1}$ and it follows that the dissociation of a neutral molecule into two ions should theoretically

Ionization Equilibria under Pressure

involve a contraction of about $-20 \text{ cm}^3 \text{ mole}^{-1}$. This is certainly of the same order as the values of ΔV^∞ given in the appendix tables. And, of course, the theory is not limited to low pressures; equation (63) can be applied at any pressure where ε is known and r can be estimated. It was used in this way by Hamann and Strauss[43,61] to estimate theoretically how the standard free energy of ionization ΔG of ammonia in water might vary with the pressure between 1 and 12,000 bars. Their results are compared in Fig. 16, curve A, with the experimental values taken from Table 8 ($\Delta G = -RT \ln K_m$). There is fair agreement, and the theoretical curve shows a decrease of (negative) slope with increasing pressure, corresponding to a decrease in the magnitude of $|\Delta V_{el}|$. This helps to explain the curvature of the plots in Fig. 15: at higher pressures the electrostriction

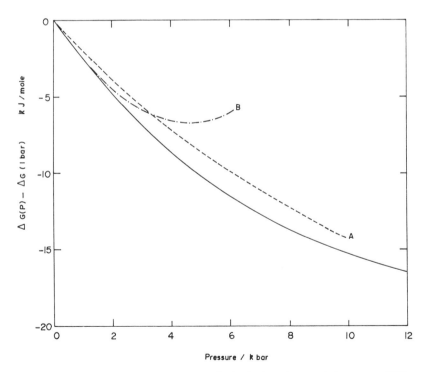

Figure 16. The standard free-energy change for ionization of ammonia in water at 45°C. The curves A and B are discussed in the text.

decreases because the solvent and ions become less compressible and a given increase in pressure produces a proportionally smaller increase in dielectric constant and decrease in ionic radius. Here it should be mentioned that a number of *empirical* equations have been proposed to describe the behavior of ΔG at high pressures. Owen and Brinkley[152] attempted to calculate its change from the value of ΔV^∞ at atmospheric pressure and the pressure dependence of ΔV^∞, which they estimated by using the Tait equation to describe the compressibilities of aqueous solutions. They calculated how pressure should change the ionization constants of water, acetic acid, and carbonic acid, but their predictions later proved to be too high for water and carbonic acid and too low for acetic acid (compare their Tables 4 and 5 with appendix Table 17). It appears that the volume and compressibility data that they used were unreliable and a more recent application of the method[141] has given much better results. Lown et al.[153] have found that the simple quadratic equation

$$RT \ln[K_m(P \text{ bar})/K_m(1 \text{ bar})]$$
$$= -\Delta V_1^\infty (P - 1) + \tfrac{1}{2} \Delta \kappa_1^\infty (P - 1)^2 \qquad (66)$$

gives a fairly good description of the behavior of acids and bases to 2000 bars. Here $\Delta \kappa_1^\infty$ is the change of "partial molar compressibility" defined by

$$-\Delta \kappa_1^\infty = \partial(\Delta V_1^\infty)/\partial P \qquad (67)$$

In these equations the subscript 1 denotes values at $P = 1$ bar. Lown et al. did not obtain $\Delta \kappa_1^\infty$ from independent measurements but simply adjusted its value to fit the experimental high-pressure data for K_m.

El'yanov and Gonikberg[154] had previously proposed a rather similar relationship

$$RT \ln[K_m(P \text{ bar})/K_m(1 \text{ bar})] = -\Delta V_1^\infty \Phi \qquad (68)$$

where Φ is an empirical function of the pressure [Φ in (68) differs from El'yanov and Gonikberg's Φ by the factor $R \ln 10$]. Interestingly, it seems to be a universal function for a range of weak electrolytes and, in terms of equation (66), this implies that $\Delta \kappa_1^\infty$ is proportional to ΔV_1^∞. Lown et al.[153] did in fact observe a linear correlation between the two quantities, with $\Delta \kappa_1^\infty \approx 2.14 \times 10^{-4} \Delta V_1^\infty$ bar^{-1}, and they suggested that a generally useful relationship for predicting pressure effects to 2000 bars might therefore be

$$RT \ln[K_m(P \text{ bar})/K_m(1 \text{ bar})] = -P\Delta V_1^\infty (1 - 1.07 \times 10^{-4} P)$$
$$= \Delta G(1 \text{ bar}) - \Delta G(P \text{ bar}) \tag{69}$$

Figure 16, curve B, shows that this is a good representation of the experimental results for the ionization of ammonia to about 2000 bars, but it is unsafe to use it to extrapolate beyond that limit.

It is significant that a linear relationship exists between the theoretical changes of volume and compressibility for electrostriction derived from Born's free-energy equation (63) assuming that ε but not r is pressure dependent. The volume change ΔV_{el} is given by (64) and the corresponding compressibility change is

$$\Delta \kappa_{el} = -\frac{\partial \Delta V_{el}}{\partial P} = -\frac{Nz^2 e^2}{r\varepsilon^3}\left(\frac{\partial \varepsilon}{\partial P}\right)^2 + \frac{Nz^2 e^2}{2r\varepsilon^2}\frac{\partial^2 \varepsilon}{\partial P^2} \tag{70}$$

$$= \Delta V_{el}\left(\frac{2}{\varepsilon}\frac{\partial \varepsilon}{\partial P} - \frac{\partial^2 \varepsilon}{\partial P^2}\frac{\partial P}{\partial \varepsilon}\right) \tag{71}$$

Inserting the values of ε, $\partial \varepsilon/\partial P$, and $\partial^2 \varepsilon/\partial P^2$ measured by Owen et al.[19] gives

$$\Delta \kappa_{el} = 1.99 \times 10^{-4} \Delta V_{el} \text{ bar}^{-1} \tag{72}$$

for water at atmospheric pressure at 25°C. The numerical factor here is remarkably close to the empirical one, 2.14×10^{-4}, and is about 60% larger than a value which Lown et al.[155] calculated neglecting the second term in (71).

The theoretical equations for electrostriction contain the derivative $(\partial \varepsilon/\partial P)/\varepsilon^2$, which varies considerably for different solvents and has the values listed in Table 2 for a few common liquids. It is several times greater for methanol than for water and for that reason it might be expected that ΔV^∞ would be more negative and the pressure effects larger for solutions in methanol than for ones in water. That prediction is borne out by the experimental values* of ΔV^∞ given in Table 10 for the two solvents.

*From here on any data that are quoted without references have been taken from the tables in the appendix. In this case they are from appendix Tables 17, 18, and 22.

Table 10

	ΔV^∞, cm^3 mole^{-1}	
	Water	Methanol
Phenol	−18.4	−38.5
4-Nitro-phenol	−10.0	−31.7
Picric acid	−10.0	−25
Piperidine	−26.8	−53
Pyridine	−24.2	−49

The equations also contain the radius r of the ions (or, more appropriately in a polyatomic ion, the radius of the particular atom or group that carries the bulk of the electric charge) in the denominator, so that ΔV^∞ should be less negative for electrolytes that form large ions than for ones that form small ones. This effect is to be seen in the experimental results for the ionization of sulfur-containing compounds on the one hand and analogous oxygen-containing compounds on the other (ΔV^∞ in cm^3 mole^{-1})

H_2S −16.3, H_2O −22.1, difference +5.8

C_6H_5SH −12.5, C_6H_5OH −18.4, difference +5.9

However, it does not always appear where it might be expected, for example, it is not apparent in the series of methylamines from ammonia to trimethylamine. There are evidently more subtle structural effects operating here than can be handled by a simple continuum theory.[242,245]

An effect which is similar to that of ion size arises from charge delocalization in ions. The carboxylic acids provide examples of this effect. If the negative charge in a carboxylate ion were located on one of the two oxygen atoms, it would be expected that ΔV^∞ for these acids would be close to the values for water and phenol, but in fact it is only about half those. The reason is that the charge is distributed equally between the two oxygen atoms and the effective radius of the charged group is larger than if the charge were wholly on one atom. Similarly, the small volume change for the ionization of HCN may be partly explained by the fact that the charge is shared about equally by the C and N atoms in the CN^- ion.[156] There are other clear examples of this effect in the following series of phenols in water (ΔV^∞ in cm^3 mole^{-1}):

water	phenol	2-nitro-phenol	4-nitro-phenol
−22.1	−18.4	−12.4	−10.0

Here the electron-withdrawing effect of the aromatic ring in the anion is greatly enhanced by the presence of a nitro group through charge transference of the type represented by the valence bond structures

$$\bar{O}-\!\!\left\langle\;\right\rangle\!\!-NO_2 \qquad O=\!\!\left\langle\;\right\rangle\!\!=\bar{N}O_2$$

The same effect of nitro groups is apparent in substituted naphthols and benzoic acids.

An alternative and in some cases a more satisfactory way of viewing charge delocalization is to split the electrostriction into parts arising from separated partial charges which act separately on their surroundings. If, in the example just considered, it were supposed that the O− and −NO$_2$ groups each carried half of the total negative charge and that their radii were effectively the same, the Born electrostrictive effect would be the sum of two terms containing $e^2/4$; that is, it would be proportional to $e^2/2$ instead of to e^2 as it would if all the charge were on one group. This approach is particularly appropriate when the charged centers are a large distance apart.

Some of the ionization reactions listed in appendix Table 17 involve specific chemical hydration as distinct from general electrostatic hydration, the distinction being that a water molecule becomes bound into the anion covalently and not just electrostatically. An example is "carbonic acid," which exists in the ionized form as HCO$_3^-$ ions. Ellis[157] has suggested that this type of hydration involves an unusually large contraction, which explains the large negative value of ΔV^∞ for the first ionization step of carbonic acid. The present author suggests that it also explains the large pressure effect for boric acid. The volume changes concerned are

$$CO_2 + H_2O \rightleftarrows HCO_3^- + H^+, \qquad \Delta V^\infty = -27.2 \text{ cm}^3 \text{ mole}^{-1}$$

$$B(OH)_3 + H_2O \rightleftarrows B(OH)_4^- + H^+, \qquad \Delta V^\infty = -31.8 \text{ cm}^3 \text{ mole}^{-1}$$

which are to be compared with

$$H_3PO_4 \rightleftarrows H_2PO_4^- + H^+, \quad \Delta V^\infty = -16.4 \text{ cm}^3 \text{ mole}^{-1}$$

$$CH_3CO_2H \rightleftarrows CH_3CO_2^- + H^+, \quad \Delta V^\infty = -11.5 \text{ cm}^3 \text{ mole}^{-1}$$

The effect is almost as great as if one water molecule (volume 18.0 cm^3 mole^{-1}) has been completely removed from the system in addition to the electrostrictive contraction of the remainder of the solvent.

The discussion so far has been oversimplified in that it has considered only the solvation or electrostriction of the ions that are formed in an ionization reaction. It is certainly not always justifiable to ignore the interactions between the parent molecules and the solvent, which may be quite strong and are by no means the same for different solute molecules. The way in which these interactions may influence ΔV^∞ can be seen from the volume changes ΔV_s that occur when liquid weak electrolytes dissolve in water in their neutral forms. For some simple molecules the values of ΔV_s and ΔV^∞ at 1 bar and 25°C are given in Table 11. At least part of the decreasing trend of $|\Delta V^\infty|$ for these molecules must be associated with the increasing compactness in water of the un-ionized states.

Table 11

	H_2O	NH_3	H_2S	HF	HCN
ΔV_s, cm^3 mole^{-1}	0	-3.3	-5.4	-7.4	-8.3
ΔV^∞, cm^3 mole^{-1}	-22.1	-28.5	-16.3	-13.7	-7.0

(ii) Ionization of Charged Molecules

The presence of the factor z^2 in the expressions for electrostriction means that the successive ionization steps of a polybasic acid should involve increasingly large decreases in volume. This effect is to be seen in appendix Tables 17 and 18, for example, in the data for phosphoric, oxalic, maleic, and citric acids and for the base ethylenediamine.

Bjerrum[158] proposed a simple model for the ionization of polybasic acids which is appropriate when the ionizing groups are structurally identical. He suggested that the relationship between the first and second ionization constants K_1 and K_2 of a dibasic acid

Ionization Equilibria under Pressure

might then be

$$K_1/K_2 = 4\exp(e^2/\varepsilon akT) \quad (73)$$

or

$$\ln K_1 - \ln K_2 - \ln 4 = e^2/\varepsilon akT \quad (74)$$

where 4 is a statistical factor (it becomes 3 for a tribasic acid) and the term in parentheses in (73) is the electrostatic free-energy change involved in bringing a second negative charge e from infinity to a distance a from the first charge in the molecule, a is the distance between the two ionizing groups; ε is the dielectric constant, k is Boltzman's constant, and T is the absolute temperature. To apply Bjerrum's model to pressure effects, (74) can be differentiated with respect to pressure to give

$$\Delta V_1^\infty - \Delta V_2^\infty = RT(e^2/\varepsilon^2 akT)\,\partial\varepsilon/\partial P \quad (75)$$

$$= RT(\ln K_1 - \ln K_2 - \ln 4)\,\partial(\ln\varepsilon)/\partial P \quad (76)$$

$$= RT(\ln 10)(pK_2 - pK_1 - \log_{10}4)\,\partial(\ln\varepsilon)/\partial P \quad (77)$$

where $pK = -\log_{10}K$. Inserting the value of $\partial(\ln\varepsilon)/\partial P$ for water at 1 bar and 25°C from Owen et al.[19] then gives the numerical result

$$\Delta V_1^\infty - \Delta V_2^\infty = 2.69(pK_2 - pK_1 - \log_{10}4) \quad \text{cm}^3\,\text{mole}^{-1} \quad (78)$$

which suggests that ΔV_2^∞ should be more negative than ΔV_1^∞ by an amount which is proportional to the difference in the two pK values for the acid. Figure 17 shows that this is more or less the case. The straight line in the diagram has the theoretical slope of 2.69 cm^3 mole^{-1}. Carbonic acid lies well below the line, but that is because its first ionization step involves specific chemical hydration (see the preceding section).

The same type of electrostatic theory can be applied to the second ionization of ions of amino acids to give zwitter ions, except that now the sign of the electrostatic free energy change is reversed and ΔV_2^∞ is less negative than it would be if the first charge were not there. The fields of the two charges cancel at a distance from the zwitter ion and z in the Born equation is zero for the whole ion. The Bjerrum theory can again be applied, but because the two ionizing groups are structurally different, the term $\log_{10}4$ no longer appears in (78) and ΔV_1^∞ and pK_1 must be replaced by ΔV_r^∞ and

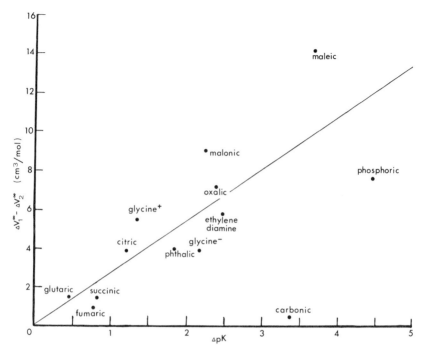

Figure 17. The difference between ΔV^∞ for the first and second ionizations of polyprotic acids and bases. The scale ΔpK is $pK_2 - pK_1 - \log_{10} 4$ for the diprotic acids and bases and $pK_2 - pK_1 - \log_{10} 3$ for phosphoric and citric acids. For the glycine ions (see text) $\Delta V_2^\infty - \Delta V_r^\infty$ has been plotted against $pK_r - pK_2$.

pK_r for a reference acid or base which is electrically neutral but is otherwise structurally similar to the acid or base which ionizes in the second stage of the amino acid. For instance, the ionization of the glycine cation may be compared with the ionization of acetylglycine:

$$NH_3{}^+CH_2CO_2H \rightarrow NH_3{}^+CH_2CO_2{}^- + H^+,$$

$$\Delta V_2^\infty = -6.8 \text{ cm}^3 \text{ mole}^{-1}$$

$$CH_3CONHCH_2CO_2H \rightarrow CH_3CONHCH_2CO_2{}^- + H^+,$$

$$\Delta V_r^\infty = -12.3 \text{ cm}^3 \text{ mole}^{-1}$$

and that of the glycine anion with that of ethyl glycine ester:

$$NH_2CH_2CO_2^- + H_2O \rightarrow NH_3^+CH_2CO_2^- + OH^-,$$
$$\Delta V_2^\infty = -23.4 \text{ cm}^3 \text{ mole}^{-1}$$

$$NH_2CH_2CO_2C_2H_5 + H_2O \rightarrow NH_3^+CH_2CO_2C_2H_5 + OH^-,$$
$$\Delta V_r^\infty = -27.3 \text{ cm}^3 \text{ mole}^{-1}$$

These results are plotted in Fig. 17.

It is apparent from the appendix tables and from Fig. 17 that for *simple* polyprotic acids and bases there are no exceptions to the general trend predicted by electrostatic theory that ΔV^∞ should become more negative for successive ionization steps. The same trend has been established by density measurements for weak polyelectrolytes. Appendix Table 19 shows that $\Delta V(b)$ becomes more negative as the degree of ionization α increases. It is therefore disturbing to find that Wall and Gill,[159] who made the first high-pressure measurements on polyelectrolytes, reported a trend in the opposite direction for polyacrylic acid. ΔV was apparently *less* negative by about 7 cm^3 mole^{-1} at $\alpha = 0.06$ than $\alpha = 0$, in contradiction both of electrostatic theory and of Begala and Strauss's[160] volumetric measurements. If their conclusion were correct, it would imply that there is something wrong with Planck's relationship (35), but in fact this is fallacious. Apart from mistakenly differentiating $\ln K_c$ instead of $\ln K_m$ with respect to pressure [see the discussion of equation (46); this error alone does not account for the discrepancy], Wall and Gill wrongly assumed that they could calculate the volume change for ionization from a supposed relationship

$$\Delta V = -\partial\{RT \ln[\alpha^2 c/(1-\alpha)]\}/\partial P \quad (79)$$

and they referred to the quantity in square brackets as an "effective" ionization constant. For a polyelectrolyte it is actually a composite quantity involving a very large number of ionization constants in a complex way. It is true that Wall and Gill found experimentally that the derivative (79) becomes less negative as α increases, and the same trend appears in the more recent results of Suzuki and Taniguchi[161] and Eldridge and Treloar,[162] but that does not mean that the volume change for ionization also becomes less negative. For a simple dibasic acid it can be shown theoretically that the derivative (79) initially becomes less negative than ΔV_1^∞ as α increases, even when

ΔV_2^∞ is more negative than ΔV_1^∞. In short, $\Delta V_1^\infty, \Delta V_2^\infty, \cdots$ can only be derived from high-pressure measurements if $K_1, K_2, \cdots$ can all be determined separately by a full analysis of all the species in the system. Differentiating the logarithm of an "effective" ionization constant gives only a fictitious "volume."

(iii) Excited States of Molecules

The electronically excited states of phenols and anilines are highly dipolar. Molecular orbital calculations[163] on 4-nitrophenol suggest that its lowest excited state is a charge-transfer one resembling the valence bond structure

$$O_2\bar{N}=\langle\ \rangle=\overset{+}{O}H$$

and this conclusion is supported by experimental evidence[164] that the dipole moment of the analogous excited state of 4-nitroaniline is about 14 Debye units, corresponding to full positive and negative charges separated by 2.9 Å. It also explains why the acid strength of the excited state is very much greater than that of the ground state; the *pK* is less by about eight units. The excited state is likely to be strongly hydrated in water and, since some delocalization of charge may occur when the anion is formed by ionic dissociation, it is to be expected that the increase of electrostriction accompanying dissociation will be small and there may even be a decrease.

There is some evidence that this is the case. The writer[165] has observed marked shifts in the long-wavelength absorption bands of nitrophenols and their anions; the frequency decreases with increasing pressure for the phenols and increases for their anions. By applying the Förster cycle to these results in the way that Weller[166] had done at atmospheric pressure, he found qualitative evidence that the values of $\Delta^* V^\infty$ for the excited states of nitrophenols are small in comparison with the values ΔV^∞ for the ground states and may even be positive instead of negative.

This conclusion was only qualitative because of the limitations of the Förster method imposed by the Franck–Condon principle. But the writer has since obtained more direct evidence by examining the fluorescence intensity of 1-naphthol (the nitrophenols do not fluoresce) in hydrochloric acid solutions in water as a function of the pH between zero and five. There is no detectable change between 1 and 2000 bars and, unless there is an unlikely cancellation of effects, this means that $|\Delta^*V^\infty|$ is less than 5 cm^3 mole^{-1}, whereas ΔV^∞ for the ground state is -18 cm^3 mole^{-1}. Also, Leiber et al.[167] have examined the fluorescence behavior of 2-naphthol and concluded that Δ^*V^∞ is slightly less negative than -6.3 cm^3 mole^{-1}, compared with $\Delta V^\infty = -17.7$ for the ground state.

The behavior of these excited phenols is analogous to that of zwitter ions, for which the loss of a proton generally involves only a small volume change, which may be positive:

$$NH_3{}^+CH_2CO_2{}^- \rightarrow NH_2CH_2CO_2{}^- + H^+,$$

$$\Delta V^\infty = +1.3 \text{ cm}^3 \text{ mole}^{-1}$$

(iv) Dissociation of Associated Ions

The general effects of compression on the dissociation of ion pairs and ion complexes are of the type to be expected from simple considerations of changes in electrostriction. The separation of two oppositely charged ions in solution involves an increase in the total field strength and in the electrostriction of solvent and so is favored by an increase of pressure. The first experimental evidence of such a change is to be found in the early measurements of Fink,[55] which showed that the electrical conductances of solutions of $ZnSO_4$ increased almost twice as rapidly with increasing pressure as the conductances of NaCl and HCl solutions. Fink's results were later confirmed by Körber[115] and subsequently interpreted by Tammann and Rohmann[131] as implying an increase in the dissociation of $ZnSO_4$ (see p. 87). More recent work has shown that the magnitude of the effect can shed light on the question of whether a particular ion pair is of the "solvent separated" or "contact" type.

The experimental results can be understood quantitatively in terms of Fuoss'[168] theory of ion-pairing reactions. Assuming that ions with opposite charges $z_1 e$ and $z_2 e$ can be considered to form ion pairs when their centres come within a contact distance a of each other, Fuoss derived the dissociation constant for ion pairs in the form

$$K_m = 3000[\exp(z_1 z_2 e^2/a\varepsilon kT)]/4\pi\rho Na^3 \qquad (80)$$

where ρ denotes the density of the solvent and the other symbols have the same meaning as in earlier parts of this section. If it is assumed that a is independent of pressure, the pressure effect can be derived in the form

$$-\frac{\partial (RT \ln K_m)}{\partial P} = \Delta V^\infty = \frac{\partial (RT \ln \rho)}{\partial P} + \frac{z_1 z_2 e^2 N}{a\varepsilon^2} \frac{\partial \varepsilon}{\partial P} \quad (81)$$

and for water at 1 bar and 25°C

$$\Delta V^\infty = 1.12 + 8.35 z_1 z_2/a(\text{Å}) \quad \text{cm}^3 \text{mole}^{-1} \quad (82)$$

There is no a priori way of determining the parameter a, but if it is derived by applying (80) to the measured value of K_m at 1 bar, then (80) and (81) can be used to predict the influence of pressure on the equilibrium. It turns out that they do so with remarkable accuracy. Hamann et al.[26] used (82) to calculate the values $\Delta V^\infty = -7.4$ and $-9.0 \text{ cm}^3 \text{mole}^{-1}$ for ion pairs of $Mg^{2+}SO_4^{2-}$ and $La^{3+}Fe(CN)_6^{3-}$, respectively, in water at 1 bar and 25°C: the experimental values[26] are -7.3 and $-8.0 \text{ cm}^3 \text{mole}^{-1}$. Moreover, (80) gives a good description of the behavior of K_m for $La^{3+}Fe(CN)_6^{3-}$ at high pressures, to at least 2 kbar.[26]

The values of ΔV^∞ quoted in the last paragraph are surprisingly similar for the two salts despite their different valence types. The reason is that a is greater by 3.5 Å for $La^{3+}Fe(CN)_6^{3-}$ than for $Mg^{2+}SO_4^{2-}$ and, in (82), this difference compensates for the increase in the factor $z_1 z_2$. The difference in a is large enough to suggest that the ion pairs of $La^{3+}Fe(CN)_6^{3-}$ contain hydrated lanthanum ions and are "solvent separated." The smallness of the values of ΔV^∞ in comparison with the corresponding values for the ionization of weak acids and bases shows that the ions are extensively hydrated in the ion-pair state. By contrast, some of the other ion pairs listed in appendix Table 20 have quite large values of $-\Delta V^\infty$ and the differences between, say, $La^{3+}SO_4^{2-}$ and $La^{3+}Fe(CN)_6^{3-}$ and between $Ce^{3+}C_2H_5CO_2^-$ and $Ce^{3+}NO_3^-$ suggest that the pressure dependence of K_m may be a useful indication of whether or not ion pairs are separated by solvent molecules.

When the associating ions have electric charges of the same sign, both terms in (81) are positive and ΔV^∞ becomes positive. The total amount of electrostriction reduces when the charges separate. This conclusion is confirmed by the results of Suzuki and Tsuchiya[169] for the dye ions listed in appendix Table 20.

The formation of micelles in solutions of ionic detergents is an extreme example of ion association. It is generally agreed that the micelles are large clusters of similarly charged ions surrounded by an almost equal number of oppositely charged ions. There is presumably some desolvation of the ions when they aggregate and the accompanying increase in volume should make it more difficult to form micelles when the solutions are compressed. In agreement with this idea, the data in appendix Table 21 (see also Fig. 11, inset) show that the critical micelle concentrations of a number of long-chain salts in water rise with increasing pressure below 1 kbar. But surprisingly, and universally, they begin to drop at higher pressures. Such inversion of behavior is most unusual. It could arise from partial solidification of the hydrocarbon chains within the micelles or simply from the fact that the hydrocarbon cores of the micelles are more compressible than the hydrocarbon chains of the free ions in water.

(v) "Complete" Ionization Constants

The treatment throughout this section has been based on conventional thermodynamics and electrostatic theory. Marshall and Quist[170] have advanced a radically different treatment of the influence of pressure on the dissociation of ion pairs and weak electrolytes, which has led them to conclude that the standard partial molar volume change for a reaction of this type "is essentially zero when the complete equilibrium is considered."[171] By complete equilibrium they mean that the solvating molecules are included explicitly in the reaction equation, for instance, in the following way for the dissociation of an ion pair in water:

$$MA_{(aq)} + kH_2O \rightleftarrows M^+_{(aq)} + A^-_{(aq)} \tag{83}$$

where k represents the net change in the "waters of solvation" of the solvated ion pair $MA_{(aq)}$. Marshall and Quist propose that the equilibrium be described by a complete equilibrium constant $K°$:

$$K° = a_{M^+_{(aq)}} a_{A^-_{(aq)}} / a_{MA_{(aq)}} a^k_{H_2O} \tag{84}$$

$$= K/a^k_{H_2O} \tag{85}$$

where the a denote activities and K is the conventional equilibrium constant. They further propose that a_{H_2O} be replaced by c_{H_2O}, the concentration of water in mole dm^{-3}. It follows that

$$\log K = \log K° + k \log c_{H_2O} \tag{86}$$

so that if $K°$ and k were independent of pressure, the logarithm of the conventional equilibrium constant would be proportional to $\log c_{H_2O}$, that is, to $\log \rho_{H_2O}$, where ρ_{H_2O} is the density of water under pressure.

Marshall and Quist found that plots of $\log K$ against $\log c_{H_2O}$ for liquid systems are remarkably linear when c_{H_2O} is changed either by altering the composition of the solvent, as in dioxan/water mixtures, or by compressing the solutions—although for a particular reaction the slopes are often completely different in the two cases. They took the linearity to mean that $K°$ is a constant, independent of dielectric constant and of pressure but dependent on temperature. That is not the only conclusion they could have drawn. A more general one would be that $\log K°$ is proportional to $\log c_{H_2O}$. In fact, however, neither conclusion is valid because there is no justification for replacing a_{H_2O} in (85) by c_{H_2O}.

The correct way to treat the full equilibrium (83) in water is to write, analogously with (32) and (33),

$$K_m° = \frac{m_{M^+_{(aq)}} m_{A^-_{(aq)}} \gamma_{M^+_{(aq)}} \gamma_{A^-_{(aq)}}}{m_{MA_{(aq)}} \gamma_{MA_{(aq)}}} \exp\left(k \frac{\mu^\infty_{H_2O} - \mu_{H_2O}}{RT}\right) \quad (87)$$

where μ_{H_2O} is the molar chemical potential of water in the mixture and $\mu^\infty_{H_2O}$ is the molar chemical potential of pure water. Differentiation of (87) gives

$$\Delta V^{\infty(0)} = -[\partial(RT \ln K_m)/\partial P] + k(V_{H_2O} - V^\infty_{H_2O}) \quad (88)$$

$$= \Delta V^\infty + k(V_{H_2O} - V^\infty_{H_2O}) \quad (89)$$

where $V^\infty_{H_2O}$ is the molar volume of pure water and V_{H_2O} is the partial molar volume of water in the mixture. The term in parentheses vanishes at zero concentration of the salt. Since most of the reported values of ΔV^∞ that have been derived from pressure measurements have been obtained by extrapolating $\partial RT \ln K_m/\partial P$ to infinite dilution, they are exactly equal to $\Delta V^{\infty(0)}$. But even at finite concentrations the second term in (89) is very small. Its value for a 1:1 salt can be estimated by combining the Debye–Hückel and Gibbs–Duhem relationships to obtain

$$V_{H_2O} - V^\infty_{H_2O} = -0.01683 m^{3/2} \quad \text{cm}^3 \text{mole}^{-1} \quad (90)$$

where m is the molality of the salt. If $m = 0.1$, this gives $V_{H_2O} - V_{H_2O}^\infty = -0.00053$ cm^3 mole^{-1}, which is entirely negligible in comparison with ΔV^∞. The complete ionization constant K_m° is very close to the conventional constant K_m, and $\Delta V^{\infty(0)}$ is similarly close to ΔV^∞; contrary to Marshall's opinion,[171] it is certainly not zero. In short, the conventional treatment of liquid-phase reactions involving hydration is correct, whereas Marshall and Quist's approach is based on gross and unnecessary approximations that lead to misleading conclusions. A recent suggestion[64] that "it is now clear that conventional constants appearing in handbook tables must be corrected, in general, for the T, P dependence of the concentration of water" is not to be taken seriously.

V. PROPERTIES OF ELECTROLYTE SOLUTIONS AT HIGH SHOCK PRESSURES

Until about 1940 virtually all high-pressure experiments had been carried out under hydrostatic conditions, where the pressure was generated in a fluid by a piston–cylinder arrangement and contained in a suitably strong pressure vessel. The attainable pressures and temperatures were limited by the weakness of the materials of construction, and the highest pressure reached in a liquid was 50 kbar, which Bridgman[172] generated in an apparatus similar to the one shown in Fig. 6. That pressure is still the practical limit for most work on truly fluid systems, although Holzapfel and Franck[75] have succeeded in reaching about 100 kbar in very small volumes of water contained in an "opposed anvil" device.

The last 30 years have seen the development of a radically different technique which uses explosive charges to produce very high shock pressures in liquids and solids. No effort is made to contain the pressure—it is generated virtually instantaneously by a shock wave and then allowed to decay as the compressed material expands into its surroundings, usually into the open air. The method is independent of the strength of materials and there is no theoretical limit to the pressures that can be reached. The highest pressure reported so far in controlled experiments is 50 Mbar in iron,[173] and Podurets et al.[174] have made an experimental estimate of the density of water compressed by a shock wave pressure of 14 Mbar, more than three times the pressure of the center of the earth.

1. Elementary Theory of Shock Waves

A shock wave is simply a compression wave with a very steep front (see Figure 19). The mathematical concept of such waves was put forward in 1848 by Stokes,[175] who showed how a wave with a discontinuous front can develop from an ordinary compression wave. He proved that such waves are stable and derived the important conditions of conservation of mass and momentum that relate the changes of pressure, density, and velocity through a shock front. For a plane-fronted shock wave moving into a material initially at rest these conditions are

$$P_1 - P_0 = Uu\rho_0 \tag{91}$$

$$\rho_1/\rho_0 = U/(U - u) \tag{92}$$

where P_0 and ρ_0 denote the pressure and density of material ahead of the shock front and P_1 and ρ_1 denote these quantities behind the front; U is the velocity of the shock front (phase velocity); and u is the velocity of flow (particle velocity) of material behind the front.

In the same paper, Stokes advanced the idea of the modern shock tube in which a section of gas, initially at rest, is suddenly expanded into another section at a lower pressure. But shortly afterwards his paper was attacked by Challis,[176] who described Stokes' waves as being an absurd concept and a physical impossibility. It seems that Challis was more upset by what he regarded as an affront to nature than by any imagined errors in Stokes' reasoning or mathematics. The fact remains that Stokes' work on this subject has almost been forgotten and his conservation relationships are now usually attributed to Rankine[177] and Hugoniot,[178] who rediscovered them later in the last century.

Hugoniot[178] added an additional relationship for the conservation of specific internal energy E:

$$E_1 - E_0 = \tfrac{1}{2}(P_1 + P_0)\left(\frac{1}{\rho_0} - \frac{1}{\rho_1}\right) \tag{93}$$

which is important because it defines all the possible P_1, ρ_1 conditions that can be reached in a single shock compression from the initial conditions P_0, ρ_0. This follows from the fact that the increase of E is related to the equation of state of the material, as well as being given by (93). If the equation of state is known, $E_1 - E_0$ can be eliminated and a single theoretical equation is obtained relating P_1 and ρ_1 to P_0 and ρ_0. The locus of points defined in this way is usually called a Hugoniot compression curve.

The three relationships (91)–(93) are exact for plane shock waves of any intensity, provided the shock front is an effective discontinuity, as it usually is.

2. Experimental Methods of Generating Strong Shock Waves

The simplest way of generating a strong shock wave in a substance is to place it on the top of a cylindrical charge of high explosive and then detonate the charge from the bottom (Fig. 18a). If the charge is long in comparison with its diameter or if it is fired by a special wave shaper,[179] the detonation front will be nearly planar and in a typical high explosive it will have a pressure of about 250 kbar and a velocity of 6–8 km sec^{-1}. When it reaches the top of the charge it will drive a strong compressive shock into the specimen.

The pressure in this shock wave depends on the resistance which the substance offers to sudden compression and this resistance is roughly proportional to the density of the substance and inversely proportional to its compressibility (more precisely, the pressure is determined by the shock impedance of the substance, defined as the product of the initial density and the shock velocity). In general an increase in shock "resistance" from one material to another will increase the shock pressure but reduce the shock temperature.

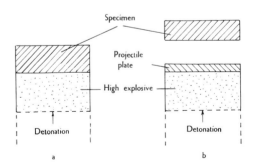

Figure 18. Arrangements for generating explosive shock waves.

Table 12

Approximate Shock Pressures and Temperatures for Materials in Contact with the Explosive Composition B(60/40 RDX/TNT)

Material	Initial density, g cm^{-3}	Initial compressibility, bar^{-1}	Shock pressure, kbar	Shock temperature, °K
Argon (gas)	0.0018	1.0	1	30,000
Water	1.00	4.5×10^{-5}	160	1,300
Copper	8.9	7×10^{-7}	500	700

These effects are illustrated by the examples in Table 12. The values of pressure and temperature listed in this table apply to newly formed shocks close to the boundary of the explosive and the specimen. As a shock advances into the material it becomes weakened by rarefaction waves and dissipative effects, and the pressure and temperature fall. It is therefore possible to cover a range of shock conditions by making measurements at different distances from the explosive.

This arrangement generates shock waves with profiles something like that shown in Fig. 19. The initial material at a pressure

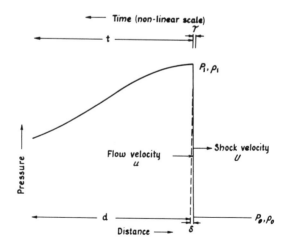

Figure 19. Schematic representation of a strong shock wave.

P_0 and density ρ_0 is rapidly compressed in a time τ and a distance δ to a state P_1, ρ_1, after which it expands relatively slowly and the pressure drops to $\frac{1}{2}(P_1 + P_0)$ in a time t and a distance d. For a 100-kbar shock wave in water initially at 1 bar and 20°C these quantities might have the following values.

Initial state: P_0, 1 bar; ρ_0, 1 g cm^3.

Shock front: δ, of the order of 100 Å (10^{-8} m); τ, of the order of 1 psec (10^{-12} sec); U, 5.05 km sec^{-1}.

State immediately behind shock front: P_1, 100 kbar; ρ_1, 1.64 g cm^{-3}; u, 1.99 km sec^{-1}.

Partially released state: d, of the order of 1 cm; t, of the order of 1 μsec (10^{-6} sec); P, 50 kbar.

The simple method described above relies on the impact of the dense detonation zone of the explosive to launch the shock wave. It is possible to reach higher pressures by making use in addition of the kinetic energy of the explosion products. This is done in the arrangement shown in Fig. 18(b), where the explosive is used to accelerate a projectile plate to a speed as great as the detonation velocity of the explosive. The impact of the plate with the specimen launches a shock wave which may be several times stronger than that produced by direct contact with the explosive. An alternative to using an explosive charge to accelerate a projectile plate is to use a specially designed high-velocity gun,[180] and there are now a number of these guns in operation.

Other, more complex, ways of producing very high shock pressures are to generate converging and colliding shock waves or to use more powerful explosive devices.[181]

3. Measurements on Shock-Compressed Materials

Whatever other properties are to be studied, it is important to know the state of a shock-compressed material in terms of its pressure, density, and temperature. The pressure and density can sometimes be measured directly by piezoelectric crystals[182] and X-ray densitometry,[183,184] respectively, but it is usually easier and more precise to derive them from measurements of the shock velocity U and flow velocity u by means of the Stokes–Rankine–Hugoniot relations (91) and (92). The velocities themselves are readily measurable by electrical or optical methods[185–187] with an accuracy sufficient to yield pressures and densities that are reliable to better than 1%. On the other hand, there is no direct way of measuring the temperature of a shocked specimen, nor is there any way of calculating it exactly from the shock propagation parameters U and u. The best that can be done is to use equation (93) to calculate the change of internal energy and combine this value with a sensible equation of state for the material to make a theoretical estimate of the temperature, which will be good or bad depending on the reliability of the equation of state. The major problem here is that the temperatures and pressures are often very high and beyond the range of any static measurements of P–V–T data, so that it is necessary to make rather long extrapolations. The kind of uncertainty that may arise in the estimated temperatures is illustrated by the following sets of values for water, calculated by different authors using different assumptions in their extrapolations:

shock pressure, kbar	0.001		100		200	
calculated shock temperature, °C	20[188,189]		570,[188]	450[189]	1297,[188]	1050[189]

In addition to the simple shock state parameters P_1, ρ_1, U, and u, it is possible to measure a remarkably wide range of properties during the brief time for which the shock pressure acts. Some of these are electrical conductivity, dielectric constant, magnetic permeability, refractive index, ultraviolet and visible absorption spectra, viscosity (indirectly), and reversible phase changes. The techniques for making these measurements have been reviewed in a number of articles[181,185–187,190,191] and a later part of this section will be concerned particularly with conductivity measurements on shock-compressed liquids.

4. Disadvantages and Advantages of Shock-Wave Methods

In comparison with static high-pressure techniques, the shock wave methods have the following fairly obvious disadvantages. (a) The duration of the shock conditions is very brief—of the order of microseconds unless a very large explosive charge is used, when it may be milliseconds. (b) The temperature of the shocked material is not directly measurable, nor is there any exact way of calculating it. Moreover, it is not easy to vary the shock temperature independently of the shock pressure, although that can be done to a limited extent by changing the initial temperature or density of the specimen or by generating interacting or reflected shock waves. (c) Although there is no theoretical upper limit to the shock pressure in a material, there *is* a theoretical limit to the increase of density which it causes. This is because the temperature increases very steeply with the pressure and the material approaches a state where the tendency for thermal expansion balances the tendency for compression as the shock pressure increases. (d) The experiments are destructive. It is usually impossible to recover the specimen, and part of the measuring equipment is often destroyed.

On the other hand, shock wave methods have the following advantages. (a) They allow the generation of much higher pressures and temperatures than can be reached in static experiments at the present time. (b) The short duration of the shock conditions is often a real advantage. For example, water at high pressures and high temperatures is strongly ionized and very corrosive; it is extremely difficult to handle it and to make reliable measurements of, say, its electrical conductivity under static conditions. But in shock conditions there is insufficient time for it to react with contaminating surfaces and it remains as pure as it began at ambient temperature and pressure.

5. Electrical Conductivities of Weak Electrolytes in Shock Waves

The static experiments summarized in Sections III and IV have shown that weak electrolytes in solution become much stronger at high pressures and that a similar change occurs in pure solvents such as water and methanol which are themselves weak electrolytes. In the case of water it might be expected that shock wave compression would have an even greater effect than static compression in enhancing ionization, since it raises the temperature as well as the pressure.

To test this prediction, the writer and co-workers[192–196] have made a series of measurements of the conductivity of pure water and aqueous solutions to shock pressures as high as 300 kbar. In the early work on pure water, David and Hamann[192] used the simple arrangement shown in Fig. 20. The best way to explain this arrangement is to describe the events during a particular experiment. The first operation consists in opening the shutter of the oscillograph camera. This automatically connects the 5-V battery to a condenser wired in series with the conductivity cell (C) and with

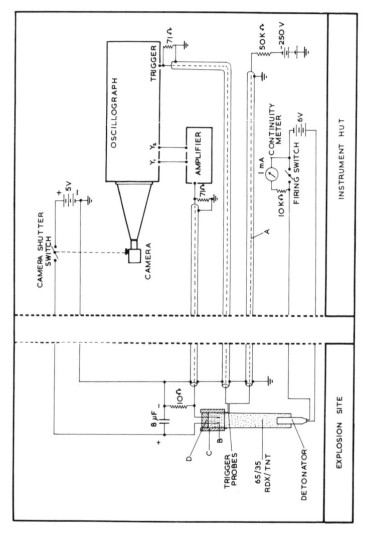

Figure 20. An arrangement for measuring the electrical conductivities of shock-compressed liquids.

a load resistor. At this stage the resistance of the cell, containing conductivity water, is about 10^6 Ω, so that no measurable voltage is developed across the small load resistor. About 1/100 sec later the firing switch is closed and the explosive detonated. A detonation wave travels up the explosive and passes a pair of trigger probes placed about 3 mm from the top. The hot ionized gases in the detonation front provide an electrical path between the probes and feed a triggering pulse from the charged transmission line (A) to the oscillograph. About $\frac{1}{3}$ μsec later the detonation reaches the top of the explosive and launches a shock wave into the water. When this reaches the electrodes (B) it causes a sudden drop in the resistance of the water between them and allows the condenser to discharge through the cell and load resistor. The resultant voltage pulse across the load is amplified and displayed as a vertical deflection of the oscillograph trace. The lower plot in Fig. 21 shows a typical oscillograph trace given by a shock in water. It will be seen that the current rises steeply as soon as the shock wave reaches the electrodes. It persists for about $\frac{1}{2}$ μsec, which is roughly the time required for the wave to travel from the bottom to the top of the electrodes.

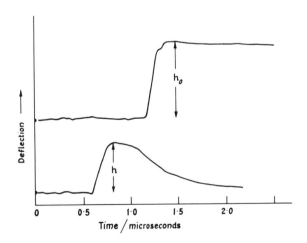

Figure 21. Typical current pulses. Upper: for a short circuit; lower: for water with the electrodes placed 2 mm away from the explosive (cf. Fig. 20).

It then begins to decay as the pressure is relieved by rarefaction waves which converge on the tail of the shock wave. The upper trace is a picture obtained by placing a sudden short circuit across a pair of electrodes in a similar circuit. The shock resistance of the water is easily derived from a comparison of the heights h and h_0 of these two pulses. The average shock velocity can be worked out from the horizontal position of the water pulse, and the temperature and pressure in the shocked state can then be estimated from Rice and Walsh's[188] tables of the thermodynamic properties of shocked water as a function of the shock velocity.

The conductivity cells (C) used in these early experiments were simply polyethylene cylinders containing platinized platinum electrodes mounted in polyethylene plugs (D), which could be pushed into the cells to vary the distance between the electrodes and the explosive charge and so vary the pressure in different experiments. The cells were completely filled with water. A serious

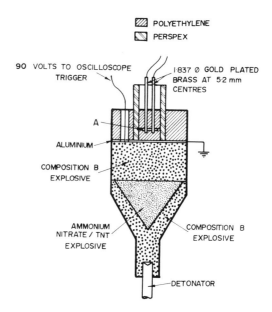

Figure 22. An arrangement for measuring electrical conductivities in plane shock waves.

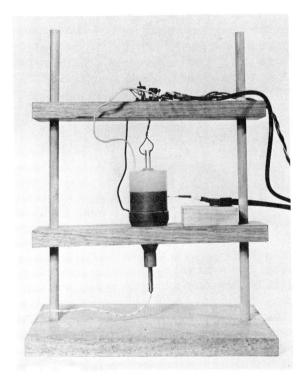

Figure 23. An arrangement for measuring the conductivities of shock-compressed liquids to 200 kbar (see Figure 22).

shortcoming of this arrangement was that the electrodes sampled the conductivity of a large volume of water, not all of which was under the same set of shock conditions at any instant. In analyzing the results, David and Hamann[192] measured the height h of the current pulse at the time when the wave front reached the top of the electrode plates, and applied an appropriate cell constant found from a previous calibration in which the plates were immersed to their tops in a 0.01 N solution of potassium chloride. Clearly, this method gave the average conductivity of the shocked water over some distance behind the shock front.

To obtain a truer estimate of the conductivity close to the shock front, it is necessary to limit the electrical path to a thin layer whose geometry is not greatly changed by the arrival of the wave. Hamann and Linton[194] later did this by sandwiching the water between two disks of polyethylene in the manner illustrated in Fig. 22, which shows the conductivity cell mounted in a polyethylene supporting ring on the top of a cylindrical charge of composition B explosive (60/40 RDX/TNT) initiated by an explosive plane wave generator.[179] The whole assembly is shown in Fig. 23. The cell itself can be used with smaller charges, without plane wave generators.[194]

The cell has parallel cylindrical electrodes mounted tightly in two polyethylene plugs, the gap (A) between the plugs contains the sample of water, and its thickness determines the cell length and hence the cell constant. The thickness can be varied in different experiments, but is usually 1 mm. The thickness of the lower polyethylene plug is also varied, to allow the pressure of the emerging shock wave to be altered and to provide a series of time/distance data that can be used to determine the velocity of shocks in polyethylene. Polyethylene was chosen as the insulating material because its shock impedance is very close to that of water and a wave can pass from one to the other without serious reflection. In this arrangement the cell constant changes in a calculable way as the shock wave traverses the water layer and compresses it, but once the wavefront reaches the top of the layer the geometry remains constant for a comparatively long time. The change of cell length caused by shock compression can be allowed for in calculating the conductivity.

Using this type of cell, Hamann and Linton[194] obtained the data shown in Table 13, where the values in the first three columns have been taken from the tables of Rice and Walsh[188] at the measured values of shock velocity U in water. It will be seen that there is a very large increase in conductivity in strong shocks, which the authors[192,194] attributed to a great enhancement of the normal ionization of water into hydrogen ions and hydroxyl ions. To estimate ionization constants from the conductivity data requires knowledge of the mobilities of the ions under shock conditions, which was obtained later in experiments described in the next section.

David and Hamann[193] found that a number of liquids other than water also become good conductors in shock waves. They are

Table 13

Values of Some Properties of Water Compressed by Shock Waves from 1 bar and 20°C

Density ρ_1, g cm^{-3}	Pressure P_1, kbar	Temperature, °C	Conductivity, Ω^{-1} cm^{-1}
1.0	0.0	20	4.5×10^{-8}
1.1	3.4	31	2.1×10^{-7}
1.2	8.6	49	1.1×10^{-6}
1.3	17.8	86	7.2×10^{-6}
1.4	32.1	156	5.4×10^{-5}
1.5	53.9	274	7.6×10^{-4}
1.6	83.2	458	4.0×10^{-2}
1.7	121.2	718	5.6×10^{-1}
1.8	167.0	1053	2.3

compounds which can ionize by autoprotolysis reactions

$$2ROH \rightleftarrows RO^- + ROH_2^+, \quad K_{auto} = [RO^-][ROH_2^+]$$

and it is significant that the order of their conductivities at 100 kbar is roughly the same as the order of their autoprotolysis constants under normal conditions (see Table 14). Glycerol, which is potentially capable of autoprotolysis, gave no sign of high conductivity, and acetone behaved in a perplexing way that has been discussed elsewhere.[181]

Table 14

Autoprotolysis Constants and Shock Conductivities of Some Liquids

Liquid[a]	K_{auto}, mole^2 dm^{-6}
Acetic acid	3×10^{-13}
Water	1×10^{-14}
Propionic acid	?
1:1 (vol.) Water–ethanol	10^{-17}
Methanol	2×10^{-17}
Ethanol	8×10^{-20}

[a]Listed in order of decreasing shock conductivity.

Finally, a few experiments have been carried out by an arrangement[193] that allowed the conductivity of water to be measured in a region that had been doubly shocked by the head-on collision of two single shocks each of 100 kbar, raising the total pressure to about 300 kbar. Under these conditions the conductivity approached $10\,\Omega^{-1}\,cm^{-1}$, which is of the same order as the conductivities of molten salts.

6. Electrical Conductivities of Solutions of Strong Electrolytes in Shock Waves

The large increase that occurs in the conductivity of water when it is shock-compressed tends to mask any changes due to dissolved salts. For example, the conductivity of pure water at a shock pressure of 130 kbar is more than ten times the conductivity of a 1 m solution of potassium chloride in water under normal conditions. Thus unless the change in the contribution of a dissolved salt is large, it is difficult to measure it except at fairly low pressures where the masking effect of the water is less troublesome. It turns out that in general the changes are not large and it has only been possible to measure them by taking great care both with the geometry of the apparatus and in interpreting the oscillograms.[195] The experimental results for KCl, KOH, and HCl are shown in Table 15, where the

Table 15

Molar Conductivities of Electrolytes in Shock-Compressed Water

Shock conditions			$\Lambda, \Omega^{-1}\,cm^2\,mole^{-1}$				
P_1 kbar	ρ_1 g cm^{-3}	Temp., °C	KCl 1.0m	KCl 0.1m	KOH 0.1m	HCl 0.1m	HCl 0.033m
0.001[a]	0.998[a]	20[a]	102[a]	117[a]	224[a]	362[a]	373[a]
7.5	1.184	45	—	150[b]	290[b]	550[b]	—
48	1.479	240	—	120 ± 30	—	—	—
59	1.519	304	—	130 ± 30	—	—	—
71	1.560	373	115 ± 20	140 ± 20	700 ± 150	900 ± 150	900 ± 150
82	1.594	444	105 ± 20	130 ± 20	800 ± 150	1000 ± 150	1100 ± 150
93	1.629	527	105 ± 20	—	900 ± 150	1200 ± 150	1200 ± 150
114	1.683	670	90 ± 20	—	800 ± 300	1300 ± 300	—
133	1.727	804	—	—	1300 ± 300	1200 ± 300	—

[a]Initial, unshocked state.
[b]From static measurements.

values of the shock pressure, density, and temperature have been taken from the tables of Rice and Walsh[188] at the measured values of shock velocity U.

It is remarkable that the molar conductivity Λ of KCl is not measurably altered by strong shock compression, in spite of the high pressures and temperatures that occur in the shocked state. This can only be interpreted as meaning that the viscosity of shocked water is close to that of normal water; the tendency of the viscosity to drop with increasing temperature at constant pressure[197] being counteracted by its tendency to rise with increasing density at constant temperature.[197] Hamann and Linton[198] have made a rough theoretical estimate of the viscosity of shocked water and concluded that it may be about 0.01 P in 100-kbar shocks and, from Walden's rule (2), the mobilities of large ions should therefore be close to their normal values. Also, the experimental results are consistent with trends that have been observed in Λ for KCl,[199] NaCl,[200] and NaBr[201] under *static* compression at high temperatures but much lower pressures. A long extrapolation of the static data indicates that Λ may approach its normal value at densities near 1.5 g cm^{-3} at temperatures in the range 300–700°C.

In other experiments the same authors[195] found that aqueous solutions of the salts NH_4Cl, KI, $MgCl_2$ and $NiCl_2$ behave in a similar manner to those of KCl, but shocked solutions of 0.1 m $MgSO_4$, $NiSO_4$, and $La_2(SO_4)_3$ show an immediate twofold increase in conductivity at 100 kbar which can be attributed to the dissociation of ion pairs. On the other hand, solutions of KI in acetone, formamide, and dimethylformamide show an immediate decrease, nearly to zero, in the case of dimethylformamide. This change could be caused either by a large increase in the viscosity of the organic solvent or by its partial freezing in shock compression, neither of which changes apparently occurs in water. Photographs taken with a fast framing camera show that the liquids do not freeze, so that the drop in conductivity can be taken as evidence of a substantial increase in viscosity.

In contrast to that of KCl, the molar conductivities of HCl and KOH in water show unmistakable jumps which increase with the intensity of the shock waves. The data[195] in Table 15 indicate that the "excess" conductivities of hydrogen and hydroxyl ions at 100 kbar are about four times their usual values. The results are consistent with Holzapfel and Frank's[75] estimate that $\Lambda_{H^+ + OH^-}$ is about $2000\,\Omega^{-1}\,cm^2\,mole^{-1}$ at 700°C at high densities, and with an extrapolation of Quist and Marshall's[202] static measurements on solutions of HBr. The persistence of the abnormally high mobilities of hydrogen and hydroxyl ions to these extreme conditions means that there must still be a large degree of hydrogen-bonded association within the water, despite its high temperature, and this conclusion is consistent with infrared evidence that supercritical water is largely hydrogen bonded even at quite low densities.[38]

7. Ionization Constant of Water at High Shock Pressures

The data presented in the last two sections can be combined to derive the ionization product K_w of shock-compressed water.

First, it may justifiably be assumed that KCl, KOH, and HCl remain fully ionized and unassociated in the shocked state[195] so that their conductivities can be combined to obtain the sum of the molar conductivities of H^+ and OH^- ions in the following way:

$$\Lambda_{H^+ + OH^-} = \Lambda_{HCl} + \Lambda_{KOH} - \Lambda_{KCl} \tag{94}$$

The fourth column of Table 16 lists results obtained from smoothed, and sometimes extrapolated, values of Λ for 0.1 m solutions (the

Table 16

Ionization Product of Water under Shock Compression

Pressure, kbar	Density,[a] g cm^{-3}	Temp., °C	$\Lambda_{H^+ + OH^-}$, Ω^{-1} cm^2 mole^{-1}	κ_{H_2O}, Ω^{-1} cm^{-1}	$m_{H^+}\,(=m_{OH^-})$, mole kg^{-1}	K'_w, mole^2 kg^{-2}
0.001	0.998	20	504	4.2×10^{-8}	—	6.8×10^{-15}
7.5	1.184	45	—	—	—	8.7×10^{-13b}
48	1.479 (1.504)	240	1250	4.5×10^{-4}	2.4×10^{-4}	5.9×10^{-8}
59	1.519 (1.535)	304	1390	1.2×10^{-3}	5.9×10^{-4}	3.5×10^{-7}
71	1.560 (1.567)	373	1540	3.4×10^{-3}	1.4×10^{-3}	2.0×10^{-6}
82	1.594 (1.584)	444	1680	1.8×10^{-2}	6.7×10^{-3}	4.6×10^{-5}
93	1.629 (1.614)	527	1820	8.8×10^{-2}	3.0×10^{-2}	8.9×10^{-4}
114	1.683 (1.646)	670	2090	4.2×10^{-1}	1.2×10^{-1}	1.4×10^{-2}
133	1.726 (1.671)	804	2330	1.2	3.0×10^{-1}	8.9×10^{-2}

[a]The densities in parentheses were calculated from equation (97).
[b]From static measurements.

accuracy of the shock wave measurements is too low to give any information on the concentration dependence of Λ). The accumulation of errors in this process may make the results uncertain to the extent of $\pm 500 \, \Omega^{-1} \, \text{cm}^2 \, \text{mole}^{-1}$, but the increase at high pressures is unmistakable.

The values of $\Lambda_{\text{H}^+ + \text{OH}^-}$ can then be combined with the conductivity data $\kappa_{\text{H}_2\text{O}}$ for pure water in Table 13 to obtain an ionization product K_w' defined in terms of the molalities of the ions

$$K_w' = m_{\text{H}^+} m_{\text{OH}^-} \tag{95}$$

so that

$$K_w' = (1000 \kappa_{\text{H}_2\text{O}} / \rho_1 \Lambda_{\text{H}^+ + \text{OH}^-})^2 \tag{96}$$

where the shock density ρ_1 is introduced to convert Λ from molar to molal units. The results are listed in Table 16 and plotted in Fig. 24,

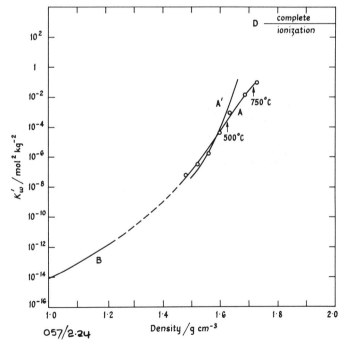

Figure 24. The ionization constant of shock-compressed water.

where the densities along curve A have been taken from Rice and Walsh's[188] equation of state for shocked water. At the low-density end the behavior of K_w' in weak shocks should theoretically follow the curve B, which is obtained by fitting some static isothermal high-pressure data[140] to an equation of the form proposed by Owen and Brinkley[152] and correcting them for the temperature jump that occurs in shock compression. The conditions near the top of curve B are listed in the second line of Table 16. The diagram also shows a line D which indicates the value that K_w' would have if the water were completely ionized into H_3O^+ and OH^- ions ($m_{H^+} = m_{OH^-} = 27.8$ moles kg^{-1}): Complete ionization into $H_9O_4^+$ ions and OH^- ions would occur at a value lower than D by a factor of 6.25.

It will be seen that K_w' increases by a factor of more than 10^{13} between normal pressure and a shock pressure of 133 kbar, and the trend is such that conversion to a completely ionized state might be expected to occur at a shock density between 1.8 and 2.0 g cm^{-3} at a pressure of 150–200 kbar. From the conductance data alone it is not easy to prove that this happens, because the high conductivity of water in that region makes it impossible any longer to estimate the value of $\Lambda_{H^+ + OH^-}$ (see the last subsection). But a transformation of this kind from a molecular fluid to one which is essentially an ionic melt, isoelectronic with ammonium fluoride (which is ionic in the liquid state[203]), should show itself in other properties than the conductivity—for example, it should be reflected in the thermodynamic and optical properties of shock-compressed water. And so it is. Some time ago Al'tshuler et al.[204] pointed out that the pressure/density plot for shock waves in water consists of two fairly distinct sections separated by a transition zone near $\rho_1 = 1.7$ g cm^{-3}. This conclusion is supported by more recent data and the "transition" can be seen clearly in the plot of measured shock velocities U against flow velocities u in Figure 25. In this diagram the lines A and B are defined by the relations

$$A: \quad U = 1.82u + 1.49 \text{ km sec}^{-1} \tag{97}$$

$$B: \quad U = 1.164u + 3.09 \text{ km sec}^{-1} \tag{98}$$

Formulas (97) and (98), together with the Stokes–Rankine–Hugoniot equations (97) and (98), yield the Hugoniot pressure-density curves shown in Fig. 26. They are sufficiently different from

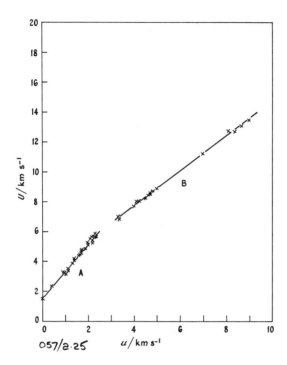

Figure 25. The relationship between shock velocity and particle velocity for water.

the earlier ones of Rice and Walsh[188] to alter the densities in Table 16 to the values in brackets and to move the curve A in Fig. 24 to A'.

The breaks in Figs. 25 and 26 are almost sharp enough to indicate a phase transition. Al'tshuler et al. originally postulated[204] that they were caused by the partial or complete freezing of water, but subsequent optical experiments[205,206] disproved this explanation and water probably remains fluid to at least 1 Mbar. It is very improbable that it undergoes an abrupt fluid–fluid phase change, and the most likely explanation is that it passes through a smooth conversion from a largely un-ionized fluid to an almost fully ionized one between 150 and 200 kbar.

The following evidence supports this theory. (a) When the results for K_w' are plotted on a revised density scale based on (97)

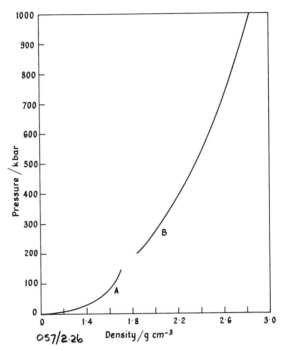

Figure 26. Hugoniot compression curves for water derived from the relationships in Fig. 25.

and (98) rather than Rice and Walsh's relations,[188] they lie on the curve A' of Fig. 24, which extrapolates to a state of full ionization at a shock density in the neighborhood of 1.6–1.8 g cm^{-3}, the density at which breaks occur in Figs. 25 and 26. (b) The coefficients 1.82 and 1.164 in formulas (97) and (98) are, respectively, close to the average values for molecular materials such as organic compounds on the one hand and ionic materials such as the alkali halides on the other hand.[207] (c) Figure 26 shows that water is more compressible above the "transition" than below it. This is understandable since water should be denser in an ionic state than in the molecular state—ammonium fluoride has a density of 1.77 g cm^{-3} at 12 kbar,[208] compared with 1.32 g cm^{-3} for ice VI at the same pressure. (d) The refractive index of water behaves anomalously in the transition region. Although it increases as the shock density increases[205,209] to $\rho_1 = 1.6$ g cm^{-3}, it appears to remain constant (within the large experimental errors) between 1.62 and 1.83 g cm^{-3}. This behavior is consistent with the fact that ammonium fluoride has a lower refractive index[208] than water and than ice if the refractive index of ice is adjusted by the Gladstone–Dale formula to the density of ammonium fluoride (1.009 g cm^{-3}). (e) The degree of solubility of water in molten glasses, in which it is extensively ionized,[210] suggests that the transition from normal water at room temperature to dense ionized water at 1100°C requires an increase in free energy of about 150 kJ mole^{-1}. A shock wave of 170 kbar raises the temperature and density of water to about 1.75 g cm^{-3} and 1100°C and increases the free energy by about 165 kJ mole^{-1}.[188]

To summarize, the indications are that water becomes fully ionized to H_3O^+ and OH^- ions at 150–200 kbar, 1.7–1.9 g cm^{-3}, and ~ 1000°C. It is then essentially a molten salt, isoelectronic with NH_4F and NaOH. At higher pressures it may ionize further to H^+ and OH^- ions or H^+ and O^{2-} ions, although these transitions require much higher energy. It is perhaps significant that recent pressure–density estimates for water at the extremely high shock pressure of 14 Mbar[174] lie close to extrapolations of the curves B in Figs. 25 and 26.

There is no reason why this kind of behavior should be limited to water, and the conductivity behavior of some alcohols and acids[193] suggests that they, too, may approach a state of full ionization at high temperatures and pressures.

ACKNOWLEDGMENTS

Preliminary work on this article was done while the author held a visiting professorship in the Chemistry Department, Kyoto University, Japan. He is grateful to Professor Jiro Osugi for his hospitality during that time and to the Japan Society for the Promotion of Science for the award of the professorship.

APPENDIX. A Critical Compilation of the Effects of Pressure on Ionization Equilibria, and of Volume Changes for the Ionization of Weak Electrolytes

1. Introduction

In 1963 the author published a collection of measured values of the volume changes for simple ionization reactions in solution, based both on volumetric determinations and on measurements of the effect of pressure on ionization constants (Ref. 127, Chapter 7.ii, Table IV). That compilation is now obsolete and is replaced by the present one.

Table 17. Volume changes for the ionization of weak acids in water at 25°C, and the effects of pressure on ionization constants. (i) Inorganic acids, (ii) carboxylic acids, (iii) aldose, (iv) phenols.

Table 18. Volume changes for the ionization of weak bases in water at 25°C, and the effects of pressure on ionization constants. (i) Inorganic bases, (ii) aliphatic bases, (iii) aromatic bases, (iv) cyclic bases.

Table 19. Volume changes for the ionization of weak polyelectrolytes in water at 25°C, and the effects of pressure on ionization. (i) Polyacids, (ii) polybases.

Table 20. Volume changes for the dissociation of ion pairs in water at 25°C, and the effects of pressure on dissociation.

Table 21. Volume changes for the formation of micelles in water, and the effects of pressure on critical micelle concentrations. (i) Acid salts, (ii) base salts.

Table 22. Volume changes for the ionization of weak acids and weak bases in nonaqueous and mixed solvents at 25°C and the effects of pressure on ionization constants. (i) Solvent: methanol; (ii) solvent: ethanol/water, 50% (wt).

Table 23. Volume changes for the dissociation of ion pairs in nonaqueous solvents, and the effects of pressure on dissociation. (i) Solvent: ethanol; (ii) solvent: 2-propanol; (iii) solvent: 2-methylpropan-1-ol; (iv) solvent: acetone.

Continued on p. 150

Appendix

Table 17
Volume Changes for the Ionization of Weak Acids in Water at 25°C, and the Effects of Pressure on Ionization Constants[a]

Acid	$K_m(1)$ at 1 bar mole kg^{-1}		$K_m(P)/K_m(1)$ P, bars: 1000	2000	3000	ΔV^∞(a), cm^3 mole^{-1}	ΔV^∞(b), cm^3 mole^{-1}
(i) Inorganic acids							
Water	1.0	−14	2.17	4.14	—	−21.8[140,142]	−22.11[211]
Hydrofluoric acid	6.5	−4	—	—	—	—	−13.7[b]
Hydrocyanic acid	6.2	−10	1.3	1.7	—	−7[c]	−7.0[d]
Hydrogen sulfide (1st ionization)	9.8	−8	1.78	2.74	—	−15.0[157]	−16.3[e]
Boric acid (1st ionization)	5.8	−10	3.68	—	—	−32.1[139e]	−31.5[f]
Carbonic acid (1st ionization)	4.2	−7	2.89	7.8	16.7	−26.6[139d,172,222]	−27.7[g]
Bicarbonate ion (2nd ionization)	4.7	−11	2.83	—	—	−25.6[139e]	−27.7[h]
Phosphoric acid (1st ionization)	7.1	−3	1.93	2.92	—	−16.6[139d,224]	−16.2[225]
Phosphate ion (2nd ionization)	6.2	−8	2.64	—	—	−24.0[139d]	−24.1[226]
Sulfurous acid (1st ionization)	1.4	−2	2.31	4.82	—	−21.9[224,227]	—
Sulfite ion (2nd ionization)	5.9	−8	—	—	—	—	−22.7[i]
Sulfate ion (2nd ionization)	1.2	−2	2.0	3.7	6.3	−21.7[j]	−21.7[230]
Adenosine triphosphate (4th ionization)	3.1	−7	2.65	—	—	−24.0[139d]	—
(ii) Carboxylic acids							
Formic acid	1.7	−4	1.39	1.86	2.38	−8.8[43,61]	−8.43[231]
Acetic acid	1.7	−5	1.54	2.20	3.06	−11.7[43,61,139a,155,221,223]	−11.50[231]
Propionic acid	1.3	−5	1.62	2.47	3.49	−12.9[61]	−14.0[k]
Butyric acid	1.5	−5	—	—	—	—	−14.22[231]
Oxalic acid (1st ionization)	5.2	−2	—	—	—	—	(−9.0)[233,234]
Oxalate ion (2nd ionization)	5.2	−5	—	—	—	—	(−16.2)[233,234]

Table 17—continued

Acid	$K_m(1)$ at 1 bar mole kg^{-1}	$K_m(P)/K_m(1)$ P, bars: 1000	2000	3000	$\Delta V^\infty(a)$, cm^3 mole^{-1}	$\Delta V^\infty(b)$, cm^3 mole^{-1}
Malonic acid (1st ionization)	1.4 −3	—	—	—	—	(−8.5)[233]
Malonate ion (2nd ionization)	2.0 −6	—	—	—	—	(−17.6)[233]
Succinic acid (1st ionization)	6.4 −5	—	—	—	—	(−11.6)[233,235]
Succinate ion (2nd ionization)	2.4 −6	—	—	—	—	(−13.1)[233,235]
Glutaric acid (1st ionization)	4.5 −5	—	—	—	—	(−12.2)[235]
Glutarate ion (2nd ionization)	3.8 −6	—	—	—	—	(−13.7)[235]
Phthalic acid (1st ionization)	1.1 −3	—	—	—	—	(−12.5)[233]
Phthalate ion (2nd ionization)	3.9 −6	—	—	—	—	(−16.5)[233]
Maleic acid (1st ionization)	1.1 −2	—	—	—	—	(−6.7)[233,235]
Maleate ion (2nd ionization)	5.7 −7	—	—	—	—	(−22.9)[233,235]
Fumaric acid (1st ionization)	9.6 −4	—	—	—	—	(−9.9)[233]
Fumarate ion (2nd ionization)	4.1 −5	1.50	2.07	2.72	−10.8[232]	(−10.9)[233]
Citric acid (1st ionization)	7.4 −4	—	—	—	—	(−8.8)[233]
Citrate ion (2nd ionization)	1.7 −5	—	—	—	—	(−12.7)[233]
Citrate ion (3rd ionization)	4.0 −7	—	—	—	—	(−18.3)[233]
4-Carboxy-pimelic acid (1st ionization)		—	—	—	—	(−13.7)[235]
4-Carboxy-pimelate ion (2nd ionization)		—	—	—	—	(−13.7)[235]
4-Carboxy-pimelate ion (3rd ionization)		—	—	—	—	(−15.6)[235]
Benzoic acid	6.2 −5	1.54	2.25	3.07	−10.9[113,236]	—
2-Nitro-benzoic acid	6.1 −3	1.52	2.09	2.91	−10.2[113]	—
3-Nitro-benzoic acid	3.2 −4	1.40	1.93	2.58	−8.7[113,236]	—
4-Nitro-benzoic acid	3.5 −4	1.44	1.95	2.58	−8.8[113,236]	—
3-Fluoro-benzoic acid	1.4 −4	1.47	2.08	2.80	−9.8[236]	—
4-Fluoro-benzoic acid	7.0 −5	1.50	2.15	2.93	−10.5[236]	—
2-Hydroxy-benzoic acid (salicylic acid)	1.0 −3	1.34	1.80	2.28	−7.2[113]	—

Appendix

Table 17—*continued*

Acid	$K_m(1)$ at 1 bar mole kg^{-1}	$K_m(P)/K_m(1)$ P, bars: 1000	2000	3000	$\Delta V^\infty(a)$, cm^3 mole^{-1}	$\Delta V^\infty(b)$, cm^3 mole^{-1}
3-Methoxy-benzoic acid	8.2 −5	1.49	2.13	2.91	−10.3[236]	—
4-Methoxy-benzoic acid	3.1 −5	1.55	2.27	3.16	−11.3[236]	—
Phenylacetic acid	4.9 −5	1.65	2.49	3.57	−12.7[236]	—
3-Nitro-phenylacetic acid	1.1 −4	1.53	2.23	3.08	−11.1[236]	—
4-Nitro-phenylacetic acid	1.2 −4	1.53	2.21	3.08	−10.9[236]	—
3-Fluoro-phenylacetic acid	7.5 −5	1.60	2.38	3.39	−12.0[236]	—
4-Fluoro-phenylacetic acid	6.0 −5	1.63	2.46	3.51	−12.5[236]	—
3-Chloro-phenylacetic acid	7.8 −5	1.58	2.36	3.37	−11.8[236]	—
4-Chloro-phenylacetic acid	6.6 −5	1.61	2.39	3.40	−12.2[236]	—
4-Methyl-phenylacetic acid	4.4 −5	1.64	2.50	3.60	−12.6[236]	—
4-Methoxy-phenylacetic acid	4.4 −5	1.62	2.49	3.53	−12.4[236]	—
3,5-Dinitro-salicylic acid	2.9 −1	1.18	1.41	—	−4.1[144]	—
Acetylglycine	2.1 −4	—	—	—	—	(−12.3)[234]
Glycine cation	4.5 −3	—	—	—	—	−6.80[231]
Alanine cation	4.5 −3	—	—	—	—	(−7.5)[234,237,238]
Valine cation	5.3 −3	—	—	—	—	(−8.5)[238]
3-Amino-butyric acid cation	2.8 −4	—	—	—	—	(−7.4)[234]
6-Amino-caproic acid cation	4.3 −5	—	—	—	—	(−13.2)[239]
Proline cation	1.1 −2	—	—	—	—	(−6.9)[234]
4-Hydroxy-proline cation	1.2 −2	—	—	—	—	(−7.4)[234]
Histidine cation	1.5 −2	—	—	—	—	(−7.7)[234]
(iii) Aldose						
Glucose	3.7 −13	—	—	—	—	(−19.1)[234]

Table 17—continued

Acid	$K_m(1)$ at 1 bar mole kg^{-1}	$K_m(P)/K_m(1)$ P, bars: 1000	2000	3000	ΔV^∞(a), cm^3 mole^{-1}	ΔV^∞(b), cm^3 mole^{-1}
(iv) Phenols						
Phenol	1.1	−10	—	—	—	−18.4[144]
2-Nitro-phenol	5.3	−8	1.65	—	—	−12.4[144]	—
3-Nitro-phenol	4.2	−9	1.67	—	—	−12.9[144]	—
4-Nitro-phenol	6.9	−8	1.49	—	—	−10.0[144]	—
2,4-Dinitro-phenol	8.9	−5	1.50	2.15	—	−10.8[144]	—
2,5-Dinitro-phenol	5.4	−6	1.56	2.26	—	−11.8[144]	—
2,6-Dinitro-phenol	1.8	−4	1.74	2.81	—	−14.3[144]	—
2,4,6-Trinitro-phenol (picric acid)	3.8	−1	1.44	1.97	—	−10.0[144]	—
4-Nitro-2-amino-phenol	7.1	−8	1.51	—	—	−9.5[144]	—
4-Nitro-2,6-dibromo-phenol	3.3	−4	1.60	2.87	—	−12.6[144]	—
2,4-Dinitro-6-amino-phenol	1.9	−5	1.36	1.75	—	−8.4[144]	—
3,5-Dinitro-salicylate ion (2nd ioniz.)	4.2	−8	2.15	—	—	−18.6[144]	—
4-Nitroso-phenol (4-benzoquinone oxime)	2.9	−7	1.53	—	—	−10.3[144,l]	—
2-Methoxy-4-formyl-phenol (vanillin)	3.2	−8	1.67	—	—	−11.3[144]	—
1-Naphthol	4.4	−10	1.91	—	—	−18.0[144]	—
2-Naphthol	2.5	−10	1.89	—	—	−17.7[144]	—
2,4-Dinitro-1-naphthol	3.6	−3	—	—	—	[−8.5][144,m]	—
1-Hydroxy-2,4-dinitro-7-naphthalene sulfonate ion (2nd ionization)(flavianate ion)	3.2	−3	1.52	2.31	—	−10.3[144]	—
Thiophenol	3.3	−7	—	—	—	—	−12.5[280]

*a*The values ΔV^∞(a) are derived from the pressure dependence of K_m, and the values ΔV^∞(b) are from volumetric measurements. The volume changes that are listed in parentheses, e.g., (−9.0), were measured at fairly high concentrations and may be 1–2 cm^3 mole^{-1} less negative than the values ΔV^∞ at infinite dilution. The temperature was not exactly 25°C in all cases, but it was close enough not to affect the values of ΔV^∞.

Appendix

[b] Calculated from $V_{F^-}^\infty = -1.2 \text{ cm}^3 \text{ mole}^{-1}$ (Ref. 213) and the densities of solutions of HF,[212] which give $V_{HF}^\infty = +12.5 \text{ cm}^3 \text{ mole}^{-1}$.

[c] Calculated from the conductivity data of Tammann and Rohmann[131] for $0.01\, M\, NH_4CN$, allowing for the effect of pressure on the ionization constants of water and ammonia.

[d] Calculated from $V_{CN^-}^\infty = +24.0 \text{ m}^3 \text{ mole}^{-1}$ (Ref. 214) and the densities of solutions of HCN,[215] which give $V_{HCN}^\infty = +31.0 \text{ cm}^3 \text{ mole}^{-1}$.

[e] Calculated from the densities of solutions of KSH[216] and of H_2S,[217] which give $V_{H_2S}^\infty = +35.5 \text{ cm}^3 \text{ mole}^{-1}$ and $V_{SH^-}^\infty = +19.2 \text{ cm}^3 \text{ mole}^{-1}$. It might be thought that ΔV^∞ for the second ionization step $SH^- \rightarrow S^{2-} + H^+$ could be derived from the value of V_{SH}^∞ and the value of $V_{S^{2-}}^\infty$ listed by Millero.[213] But there is now strong evidence that the ion S^{2-} only exists in solution under extremely alkaline conditions[218] and that the supposed value of $V_{S^{2-}}^\infty$ really relates to a mixture of SH^- and OH^- ions.

[f] Calculated from $V_{Na_2B_4O_7}^\infty = +4.5 \text{ cm}^3 \text{ mole}^{-1}$ (from density data in Refs. 219 and 220 and densities measured by the author) and from $V_{H_3BO_3}^\infty = +40.0 \text{ cm}^3 \text{ mole}^{-1}$,[221] assuming that borax dissolves according to the formula

$$\tfrac{1}{2}Na_2B_4O_7 + \tfrac{7}{2}H_2O \rightarrow Na^+ + H_3BO_3 + B(OH)_4^-$$

[g] Calculated from $V_{HCO_3^-}^\infty = +23.4 \text{ cm}^3 \text{ mole}^{-1}$ (Ref. 213) and the densities of solutions of CO_2,[223] which give $V_{CO_2}^\infty = 33.1 \text{ cm}^3 \text{ mole}^{-1}$.

[h] Calculated from $V_{CO_3^{2-}}^\infty = -4.3 \text{ cm}^3 \text{ mole}^{-1}$ and $V_{HCO_3^-}^\infty = +23.4 \text{ cm}^3 \text{ mole}^{-1}$.[213]

[i] Calculated from $V_{SO_3^{2-}}^\infty = +8.9 \text{ cm}^3 \text{ mole}^{-1}$,[213] together with $V_{SO_2}^\infty = +35.5 \text{ cm}^3 \text{ mole}^{-1}$ (derived from density measurements by Blair and Quinn[223]) and the value $\Delta V^\infty = -21.9 \text{ cm}^3 \text{ mole}^{-1}$ for the first ionization step.

[j] From Ref. 228. The results of Ref. 229 are clearly wrong.

[k] From Ref. 232, corrected by using the most recent values for $V_{Na^+}^\infty$, $V_{OH^-}^\infty$, and $V_{Cl^-}^\infty$.[213]

[l] This compound exists in water mainly in the oxime form, rather than the tautomeric phenol form (see Norris and Sternhell[240]).

[m] The results for 2,4-dinitro-1-naphthol are less accurate than those for the other phenols. Its solubility in water is very low, and the presence of overlapping peaks in its ultraviolet spectrum reduces the accuracy of the optical method that was used for the phenols.[144]

Table 18
Volume Changes for the Ionization of Weak Bases in Water at 25°C, and the Effects of Pressure on Ionization Constants[a]

Base	$K_m(1)$ at 1 bar mole kg^{-1}	$K_m(P)/K_m(1)$ P, bars: 1000	2000	3000	$\Delta V^\infty(a)$, cm^3 mole^{-1}	$\Delta V^\infty(b)$, cm^3 mole^{-1}
(i) Inorganic base						
Ammonia	1.7 −5	2.92	6.89	13.7	−28.9[65]	−28.5[226,b,c]
(ii) Aliphatic bases						
Methylamine	4.2 −4	2.80	6.06	11.2	−26.4[65]	−26.2[242,c]
Dimethylamine	5.9 −4	2.73	6.14	12.7	−27.2[43]	−26.0[242,c]
Trimethylamine	6.3 −5	2.85	6.81	14.4	−28.1[43]	−27.0[242,c]
Triethylamine	5.7 −4	—	—	—	—	−21.1[242]
Ethylenediamine (1st ionization)	8.5 −5	—	—	—	—	−26.8[233,243]
Ethylenediamine (2nd ionization)	7.1 −8	—	—	—	—	−32.6[233,243]
Diethylenetriamine (2nd ionization)	5.5 −6	—	—	—	—	−30.8[243]
Triethylenetetramine (2nd ionization)	2.0 −5	—	—	—	—	−25.8[243]
Tetraethylenepentamine (2nd ionization)	1.2 −5	—	—	—	—	−33.6[243]
Ethyl glycine ester	4.4 −7	—	—	—	—	−27.3[226,234]
Glycine anion	6.0 −5	—	—	—	—	−23.4[226,237−239]
Alanine anion	7.5 −5	—	—	—	—	−22.3[234,237]
Valine anion	5.3 −5	—	—	—	—	−21.5[237]
3-Amino-butyrate ion	1.4 −4	—	—	—	—	−22.5[234]
6-Amino-caproate ion	6.4 −4	—	—	—	—	−25.0[239]
(iii) Aromatic bases						
Aniline	4.1 −10	—	—	—	—	−27.1[234,c]
2-Nitro-aniline	7.0 −15	2.49	5.32	—	−26.3[144]	—

Appendix

Table 18—*continued*

Base	$K_m(1)$ at 1 bar mole kg^{-1}	$K_m(P)/K_m(1)$ P, bars: 1000	2000	3000	ΔV^∞(a), cm^3 mole^{-1}	ΔV^∞(b), cm^3 mole^{-1}
3-Nitro-aniline	2.6 −12	2.71	6.04	—	−28.5[144]	—
4-Nitro-aniline	1.0 −13	2.72	6.19	—	−28.5[144]	—
4-Methyl-2-nitro-aniline	2.9 −14	2.46	5.25	—	−25.8[144]	—
4-Methyl-3-nitro-aniline	7.0 −12	2.40	5.15	—	−25.3[144]	—
2-Methyl-4-nitro-aniline	9.1 −14	2.52	5.44	—	−26.4[144]	—
2-Methyl-5-nitro-aniline	2.1 −12	2.61	5.74	—	−27.5[144]	—
2-Hydroxy-5-nitro-aniline	5.7 −12	2.35	4.86	—	−24.5[144]	—
(iv) Cyclic bases						
Aziridine	1.0 −6	—	—	—	—	−26.8[244]
Azetidine	2.0 −3	—	—	—	—	−27.1[244]
Pyrrolidine	1.9 −3	—	—	—	—	−26.0[244]
Piperidine	1.3 −3	—	—	—	—	−24.2[244,245,c]
Hexamethylene imine		—	—	—	—	−24.6[244]
Heptamethylene imine	—	—	—	—	—	−25.1[244]
1-Methyl-pyrrolidine	2.3 −4	—	—	—	—	−26.6[244]
1-Methyl-piperidine	1.4 −4	—	—	—	—	−24.6[244,245]
Pyridine	1.6 −9	—	—	—	—	−26.8[234,246,c]
2-Methyl-pyridine	8.6 −9	—	—	—	—	−25.2[246]
2,6-Dimethyl-pyridine	4.5 −8	—	—	—	—	−23.7[246]
Imidazole (1st ionization)	9.0 −8	—	—	—	—	−22.8[233,234,238,247]

[a] The values of ΔV^∞(a) are derived from the pressure dependence of K_m, and the values of ΔV^∞(b) from volumetric measurements. The values of K_m and of ΔV^∞ are those for the basic ionization reaction $B + H_2O \rightleftharpoons BH^+ + OH^-$. In cases where the measurements were made on the acidic ionization reaction $BH^+ \rightleftharpoons B + H^+$ the results have been converted using the values of K_m for water listed in Table 17 and the value $\Delta V^\infty = -22.1$ cm^3 mole^{-1} for water (Ref. 211).

[b] Calculated from $V^\infty_{OH^-} = -4.0$, $V^\infty_{NH_4^+} = +17.9$ cm^3 mole^{-1},[213] and $V^\infty_{NH_3} = +24.8$ cm^3 mole^{-1} (from density data in Refs. 241).

[c] See footnote k, Table 17.

Table 19

Volume Changes for the Ionization of Weak Polyelectrolytes in Water at 25°C and the Effects of Pressure on Ionization[a]

Polyelectrolyte	$\alpha(1)$ at 1 bar	$\alpha(P)/\alpha(1)$ P, bars: 1000	2000	3000	$\Delta V(a)$, cm³ mole⁻¹	$\Delta V(b)$, cm³ mole⁻¹
(i) Polyacids						
Poly(acrylic acid)	0	—	—	—	—	−12.3[235]
	1	—	—	—	—	−24.6[235,248]
Poly(methacrylic acid)	0.026	0.032	—	—	−13.9[b]	—
	>0.5	—	—	—	—	−12.5[c]
	1	—	—	—	—	−27[281]
Poly(ethylene/maleic acid)	0.015	0.021	0.025	0.031	−15[c]	−9.1[235]
	0–0.5	—	—	—	—	−34.5[235]
Poly(methyl vinyl ether/maleic acid)	0.5–1	—	—	—	—	−8.8[235]
	0–0.5	—	—	—	—	−37.4[235]
	0.5–1	—	—	—	—	
Poly(D-glutamic acid)	0.030	0.034	0.040	0.046	−9[d]	—
Poly(L-glutamic acid)	1	—	—	—	—	−12.0[251,252]
Poly(S-carboxymethyl-L-cysteine)	1	—	—	—	—	−12.7[252]

Appendix

Table 19—*continued*

Polyelectrolyte	$\alpha(1)$ at 1 bar	$\alpha(P)/\alpha(1)$ P, bars: 1000	2000	3000	$\Delta V(\text{a})$, cm³ mole⁻¹	$\Delta V(\text{b})$, cm³ mole⁻¹
(ii) Polybases						
Poly(ethyleneimine)	0.5	—	—	—	—	−24[243,248]
Poly(L-lysine)	0–0.5	—	—	—	—	−25[253]

[a] The values of ΔV are all for finite concentrations of the polyelectrolytes. The values $\Delta V(\text{a})$ were derived from the relationship

$$\Delta V(\text{a}) = \left[-RT\,\partial\left(\ln\frac{\alpha^2 c}{1-\alpha}\right)\Big/\partial P\right]_{1\,\text{bar}}$$

(c = molal concentration of ionizing groups), which, for polyelectrolytes, is not a thermodynamic one. They are therefore only apparent volume changes, and in contrast to the real volume changes $\Delta V(\text{b})$, they show an apparent but false tendency to become less negative as α increases (as c decreases—see **Ref.** 159). The values $\Delta V(\text{b})$ are from volumetric measurements.

[b] The value of ΔV has been recalculated from the results of Ref. 159 using a simpler method than the authors used of extrapolating to $\alpha = 0$, and omitting the term $\partial(\ln c)/\partial P$, which they wrongly included in their expression for ΔV.

[c] Calculated from the results of Ref. 250.

[d] Calculated from the results of Ref. 250.

Table 20

Volume Changes for the Dissociation of Ion Pairs in Water at 25°C, and the Effects of Pressure on Dissociation[a]

Ion pair	Valence type	$K_m(1)$, at 1 bar mole kg^{-1}	$K_m(P)/K_m(1)$ P, bars: 1000	2000	3000	ΔV^∞(a), cm^3 mole^{-1}	ΔV^∞(b), cm^3 mole^{-1}
Na$^+$ SO$_4^{2-}$	1:2	2.9 −1	2.00	—	—	−15.8[254]	—
Mg^{2+} SO$_4^{2-}$	2:2	4.7 −3	1.27	1.93	—	−8.5[81]	—
Ca^{2+} SO$_4^{2-}$	2:2	4.9 −3	1.48	—	—	−10.2[255]	—
Mn^{2+} SO$_4^{2-}$	2:2	4.4 −3	1.30	1.65	—	−7.48[6a]	—
Co^{2+} SO$_4^{2-}$	2:2	2.5 −3	1.57	—	—	−10.5[278]	—
Ni^{2+} SO$_4^{2-}$	2:2	3.8 −3	1.68	—	—	−11.6[87]	—
Co(NH$_3$)$_5$NO$_2^{2+}$ SO$_4^{2-}$	2:2	2.0 −3	1.49	2.13	2.58	−10.5[279]	—
La^{3+} SO$_4^{2-}$	3:2	2.2 −4	2.13	3.64	—	−22.58[6b]	—
Ce^{3+} SO$_4^{2-}$	3:2	4.0 −4	—	—	—	—	−15.1[256]
Eu^{3+} SO$_4^{2-}$	3:2	2.1 −4	2.34	4.51	—	−25.6[282]	—
Ce^{3+} C$_2$H$_5$CO$_2^-$	3:1	2.1 −2	—	—	—	—	−23.6[256]
Ce^{3+} Cl$^-$	3:1	9 −1	—	—	—	—	−1[256]
Ce^{3+} NO$_3^-$	3:1	3.4 −1	—	—	—	—	−3.4[256]
Nd^{3+} NO$_3^-$	3:1	5.9 −2	—	—	—	—	−11[257]
Eu^{3+} NO$_3^-$	3:1	4.1 −1	—	—	—	—	−4.2[256]
Gd^{3+} NO$_3^-$	3:1	6.7 −2	—	—	—	—	−7.7[257]
Cr^{3+} OH$^-$	3:1	1.4 −10	1.65	—	—	−13[b]	—
Fe^{3+} SCN$^-$	3:1	6.9 −3	2.01	2.87	—	−17[262]	−17[262]
Co(NH$_3$)$_6^{3+}$ SO$_4^{2-}$	3:2	2.1 −4	1.26	1.55	1.81	−6.8[76]	—
La^{3+} Fe(CN)$_6^{3-}$	3:3	1.6 −4	1.34	1.66	—	−8.0[259]	—

Appendix

Table 20—*continued*

Ion pair	Valence type	$K_m(1)$, at 1 bar mole kg^{-1}	$K_m(P)/K_m(1)$			$\Delta V^\infty(a)$, cm^3 mole^{-1}	$\Delta V^\infty(b)$, cm^3 mole^{-1}
			P, bars: 1000	2000	3000		
Mg^{2+} (adenosine triphosphate)$^{4-}$	2:4	3 $\times 10^{-4}$	—	—	—	—	$-27.7^{261,c}$
(rhodamine B)$^+$ (rhodamine B)$^+$	—	$1.7 \times 10^{+2}$	0.79	0.58	0.39	$+8^{169}$	—
(methylene blue)$^+$ (methylene blue)$^+$	—	$1.8 \times 10^{+3}$	0.63	0.50	0.35	$+9^{169}$	—
Mg^{2+} (polyacrylate)	—	—	—	—	—	—	-26.6^{235}
Ba^{2+} (polyacrylate)	—	—	—	—	—	—	-33.2^{235}
Ag$^+$ (polyacrylate)	—	—	—	—	—	—	-20.8^{235}
Li$^+$ (polyphosphate)	—	—	—	—	—	—	-14.2^{263}
Na$^+$ (polyphosphate)	—	—	—	—	—	—	-13.1^{263}
Mg^{2+} (polyphosphate)	—	—	—	—	—	—	-47.6^{263}

*The values of $\Delta V^\infty(a)$ are derived from the pressure dependence of K_m, and the values of $\Delta V^\infty(b)$ from volumetric measurements. The values of K_m relate to the *dissociation* of ion pairs into free ions.
*Calculated from "hydrolysis" constants^{258} and the high-pressure values of K_w (Table 17).
*Calculated from the results of Rainford *et al.*,^{260} who reported that $\Delta V^\infty_{apparent} = -21.9$ cm^3 mole^{-1}. This value has been corrected for the increased ionization of ATPH^{3-} to ATP^{4-} using $\Delta V^\infty = -24.0$ cm^3 mole^{-1} (see entry in Table 17).

Table 21
Volume Changes for the Formation of Micelles in Water, and the Effects of Pressure on Critical Micelle Concentrations[a]

Salt	Temp., °C	CMC(1) at 1 bar, mole kg^{-1}		CMC(P)/CMC(1) P, bars:			ΔV(a), cm^3 mole^{-1}	ΔV(b), cm^3 mole^{-1}
				1000	2000	3000		
(i) *Acid salts*								
H$^+$ C$_{12}$H$_{25}$SO$_3^-$	25	4.4	-3	—	—	—	—	$+11^{264}$
Na$^+$ C$_{10}$H$_{21}$SO$_4^-$	25	3.3	-2	—	—	—	—	$+8.4^{265}$
Na$^+$ C$_{12}$H$_{25}$SO$_4^-$	25	8.3	-3	1.10	1.10	—	$+11^{133}$	$+11.3^{264,266}$
Na$^+$ C$_{14}$H$_{29}$SO$_4^-$	26	2.2	-3	—	—	—	—	$+15.3^{264,265}$
Na$^+$ C$_{12}$H$_{25}$SO$_3^-$	31.5	6.9	-3	—	—	—	—	$+10^{264}$
Na$^+$ C$_{14}$H$_{29}$SO$_3^-$	39	2.0	-3	—	—	—	—	$+11.5^{264}$
Na$^+$ C$_8$H$_{17}$C$_6$H$_4$SO$_3^-$	25	1.1	-2	—	—	—	—	$+16^{264}$
Na$^+$ [poly(D-glutamate)]	30	8.6	-3	1.06	1.07	1.05	$+5^{267}$	
K$^+$ C$_7$H$_{15}$CO$_2^-$	25	4.7	-1	—	—	—	—	$+10^{264}$
K$^+$ C$_{11}$H$_{23}$CO$_2^-$	25	2.4	-2	—	—	—	—	$+15.3^{267}$
(ii) *Base salts*								
C$_{12}$H$_{25}$NH$_3^+$ Cl$^-$	25	1.3	-2	1.04	1.05	—	$+2.6^{269}$	—
C$_{12}$H$_{25}$N(CH$_3$)$_3^+$ Cl$^-$	25	2.0	-2	1.03	0.98	0.92	$+2.9^{269}$	—
C$_{12}$H$_{25}$N(CH$_3$)$_2$(C$_7$H$_7$)$^+$ Cl$^-$	25	7.8	-3	1.06	1.04	0.97	$+4.4^{269}$	—

Appendix

Table 21—*continued*

Salt	Temp., °C	CMC(1) at 1 bar, mole kg^{-1}	CMC(P)/CMC(1) P, bars: 1000	2000	3000	ΔV(a), cm^3 mole^{-1}	ΔV(b), cm^3 mole^{-1}
C$_8$H$_{17}$NH$_3^+$ Br$^-$	25	2.3 −1	—	—	—	—	+6.0[270]
C$_8$H$_{17}$N(CH$_3$)$_3^+$ Br$^-$	25	1.4 −1	—	—	—	—	+3.8[265]
C$_{10}$H$_{21}$N(CH$_3$)$_3^+$ Br$^-$	25	6.5 −2	1.04	0.96	0.86	+4.1[271]	+6.9[265]
C$_{12}$H$_{25}$N(CH$_3$)$_3^+$ Br$^-$	25	1.6 −2	1.03	0.94	0.81	+3[271]	+8.5[265]
C$_{14}$H$_{29}$N(CH$_3$)$_3^+$ Br$^-$	25	3.2 −3	—	—	—	—	+10.6[265]

^{a}The values of ΔV(a) were derived from the critical micelle concentration CMC by the relationship

$$\Delta V(a) = [kRT\, \partial(\ln \text{CMC})/\partial P]_{1\,\text{bar}}$$

where k is a number near 1.8 (see Phillips[283]). The values of both ΔV(a) and ΔV(b) relate to micelle formation at the critical micelle concentration. The values of ΔV(b) are from volumetric measurements.

Table 22
Volume Changes for the Ionization of Weak Acids and Weak Bases in Nonaqueous and Mixed Solvents at 25°C, and the Effects of Pressure on Ionization Constants[a]

Weak acid or base	$K_m(1)$ at 1 bar, mole kg^{-1}		$K_m(P)/K_m(1)$ P, bars:			$\Delta V^\infty(a)$, cm^3 mole^{-1}	$\Delta V^\infty(b)$, cm^3 mole^{-1}
			1000	2000	3000		
(i) Solvent: methanol							
Methanol	2.0	−17	—	—	—	—	−45[b]
phenol	8.0	−15	—	—	—	—	−38.5[272]
4-Nitro-phenol			—	—	—	—	−31.7[272]
4-Bromo-phenol			—	—	—	—	−35.6[272]
4-Formyl-phenol			—	—	—	—	−30.4[272]
4-*Tert*-butyl-phenol			—	—	—	—	−37.4[272]
2,6-Di(*tert*-butyl)-4-nitro-phenol			—	—	—	—	−28.0[272]
2,6-Di(*tert*-butyl)-4-formyl-phenol			—	—	—	—	−31.8[272]
3,5-Di(*tert*-butyl)-phenol			—	—	—	—	−39.3[272]
2,4,6-Trinitro-phenol (picric acid)	2.6	−4	2.48	4.9	8.4	−25[97]	—
Piperidine	6.1	−6	3.56	9.1	20.3	−45[273]	−49[c]
Pyridine			—	—	—	—	−53[c]
2,6-Dimethyl-pyridine			—	—	—	—	−53[272]
2,6-Diethyl-pyridine			—	—	—	—	−55[277]

Table 22—*continued*

Weak acid or base	$K_m(1)$ at 1 bar, mole kg^{-1}		$K_m(P)/K_m(1)$ P, bars:			ΔV^∞(a), cm^3 mole^{-1}	ΔV^∞(b), cm^3 mole^{-1}
			1000	2000	3000		
2,6-Di(*iso*-propyl)-pyridine	—		—	—	—	—	−56[277]
2,6-Di(*tert*-butyl)-pyridine	—		—	—	—	—	−67[277]
(ii) *Solvent: ethanol/water, 50% (wt.)*							
Benzoic acid	2.2	−6	1.80	3.1	4.8	−16.1[113]	—
2-Hydroxy-benzoic acid (salicylic acid)	1.9	−4	1.67	2.6	3.6	−12.7[113]	—
4-Methyl-benzoic acid	2.3	−6	1.82	3.2	5.0	−16.1[113]	—

[a] The values ΔV^∞(a) are derived from the pressure dependence of K_m, and ΔV^∞(b) from volumetric measurements.
[b] Calculated from partial molar volume data given in Refs. 213 and 232.
[c] See Table 17, footnote *k*.

Table 23
Volume Changes for the Dissociation of Ion Pairs in Nonaqueous Solvents, and the Effects of Pressure on Dissociation[a]

Salt	Temp., °C	$K_m(1)$ at 1 bar, mole kg^{-1}	$K_m(P)/K_m(1)$ P, bars: 1000	2000	3000	$\Delta V^\infty(a)$, cm^3 mole^{-1}	$\Delta V^\infty(b)$, cm^3 mole^{-1}
(i) Solvent: ethanol							
K^+I^-	20	1.0 −2	1.8	2.5	—	-15^{274}	—
$N(CH_3)_4{}^+Br^-$	30	6.4 −3	1.4	1.7	2.0	-9^{275}	—
$N(C_4H_9)_4{}^+Br^-$	30	7.5 −3	1.3	1.6	1.8	-8^{275}	—
(ii) Solvent: 2-propanol (iso-propyl alcohol)							
K^+I^-	30	8.2 −4	2.4	5.0	4.8	-26^b	—
$N(CH_3)_4{}^+Br^-$	30	7.0 −4	2.3	3.9	6.3	-26^b	—
$N(CH_3)_4{}^+$ picrate$^-$	30	1.9 −3	2.2	3.1	3.3	-20^b	—
$N(C_4H_9)_4{}^+Br^-$	30	1.5 −3	1.7	2.6	2.8	-18^b	—
$N(C_4H_9)_4{}^+$ picrate$^-$	30	1.5 −3	1.8	3.7	4.6	-24^b	—
(iii) Solvent: 2-methyl-propan-1-ol (iso-butyl alcohol)							
Na^+I^-	25	1.4 −3	1.8	2.7	3.4	-17^{103}	—
MMP^+I^{-c}	30	2.4 −4	1.8	2.1	—	-16^{103}	—

Appendix

Table 23—*continued*

Salt	Temp., °C	$K_m(1)$ at 1 bar, mole kg^{-1}	$K_m(P)/K_m(1)$ P, bars: 1000	2000	3000	ΔV^∞(a), cm^3 mole^{-1}	ΔV^∞(b), cm^3 mole^{-1}
(iv) *Solvent: acetone*							
K^+I^-	20	8.4 -3	2.3	5.4	—	-20^{274}	—
$N(CH_3)_4{}^+I^-$	26.6	3.6 -3	2.15	—	—	-27^{102}	—
$N(C_2H_5)_4{}^+I^-$	26.6	7.7 -3	1.54	—	—	-17^{102}	—
$N(C_3H_7)_4{}^+I^-$	26.6	8.2 -3	1.38	—	—	-15^{102}	—
MMP^+I^- [c]	20	2.1 -3	2.3	3.3	3.3	-20^{102}	—

[a] The values of ΔV^∞(a) are derived from the pressure dependence of K_m. The accuracy of the values of K_m and ΔV^∞ is considerably less here than in the other tables. The errors in ΔV^∞ may be as great as ± 5 cm^3 mole^{-1}.
[b] Recalculated from the data of Ref. 276.
[c] MMP = 4-methoxycarbonyl-*N*-methyl-pyridinium.

2. Symbols and Units

K_m Ionization constant on the molality scale (moles of solute species per kilogram of solvent). In most cases the constants have been corrected by activity coefficients or extrapolated to infinite dilution. In the tables an ionization constant of, say, 6.2×10^{-8} mole kg^{-1} is shown as $6.2 - 8$.

α Degree of ionization.

P Pressure in bars (1 bar = 10^5 Pa $\equiv 0.9869$ atm).

V_B^∞ Partial molar volume of the solute species B at infinite dilution, in the units cm^3 mole^{-1}.

ΔV^∞(a)   Volume change for ionization at 1 bar and at infinite dilution, derived from the relationship

$$\Delta V^\infty(a) = -RT(\partial \ln K_m/\partial P)_{1\,\text{bar}}$$

where R denotes the gas constant and T the absolute temperature. The units are cm^3 mole^{-1}.

ΔV^∞(b)   Volume change for ionization at 1 bar and infinite dilution, derived from measurements of the partial molar volumes of the ionized and un-ionized species or from measurements of the volume change accompanying neutralization. The units are cm^3 mole^{-1}.

REFERENCES

[1] G. J. Hills, "Pressure coefficients of electrode processes," in *Advances in High Pressure Research*, Ed. by R. S. Bradley, Academic Press, London, 1969, Volume 2, p. 225; *Rev. Pure and App. Chem.* **18** (1968) 153.

[2] G. J. Hills and P. J. Ovenden, "Electrochemistry at high pressures," in *Advances in Electrochemistry and Electrochemical Engineering*, Ed. by P. Delahay, Interscience, New York, 1966, Vol. 4, 185.

[3] E. U. Franck, *Water and aqueous solutions at high pressures and temperatures*, XXII *International Congress of Pure and Applied Chemistry* (Sydney, 1969), Butterworths, London, 1970, p. 13; *Pure App. Chem.* **24** (1970) 13; *J. Chim. Phys.* Special Number (October 1969), p. 9.

[4] W. L. Marshall, *Rev. Pure Appl. Chem.* **18** (1968) 167.

References

[5] J. A. Barker and R. O. Watts, *Chem. Phys. Lett.* **3** (1969) 144.
[6] A. Rahman and F. H. Stillinger, *J. Chem. Phys.* **55** (1971) 3336; **57** (1972) 1281.
[7] T. Grindley and J. E. Lind, *J. Chem. Phys.* **54** (1971) 3983.
[8] G. S. Kell and E. Whalley, *Phil. Trans. Roy. Soc.* **A258** (1965) 565.
[9] G. C. Kennedy and W. T. Holser, *Handbook of Physical Constants; Geol. Soc. Am. Mem.* **97** (1966) 374.
[10] J. Jůza, *An Equation of State for Steam and Water—Steam Tables in the Critical Region and in the Range from 1000 to 100,000 Bars*, Naklad. Česk. Akad. Věd, Prague, 1966.
[11] W. E. Sharp, *Thermodynamic Functions for Water in the Range -10 to $1000°C$ and 1 to 250,000 Bars, UCRL Report* 7118, Livermore, California, 1962.
[12] H. Köster and E. U. Franck, *Ber. Bunsenges.* **73** (1969) 716.
[13] P. W. Bridgman, *Proc. Am. Acad. Arts Sci.* **47** (1912) 441; see also N. E. Dorsey, *Properties of Ordinary Water Substance*, Reinhold, New York, 1940, p. 214.
[14] K. E. Bett and J. B. Cappi, *Nature* **207** (1965) 620; J. B. Cappi, *Thesis*, Univ. of London, 1964.
[15] J. Wonham, *Nature* **215** (1967) 1053.
[16] N. A. Agaev and A. D. Yusibova, *Soviet Phys.—Doklady* **13** (1968) 472.
[17] E. M. Stanley and R. C. Batten, *J. Phys. Chem.* **73** (1969) 1187.
[18] V. Singh, S. K. Deb, and A. K. Barua, *J. Chem. Phys.* **46** (1967) 4036.
[19] B. B. Owen, R. C. Miller, C. E. Milner, and H. L. Cogan, *J. Phys. Chem.* **65** (1961) 2065.
[20] L. A. Dunn and R. H. Stokes, *Trans. Faraday Soc.* **65** (1969) 2906.
[21] H. Kyropoulos, *Z. Physik.* **40** (1926) 507.
[22] B. K. P. Scaife, *Proc. Phys. Soc.* **B68** (1955) 790.
[23] E. Schadow and R. Steiner, *Z. Phys. Chem. NF* **66** (1969) 105.
[24] B. B. Owen and S. R. Brinkley, *Phys. Rev.* **64** (1943) 32.
[25] H. S. Harned and B. B. Owen, *Physical Chemistry of Electrolytic Solutions*, Reinhold, New York (1958), Chapter 5.
[26] S. D. Hamann, P. J. Pearce, and W. Strauss, *J. Phys. Chem.* **68** (1964) 375.
[27] M. Nakahara, K. Shimizu, and J. Osugi, *Rev. Phys. Chem. Japan* **42** (1972) 12.
[28] O. F. Hale and F. H. Spedding, *J. Phys. Chem.* **76** (1972) 2925.
[29] W. L. Lees, Dissertation, Harvard Univ., 1949.
[30] E. U. Franck, *Ber. Bunsenges.* **76** (1972) 341.
[31] R. M. Waxler, C. E. Weir, and H. W. Schamp, *J. Res. Nat. Bur. Std.* **68A** (1964) 489.
[32] K. Tödheide, in *Water—A Comprehensive Treatise*, Ed. by F. Franks, Plenum, New York, 1972, Chapter 13.
[33] Ya. B. Zel'dovich, S. B. Kormer, M. V. Sinitsyn, and K. B. Yushko, *Soviet Phys.—Doklady* **6** (1961) 494.
[34] G. B. Benedek and E. M. Purcell, *J. Chem. Phys.* **22** (1954) 2003.
[35] H. G. Hertz and C. Rädle, *Z. Phys. Chem. NF* **68** (1969) 324.
[36] G. E. Walrafen, *J. Chem. Phys.* **48** (1968) 244; **52** (1970) 4176; **55** (1971) 768, 5137.
[37] D. Eisenberg and W. Kauzmann, *The Structure and Properties of Water*, Oxford, 1969, p. 236.
[38] E. U. Franck and K. Roth, *Disc. Faraday Soc.* **43** (1967) 108.
[39] J. A. Pople, *Proc. Roy. Soc.* **A205** (1951) 163.
[40] P. W. Bridgman, *Proc. Am. Acad. Arts and Sci.* **47** (1912) 441; *J. Chem. Phys.* **5** (1937) 964.
[41] E. Whalley, J. B. R. Heath, and D. W. Davidson, *J. Chem. Phys.* **48** (1968) 2362.
[42] E. Whalley, *Ann. Rev. Phys. Chem.* **18** (1967) 224.
[43] S. D. Hamann and W. Strauss, *Trans. Faraday Soc.* **51** (1955) 1684.
[44] S. D. Hamann and M. Linton, *Trans. Faraday Soc.* **65** (1969) 2186.

[45] S. D. Hamann and W. Strauss, *Disc. Faraday Soc.* **22** (1956) 70.
[46] S. D. Hamann and F. Smith, *Aust. J. Chem.* **24** (1971) 2431.
[47] J. E. Stutchbury, *Aust. J. Chem.* **9** (1956) 536.
[48] P. W. Bridgman, *Physics of High Pressure*, Bell and Sons, London, 1958.
[49] P. W. Bridgman, *Proc. Am. Acad. Arts and Sci.* **48** (1912) 309; **66** (1931) 185; **74** (1942) 399.
[50] P. W. Bridgman, *Proc. Am. Acad. Arts and Sci.* **61** (1926) 57; **77** (1949) 115.
[51] J. F. Skinner, E. L. Cussler, and R. M. Fuoss, *J. Phys. Chem.* **72** (1968) 1057.
[52] H. Hartmann, A. Neumann, and G. Rinck, *Z. Phys. Chem. NF* **44** (1965) 204; H. Hartmann and A. P. Schmidt, *Ber. Bunsenges.* **72** (1968) 875; H. Hartmann, R. Engelmann, and A. Neumann, *Z. Phys. Chem. NF* **66** (1969) 298.
[53] D. Colladon and C. Sturm, *Ann. Chim. Phys.* **36** (1827) 231.
[54] H. Herwig, *Ann. Phys. Chem.* **160** (1877) 110.
[55] J. Fink, *Ann. Phys. (Leipzig)* **26** (1885) 481.
[56] I. Fanjung, *Z. Phys. Chem.* **14** (1894) 673.
[57] W. Ostwald, *J. Prakt. Chem.* **18** (1878) 328.
[58] G. Tammann, *Z. Elektrochem.* **16** (1910) 592.
[59] E. Brander, *Comm. Phys.-Math., Helsingf.* **6**(8) (1932) 21, 42.
[60] E. Cohen and W. Schut, *Piezochemie kondensierte Systeme*, Akad. Verlag., Leipzig, 1919.
[61] S. D. Hamann, *Physico-Chemical Effects of Pressure*, Butterworths, London, 1957.
[62] W. Strauss, *Univ. of Sheffield Fuel. Soc. J.* **8** (1957) 1.
[63] R. A. Horne, "The effect of pressure on electrical conductivity of aqueous solutions," in *Advances in High Pressure Research*, Ed. by R. S. Bradley, Academic Press, New York, 1969, Vol. 2, p. 169.
[64] S. B. Brummer and A. B. Gancy, "Aqueous solutions under extreme conditions," in *Water and Aqueous Solutions*, Ed. by R. A. Horne, Wiley—Interscience, New York, 1972, p. 745.
[65] J. Buchanan and S. D. Hamann, *Trans. Faraday Soc.* **49** (1953) 1425.
[66] H. J. Welbergen, *J. Sci. Instr.* **10** (1933) 247.
[67] A. B. Gancy and S. B. Brummer, *J. Electrochem. Soc.* **115** (1968) 804.
[68] J. C. Jamieson, *J. Chem. Phys.* **21** (1953) 1385.
[69] C. E. Weir, *J. Res. Nat. Bur. Std.* **45** (1950) 465; **50** (1953) 95; **53** (1954) 245.
[70] C. W. F. T. Pistorius, *Polymer* **5** (1964) 315.
[71] R. I. Beecroft and C. A. Swenson, *J. Appl. Phys.* **30** (1959) 1793.
[72] A. J. Ellis, *J. Chem. Soc. London* (1959) 3689.
[73] S. D. Hamann, *Aust. J. Chem.* **11** (1958) 391.
[74] K. Mangold and E. U. Franck, *Ber. Bunsenges.* **73** (1969) 21, and earlier references given there.
[75] W. Holzapfel and E. U. Franck, *Ber. Bunsenges.* **70** (1966) 1105.
[76] M. Nakahara, K. Shimizu, and J. Osugi, *Rev. Phys. Chem. Japan* **40** (1970) 12.
[77] W. A. Zisman, *Phys. Rev.* **39** (1932) 151.
[78] R. A. Horne, B. R. Myers, and G. R. Frysinger, *J. Chem. Phys.* **39** (1963) 2666.
[79] A. B. Gancy and S. B. Brummer, *J. Phys. Chem.* **73** (1969) 2429; *J. Chem. Eng. Data* **16** (1971) 385.
[80] P. J. Ovenden, quoted in Ref. 2, p. 225.
[81] F. H. Fisher, *J. Phys. Chem.* **66** (1962) 1607.
[82] B. B. Owen and H. Zeldes, *J. Chem. Phys.* **18** (1950) 1083.
[83] F. T. Wall and J. Berkowitz, *J. Phys. Chem.* **62** (1958) 87.
[84] R. L. Kay, K. S. Pribadi, and B. Watson, *J. Phys. Chem.* **74** (1970) 2724.

References

[85] Y. Matsubara, K. Shimizu, and J. Osugi, *Rev. Phys. Chem. Japan* **43** (1973) (to be published).

[86a] F. H. Fisher and D. F. Davis, *J. Phys. Chem.* **69** (1965) 2595.

[86b] F. H. Fisher and D. F. Davis, *J. Phys. Chem.* **71** (1967) 819.

[87] E. Inada, K. Shimizu, and J. Osugi, paper presented at the 14th High Pressure Conference of Japan, Osaka, Japan, October 23–25, 1972.

[88] R. A. Horne, *Nature* **200** (1963) 418.

[89] R. A. Horne, *Structure Makers and Breakers in Water: Pressure-Induced Changes in the Hydration Atmospheres of Ions in Solution*, Office of Naval Research Report AD 449831, Defence Documentation Center, Virginia, 1964; *J. Geophys. Res.* **69** (1964) 1971.

[90] R. A. Horne, "The structure of water and aqueous solutions," in *Survey of Progress in Chemistry*, Ed. by A. F. Scott, Academic Press, New York, 1968, Vol. 4, p. 1.

[91] F. Hensel and E. U. Franck, *Z. Naturforsch.* **19a** (1964) 127.

[92] G. R. Hills, P. J. Ovenden, and D. R. Whitehouse, *Disc. Faraday Soc.* **39** (1965) 207.

[93] B. E. Conway, J. O'M. Bockris, and H. Linton, *J. Chem. Phys.* **24** (1956) 834.

[94] M. Eigen and L. de Maeyer, *Proc. Roy. Soc. London* **A247** (1958) 505.

[95] B. E. Conway, *Disc. Faraday Soc.* **39** (1965) 219.

[96] E. W. Schmidt, *Z. Phys. Chem.* **75** (1911) 305.

[97] W. Strauss, *Aust. J. Chem.* **10** (1957) 277.

[98] J. F. Skinner and R. M. Fuoss, *J. Phys. Chem.* **69** (1965) 1437; **70** (1966) 1426.

[99] S. B. Brummer and G. J. Hills, *Trans. Faraday Soc.* **57** (1961) 1823.

[100] E. L. Cussler and R. M. Fuoss, *J. Phys. Chem.* **71** (1967) 4459.

[101] S. P. Sawin and C. A. Eckert, *J. Chem. Eng. Data* **16** (1971) 476.

[102] W. A. Adams and K. J. Laidler, *Can. J. Chem.* **46** (1968) 1977, 2005.

[103] A. H. Ewald and J. A. Scudder, *Aust. J. Chem.* **23** (1970) 1939.

[104] F. Barreira and G. J. Hills, *Trans. Faraday Soc.* **64** (1968) 1359.

[105] S. B. Brummer, *J. Chem. Phys.* **42** (1965) 1636.

[106] C. M. Apt, F. F. Margosian, I. Simon, J. H. Vreeland, and R. M. Fuoss, *J. Phys. Chem.* **66** (1962) 1210.

[107] J. F. Skinner, E. L. Cussler, and R. M. Fuoss, *J. Phys. Chem.* **71** (1967) 4455.

[108] P. J. Pearce and W. Strauss, *Aust. J. Chem.* **23** (1970) 905.

[109] R. A. Horne, D. S. Johnson, and R. P. Young, *J. Phys. Chem.* **72** (1968) 866.

[110] R. A. Robinson and R. H. Stokes, *Electrolyte Solutions*, Butterworths, London, 1959.

[111] R. M. Fuoss and L. Onsager, *J. Phys. Chem.* **61** (1957) 668.

[112] W. G. Davies, R. J. Otter, and J. E. Prue, *Disc. Faraday Soc.* **24** (1957) 103.

[113] R. J. H. Clark and A. J. Ellis, *J. Chem. Soc. London* (1960) 247.

[114] J. Molenat, *J. Chim. Phys.* **67** (1970) 368.

[115] F. Körber, *Z. Phys. Chem.* **67** (1909) 212.

[116] L. H. Adams and R. E. Hall, *J. Phys. Chem.* **35** (1931) 2145.

[117] V. I. Sergeevich, T. P. Zhuze, and A. I. Chestnov, *Izv. Akad. Nauk SSSR, Otd. Tekh. Nauk* (1953) 896.

[118] G. Tammann, *Über die Beziehungen zwischen den innern Kräften und Eigenschaften der Lösungen*, Voss, Leipzig, 1907, p. 36.

[119] B. Cleaver, S. I. Smedley, and P. N. Spencer, *J. Chem. Soc. Faraday II* **68** (1972) 1720; A. F. M. Barton, B. Cleaver, and G. J. Hills, *Trans. Faraday Soc.* **64** (1968) 208.

[120] H. Le Chatelier, *Compt. Rend.* **99** (1884) 786.

[121] E. Braun, *Ann. Phys. Chem.* **30** (1887) 250.
[122] M. Planck, *Ann. Phys. Chem.* **32** (1887) 462.
[123] P. Drude and W. Nernst, *Z. Phys. Chem.* **15** (1894) 79.
[124] M. Born, *Z. Phys.* **1** (1920) 45.
[125] E. A. Guggenheim, *Thermodynamics*, North-Holland, Amsterdam, 1950, p. 205.
[126] E. A. Guggenheim, *Trans. Faraday Soc.* **33** (1937) 607.
[127] S. D. Hamann, "Chemical equilibria in condensed systems," in *High Pressure Physics and Chemistry*, Ed. by R. S. Bradley, Academic Press, New York, 1963, Vol. 2, p. 131.
[128] C. W. Davies, *J. Phys. Chem.* **29** (1925) 977.
[129] M. S. Sherrill and A. A. Noyes, *J. Am. Chem. Soc.* **48** (1926) 1861.
[130] D. A. MacInnes and T. Shedlovsky, *J. Am. Chem. Soc.* **54** (1932) 1429.
[131] G. Tammann and A. Rohmann, *Z. Anorg. Allg. Chem.* **183** (1929) 1.
[132] W. Hallwachs, *Ann. Phys. Chem.* **53** (1894) 1, 14.
[133] S. D. Hamann, *J. Phys. Chem.* **66** (1962) 1359.
[134] W. R. Hainsworth, H. J. Rowley, and D. A. MacInnes, *J. Am. Chem. Soc.* **46** (1924) 1437.
[135] G. J. Hills and D. R. Kinnibrugh, quoted by G. J. Hills and P. J. Ovenden in Ref. 2; *J. Electrochem. Soc.* **113** (1966) 1111.
[136] E. W. Tiepel and K. E. Gubbins, *J. Phys. Chem.* **76** (1972) 3044.
[137] I. R. Krichevskiĭ and J. S. Kazarnovskiĭ, *J. Am. Chem. Soc.* **57** (1935) 2168.
[138] K. E. Heusler and L. Gaiser, *Ber. Bunsenges.* **73** (1969) 1059.
[139a] A. Distèche, *Rev. Sci. Instr.* **30** (1959) 474.
[139b] A. Distèche and M. Dubuisson, *Bull. Inst. Oceanog. (Monaco)* **75** (1960) No. 1174.
[139c] A. Distèche, *J. Electrochem. Soc.* **109** (1962) 1084.
[139d] A. Distèche and S. Distèche, *J. Electrochem. Soc.* **112** (1965) 530.
[139e] A. Distèche and S. Distèche, *J. Electrochem. Soc.* **114** (1967) 330.
[140] S. D. Hamann, *J. Phys. Chem.* **67** (1963) 2233.
[141] E. R. Kearns, *Compressibilities of Some Dilute Aqueous Solutions*, Dissertation, Yale Univ., 1966.
[142] M. Whitfield, *J. Chem. Eng. Data* **17** (1972) 124.
[143] E. D. Linov and P. A. Kryukov, quoted by B. S. El'yanov and M. G. Gonikberg, Ref. 154.
[144] S. D. Hamann and M. Linton (to be published).
[145] A. H. Ewald and S. D. Hamann, *Aust. J. Chem.* **9** (1956) 54.
[146] H.-D. Lüdemann and E. U. Franck, *Ber. Bunsenges.* **71** (1967) 455; **72** (1968) 514.
[147] B. Scholz, H.-D. Lüdemann, and E. U. Franck, *Ber. Bunsenges.* **76** (1972) 406.
[148] E. Whalley, *J. Chem. Phys.* **38** (1963) 1400.
[149] S. W. Benson and C. S. Copeland, *J. Phys. Chem.* **67** (1963) 1194.
[150] J. E. Desnoyers, R. E. Verrall, and B. E. Conway, *J. Chem. Phys.* **43** (1965) 243.
[151] I. R. Krichevskiĭ, *Acta Phys. Chem. USSR* **8** (1938) 181.
[152] B. B. Owen and S. R. Brinkley, *Chem. Rev.* **29** (1941) 461.
[153] D. A. Lown, H. R. Thirsk, and Lord Wynne-Jones, *Trans. Faraday Soc.* **64** (1968) 2073.
[154] B. S. El'yanov and M. G. Gonikberg, *Izv. Akad. Nauk SSSR, Ser. Khim.* (1967) 1044; *Zh. Fiz. Khim.* **46** (1972) 1494.
[155] D. A. Lown, H. R. Thirsk, and Lord Wynne-Jones, *Trans. Faraday Soc.* **66** (1970) 51.
[156] R. Bonaccorsi, C. Petrongolo, E. Scrocco, and J. Tomasi, *J. Chem. Phys.* **48** (1968) 1500.
[157] A. J. Ellis, *J. Chem. Soc. Lond.* (1959) 3639; (1961) 4678.

References

[158] N. Bjerrum, *Z. Phys. Chem.* **106** (1923) 219.
[159] F. T. Wall and S. J. Gill, *J. Phys. Chem.* **58** (1964) 740.
[160] A. J. Begala and U. P. Strauss, *J. Phys. Chem.* **76** (1972) 254.
[161] K. Suzuki and Y. Taniguchi, *J. Polym. Sci.* (A-2) **8** (1970) 1679; *Biopolymers* **6** (1968) 215.
[162] R. J. Eldridge and F. E. Treloar (to be published); R. J. Eldridge, *Thesis*, Univ. of Melbourne. 1972.
[163] S. Nagakura, *J. Chem. Phys.* **23** (1955) 1441; J. N. Murrell, *The Theory of the Electronic Spectra of Organic Molecules*, Wiley, New York, 1963, Chapter 10.
[164] J. Czekalla and G. Wick, *Z. Elektrochem.* **65** (1961) 727.
[165] S. D. Hamann, *J. Phys. Chem.* **70** (1966) 2418.
[166] A. Weller, in *Progress in Reaction Kinetics*, Vol. 1, Ed. by G. Porter, Pergamon, New York, 1961, p. 189.
[167] C. O. Leiber, D. Rehm, and A. Weller, *Ber. Bunsenges.* **70** (1966) 1086.
[168] R. M. Fuoss, *J. Am. Chem. Soc.* **80** (1958) 5059.
[169] K. Suzuki and M. Tsuchiya, *Bull. Chem. Soc. Japan* **44** (1971) 967.
[170] W. L. Marshall and A. S. Quist, *Proc. Nat. Acad. Sci.* **58** (1967) 901; A. S. Quist and W. L. Marshall, *J. Phys. Chem.* **72** (1968) 1536.
[171] W. K. Marshall, *Rec. Chem. Prog.* **30** (1969) 61; *J. Phys. Chem.* **74** (1970) 346.
[172] P. W. Bridgman, *Proc. Am. Acad. Arts and Sci.* **74** (1942) 21.
[173] R. F. Trunin, M. A. Podurets, G. V. Simakov, L. V. Popov, and B. N. Moiseyev, *Zh. Eksperim. i Teor. Fiz.* **62** (1972) 1043.
[174] M. A. Podurets, G. V. Simakov, R. F. Trunin, L. V. Popov, and B. N. Moiseyev, *Zh. Eksperim. i Teor. Fiz.* **62** (1972) 710.
[175] G. G. Stokes, *Phil. Mag.* Ser. 3, **33** (1848) 349.
[176] J. Challis, *Phil. Mag.* Ser. 3, **33** (1848) 462.
[177] W. J. M. Rankine, *Phil. Trans. Roy. Soc.* **160** (1870) 277.
[178] H. Hugoniot, *J. École Polyt. Paris* **57** (1887) 1; **58** (1889) 1.
[179] J. H. Cook, *Research* **1** (1948) 474.
[180] A. H. Jones, W. M. Isbell, and C. J. Maiden, *J. Appl. Phys.* **37** (1966) 3493.
[181] S. D. Hamann, "Effects of intense shock waves," in *Advances in High Pressure Research*, Ed. by R. S. Bradley, Academic Press, New York, 1966, Vol. 1, p. 85.
[182] R. W. Goranson, D. Bancroft, B. L. Burton, T. Blechar, E. E. Houston, E. F. Gittings, and S. A. Landeen, *J. Appl. Phys.* **26** (1955) 1472.
[183] R. Schall, *Z. Angew. Phys.* **2** (1950) 252.
[184] L. V. Al'tshuler and A. P. Petrunin, *Soviet Phys.—Tech. Phys.* **6** (1961) 516.
[185] L. V. Al'tshuler, *Soviet Phys.—Uspekhi* **8** (1965) 52.
[186] G. E. Duvall and G. R. Fowles, "Shock Waves," in *High Pressure Physics and Chemistry*, Ed. by R. S. Bradley, Academic Press, New York (1963), Vol. 2, p. 209.
[187] R. N. Keeler, in *Proc. Intern. School Phys. Enrico Fermi* **48** (1971) 51, 138; and other articles in the same volume.
[188] M. H. Rice and J. M. Walsh, *J. Chem. Phys.* **26** (1957) 824.
[189] M. Cowperthwaite and R. Shaw, *J. Chem. Phys.* **53** (1970) 555.
[190] S. D. Hamann, *Rev. Pure Appl. Chem.* **10** (1960) 139.
[191] S. B. Kormer, *Soviet Phys.—Uspekhi* **11** (1968) 229.
[192] H. G. David and S. D. Hamann, *Trans. Faraday Soc.* **55** (1959) 72.
[193] H. G. David and S. D. Hamann, *Trans. Faraday Soc.* **56** (1960) 1043.
[194] S. D. Hamann and M. Linton, *Trans. Faraday Soc.* **62** (1966) 2234.
[195] S. D. Hamann and M. Linton, *Trans. Faraday Soc.* **65** (1969) 2186.
[196] S. D. Hamann, "Reactions at ultrahigh pressures," in *Proc. the XVII IUPAC Congress, Munich, 1959*, Butterworths, London, 1960.

[197] K. H. Dudziak and E. U. Franck, *Ber. Bunsenges.* **70** (1966) 1120.
[198] S. D. Hamann and M. Lintern, *J. Appl. Phys.* **40** (1969) 913.
[199] G. Rizert and E. U. Franck, *Ber. Bunsenges.* **72** (1968) 798.
[200] A. S. Quist and W. L. Marshall, *J. Phys. Chem.* **72** (1968) 798.
[201] A. S. Quist and W. L. Marshall, *J. Phys. Chem.* **72** (1968) 2100.
[202] A. S. Quist and W. L. Marshall, *J. Phys. Chem.* **72** (1968) 1545.
[203] S. D. Hamann and M. Linton, *J. Chem. Phys.* **51** (1969) 1660.
[204] L. V. Al'tshuler, A. A. Bakanova, and R. F. Trunin, *Soviet Phys.—Doklady* **3** (1958) 761.
[205] Ya. B. Zel'dovich, S. B. Kormer, M. V. Sinitsyn, and K. B. Yushko, *Soviet Phys.—Doklady* **6** (1961) 494.
[206] S. B. Kormer, K. B. Yushko, and G. V. Krishkevich, *Zh. Eksperim. i Teor. Fiz.* **54** (1968) 1640.
[207] M. van Thiel (Ed.), *Compendium of Shock Wave Data*, UCRL Report 50108, Lawrence Radiation Laboratory, Livermore, California, 1967.
[208] P. Wulff and H. K. Cameron, *Z. Phys. Chem.* **B10** (1930) 347; C. A. Swenson and J. R. Tedeschi, *J. Chem. Phys.* **40** (1964) 1141; R. Stevenson, *J. Chem. Phys.* **34** (1961) 346.
[209] T. J. Ahrens and M. H. Ruderman, *J. Appl. Phys.* **37** (1960) 346.
[210] C. R. Kurkjian and L. E. Russell, *J. Soc. Glass Tech.* **42** (1958) 130 T; H. Franz, *J. Am. Ceram. Soc.* **49** (1966) 473; L. Němec and J. Götz, *J. Am. Ceram. Soc.* **53** (1970) 526; A. G. Turnbull, *Aust. J. Chem.* **24** (1971) 2213.
[211] L. A. Dunn, R. H. Stokes, and L. G. Hepler, *J. Phys. Chem.* **69** (1965) 2808.
[212] *Landolt-Börnstein Tab.* **1** (1923) 390–1, 411–2; *Erg. 3* (1935) 254.
[213] J. Millero, *Chem. Rev.* **71** (1971) 147.
[214] J. G. Mathieson, Ph.D. Thesis, Univ. of Newcastle (1970)
[215] J. Timmermans, *Physico-Chemical Constants of Binary Systems in Concentrated Solutions*, Interscience, New York, 1940, Vol. 4, p. 61.
[216] *Landolt-Börnstein Tab. Erg. 2* (1931) 260.
[217] F. T. Selleck, L. T. Carmichael, and B. H. Sage, *Ind. Eng. Chem.* **44** (1952) 2219.
[218] A. J. Ellis and W. Giggenbach, *Geochim. Cosmochim. Acta* **35** (1971) 242.
[219] *Landolt-Börnstein Tab. 1* (1923) 411.
[220] *Landolt-Börnstein Tab. Erg. 1* (1927) 407.
[221] J. A. Ellis, *Chem. Commun.* (1966) 802.
[222] E. Brander, *Soc. Sci. Fennica Comm. Phys.—Math.* **6**(8) (1932) 42.
[223] L. M. Blair and J. A. Quinn, *Rev. Sci. Instr.* **39** (1968) 75.
[224] A. J. Ellis and D. W. Anderson, *J. Chem. Soc. Lond.* (1961) 1765.
[225] J. S. Smith, Dissertation, Yale Univ. (1943), quoted by H. S. Harned and B. B. Owen, Ref. 25, p. 405.
[226] R. Linnerstrøm-Lang and C. F. Jacobsen, *Compt. Rend. Trav. Lab. Carlsberg, Ser. Chim.* **24**(1) (1941) 1.
[227] H.-D. Lüdemann and E. U. Franck, *Ber. Bunsenges. phys. Chem.* **72** (1968) 523.
[228] E. U. Franck, D. Hartmann, and F. Hensel, *Disc. Faraday Soc.* **39** (1965) 200.
[229] R. A. Horne, R. A. Courant, and D. Frysinger, *J. Chem. Soc.* (1964) 1515.
[230] R. E. Lindstrom and A. E. Wirth, *J. Phys. Chem.* **73** (1969) 218.
[231] E. J. King, *J. Phys. Chem.* **73** (1969) 1220.
[232] S. D. Hamann and S. C. Lim, *Aust. J. Chem.* **7** (1954) 329.
[233] W. Kauzmann, A. Bodanszky, and J. Rasper, *J. Am. Chem. Soc.* **84** (1962) 1777.
[234] H. H. Weber, *Biochem. Z.* **218** (1930) 1.
[235] A. J. Begala and U. P. Strauss, *J. Phys. Chem.* **76** (1972) 254.
[236] A. Fischer, B. R. Mann, and J. Vaughan, *J. Chem. Soc. Lond.* (1961) 1093.

References

[237] E. J. Cohn, T. L. McMeekin, J. T. Edsall, and M. H. Blanchard, *J. Am. Chem. Soc.* **56** (1934) 784.
[238] S. Katz and J. E. Miller, *J. Phys. Chem.* **75** (1971) 1120.
[239] J. Daniel and E. J. Cohn, *J. Am. Chem. Soc.* **58** (1936) 415.
[240] R. K. Norris and S. Sternhell, *Aust. J. Chem.* **19** (1966) 841.
[241] *Landolt–Börnstein Tab* **1** (1923) 871; *Erg.* **1** (1927) 203; *Erg.* **2** (1931) 254; H. H. King, J. L. Hall, and G. C. Ware, *J. Am. Chem. Soc.* **52** (1930) 5128].
[242] R. E. Verrall and B. E. Conway, *J. Phys. Chem.* **70** (1966) 3961.
[243] J. Lawrence and B. E. Conway, *J. Phys. Chem.* **75** (1971) 2353.
[244] S. Cabani, G. Conti, and L. Lepori, *J. Phys. Chem.* **76** (1972) 1338, 1343.
[245] L. H. Laliberté and B. E. Conway, *J. Phys. Chem.* **74** (1970) 4116.
[246] B. E. Conway and L. H. Laliberté, in *Hydrogen Bonded Solvent Systems*, Ed. by A. K. Covington and P. Jones, Taylor and Francis, London, 1968, p. 139.
[247] L. M. Krausz, *J. Am. Chem. Soc.* **92** (1970) 1771.
[248] N. Ise and T. Okubo, *J. Am. Chem. Soc.* **90** (1968) 4527.
[249] K. Suzuki and Y. Taniguchi, *J. Polym. Sci.* (A-2) **8** (1970) 1679.
[250] K. Suzuki and Y. Taniguchi, *Biopolymers* **6** (1968) 215.
[251] H. Noguchi and J. T. Yang, *Biopolymers* **1** (1963) 359.
[252] S. Makino and H. Noguchi, *Biopolymers* **10** (1971) 1253.
[253] H. Noguchi, *Biopolymers* **4** (1966) 1105.
[254] D. R. Kester and R. M. Pytkowicz, *Geochim. Cosmochim. Acta* **34** (1970) 1039.
[255] E. Inada, K. Shimizu, and J. Osugi, *Nippon Kagaku Zasshi* **92** (1971) 1096; *Rev. Phys. Chem. Japan* **42** (1972) 1.
[256] T. G. Spiro, A. Revesz, and J. Lee, *J. Am. Chem. Soc.* **90** (1968) 4000.
[257] R. Garnsey and D. W. Ebdon, *J. Am. Chem. Soc.* **91** (1969) 50.
[258] T. W. Swaddle and Pi-Chang Kong, *Can. J. Chem.* **48** (1970) 3223.
[259] S. D. Hamann, P. J. Pearce, and W. Strauss, *J. Phys. Chem.* **68** (1964) 375.
[260] P. Rainford, H. Noguchi, and M. Morales, *Biochemistry* **4** (1965) 1958.
[261] T. Ikkai, T. Ooi, and H. Noguchi, *Science* **152** (1966) 1756.
[262] K. R. Brower, *J. Am. Chem. Soc.* **90** (1968) 5401.
[263] U. P. Strauss and Y. P. Leung, *J. Am. Chem. Soc.* **87** (1965) 1476.
[264] K. Shinoda and T. Soda, *J. Phys. Chem.* **67** (1963) 2072.
[265] J. M. Corkill, J. F. Goodman, and T. Walker, *Trans. Faraday Soc.* **63** (1967) 768.
[266] L. M. Kuchner, B. C. Duncan, and J. I. Hoffman, *J. Res. Nat. Bur. Std.* **49** (1952) 85.
[267] K. Suzuki and Y. Taniguchi, *Bull. Chem. Soc. Japan* **40** (1967) 1004.
[268] H. Lal, *J. Colloid Sci.* **8** (1953) 614.
[269] J. Osugi, M. Sato, and N. Ifuku, *Rev. Phys. Chem. Japan* **35** (1965) 32.
[270] J. E. Desnoyers and M. Arel, *Can. J. Chem.* **47** (1967) 359.
[271] R. F. Tuddenham and A. E. Alexander, *J. Phys. Chem.* **66** (1962) 1839.
[272] C. H. Rochester and B. Rossall, *Trans. Faraday Soc.* **65** (1969) 992.
[273] S. D. Hamann and W. Strauss, *Disc. Faraday Soc.* **22** (1956) 70.
[274] Yu. A. Ershov, M. G. Gonikberg, M. B. Neiman, and A. A. Opekunov, *Proc. Acad. Sci. USSR (Phys. Chem. Sect. Trans.)* **129** (1959) 803.
[275] E. L. Cussler and R. M. Fuoss, *J. Phys. Chem.* **71** (1967) 4459.
[276] C. G. Seefried, Ph.D. Dissertation, Yale, 1969.
[277] T. Asano and W. J. le Noble, Paper presented at the International Regional Meeting of AIRAPT (International Association for the Advancement of High Pressure Science and Technology), Kyoto, Japan, October 26, 1972, to be published in *Rev. Phys. Chem. Japan*.
[278] K. Shimizu and T. Okamoto, Paper presented at the 14th High Pressure Conference of Japan, Osaka, Japan, October 1972.

[279] M. Ueno, J. Osugi and K. Shimizu, Paper presented at the 14th High Pressure Conference of Japan, Osaka, Japan, October 1972, to be published in *Rev. Phys. Chem. Japan*.
[280] H. P. Hopkins, C. L. Liotta, and E. M. Perdue, *Abstracts, American Chemical Society 164th Meeting*, New York City, N.Y., 1972.
[281] C. Tondre and R. Zana, *J. Phys. Chem.* **76** (1972) 3451.
[282] C. F. Hale and F. H. Spedding, *J. Phys. Chem.* **76** (1972) 2925.
[283] J. N. Phillips, *Trans. Faraday Soc.* **51** (1955) 561.

3

Models for Molten Salts

H. Bloom and I. K. Snook

*Chemistry Department, The University of Tasmania, Hobart
Tasmania, Australia*

*Applied Physics Department, Royal Melbourne Institute of Technology
Victoria, Australia*

I. INTRODUCTION

1. Models

After more than 100 years of experimental work to determine the physical properties of molten salts and their mixtures to reasonably acceptable standards of accuracy, for the last 20 or so years much effort has gone toward the development of models for those systems capable of mathematical interpretation.

Many authors have discussed the interpretation of models of molten salts. Some recent contributors have been Stillinger,[1] Bloom and Bockris,[2] Blander,[3] Dampier and Janz,[4] and Angell.[5]

With systems as complicated as molten electrolytes, different models and different mathematical interpretations may be needed to explain different types of properties, e.g., equilibrium on the one hand and transport on the other. Some reasons for having models are to give an intuitively reasonable picture of the physical structure of molten salts; to explain the widely different properties of the various molten salts; and to assist prediction of properties that cannot easily be measured. Models need not be completely plausible as physical pictures of the systems described, but may yet enable prediction of properties and be used to rationalize the comparison between properties of various different systems. In the present state of development of the theory greater emphasis should perhaps be placed on the success of the model in enabling calculations of properties rather than the mathematical or physical plausibility of the model, even though this appears to contradict the first of the above reasons for setting up the model.

Liquids lie in structure between solids and gases. They do not have the advantages of the regularity of order characteristic of the solid state or of the complete disorder of the perfect gas. Their properties are, however, closer to those of the solid than those of the gaseous state [e.g., for NaCl the molar volume at the melting point, 1074°K, is 30.0 cm^3 mole^{-1} for the solid and 37.5 for the liquid, but is $\sim$138,336 for the vapor at 1686°K (the boiling point)]. Rigorous theories for liquids are at present only available for the noble gases,[44-49] although some progress has been made with liquids consisting of simple molecular species.[45,46] For molten salts general theories are even more difficult to establish than for, say, liquid argon, due partly to the fact that at least two types of particles, usually of different size, are involved—and to make for even greater complication, they have opposite electric charges, so that additional structuring occurs.

2. Radial Distribution Functions (RDF)

As a result of lack of knowledge, for a liquid of fixed structure extending over macroscopic regions of space for finite times, only approximate average properties can be specified. If an element of volume in the liquid is chosen which is small compared with the volume of a single particle, which in this case is one of the ions, it will sometimes contain the center of the particle, but at other times will not. The only statement that can be made is that the probability of the center of the particle being in the particular volume element is proportional to (a) the volume of the element and (b) the mean particle density of the liquid. If there are two volume elements separated by a distance large compared with the particle diameter, then the occupancy or vacancy of one is usually assumed not to affect that of the other.

In Fig. 1 the situation as it applies to a liquid can be specified by vectors $\mathbf{r}_1, \mathbf{r}_2$, etc. from an arbitrary origin, with elements of space at the ends of these vectors being specified by their volumes $d\mathbf{r}_1(= dx_1\, dy_1\, dz_1)$ and $d\mathbf{r}_2(= dx_2\, dy_2\, dz_2)$, which are not vectors. Distribution functions can now be defined as the probability of

Introduction

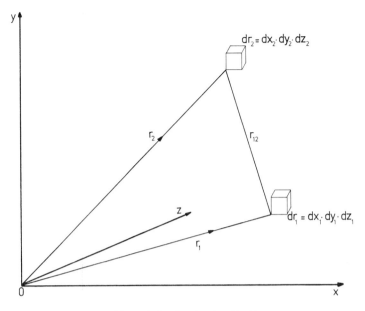

Figure 1. Lattice sites in a liquid.

finding particles in one or more volume elements. These probabilities can also be interpreted as the fraction of time that the elements of volume are occupied.

The probability of finding one particle in $d\mathbf{r}_1$ is defined to be $n^{(1)}(\mathbf{r}_1)\,d\mathbf{r}_1$, and is thus a dimensionless quantity lying between zero and one. Thus $n^{(1)}(\mathbf{r}_1)$ has the dimensions of reciprocal volume, or number density. Thus in an isotropic melt at equilibrium it is the total number of particles N divided by the total volume V and is independent of the position of $\mathbf{r}_1$. Thus

$$n^{(1)}(\mathbf{r}_1) = n \equiv N/V \tag{1}$$

If there are two volume elements as shown in Fig. 1 with scalar separation r_{12}, the probability of simultaneously finding particles in each element is denoted by $n^{(2)}(\mathbf{r}_1, \mathbf{r}_2)\,d\mathbf{r}_1\,d\mathbf{r}_2$. Since the liquid is isotropic, $n^{(2)}$ is a function only of r_{12} and not the absolute positions of $\mathbf{r}_1$ and $\mathbf{r}_2$. Since there is no long-range order in a liquid, the probability of simultaneous occupancy of the two volume elements is the product of the probability of independent occupancy, provided r_{12} is large, i.e.,

$$\lim_{r_{12} \to \infty} n^{(2)}(r_{12})\, d\mathbf{r}_1\, d\mathbf{r}_2 = (n^{(1)}\, d\mathbf{r}_1)(n^{(1)}\, d\mathbf{r}_2) = n^2\, d\mathbf{r}_1\, d\mathbf{r}_2 \qquad (2)$$

In order to aid the description of the structure, a dimensionless number $g^{(2)}(r_{12})$ is introduced called the radial (or pair) distribution function (RDF), defined by the equation

$$n^2 g^{(2)}(r_{12}) = n^{(2)}(r_{12}) \qquad (3)$$

so that $g^{(2)}(r_{12})$ is normalized to unity at large r_{12}. It can be seen from (3) that $g^{(2)}(r_{12})$ tends to zero as r_{12} becomes zero, this being the consequence of repulsive forces between particles at short distances preventing the occupation of the space by two particles simultaneously. The RDF is usually simply denoted by $g(r)$. If $g(r)$ is known, this leads to a knowledge of the time-averaging environment of any particle. The chance of finding another at distance r is greater than random for $g(r) > 1$ and less for $g(r) < 1$.

The general shape of $g(r)$ as determined by Levy et al.[6] for the molten salts KCl and LiBr is shown in Fig. 2, from which it can be seen that each ion tends to be surrounded by a shell of its neighbors and this tendency falls off rapidly with distance from the arbitrary origin (i.e., the reference ion). In contrast, similar diagrams for crystals show that $g(r)$ along a crystal axis consists of a set of sharp, narrow peaks separated by regions in which $g(r) = 0$ and for which the range is essentially infinite.

To measure structure, X-ray and neutron diffraction methods are commonly used for crystalline solids. Since liquids have some structure (in contrast to the perfect gas), $g(r) \neq 1$ for all $\mathbf{r}$, hence molten salts diffract X rays and neutrons, giving patterns which are diffuse compared with those of the crystalline salts. This diffraction arises from the correlations existing between pairs of ions. If there were no correlation, then $g(r)$ would equal unity and there would be no diffraction pattern (for monatomic ions). It is useful to define a pair correlation function $h(r)$ (sometimes called the total correlation function[97]) which measures the departure of $g(r)$ from its random value of unity. Thus

$$h(r) = g(r) - 1 \qquad (4)$$

Introduction

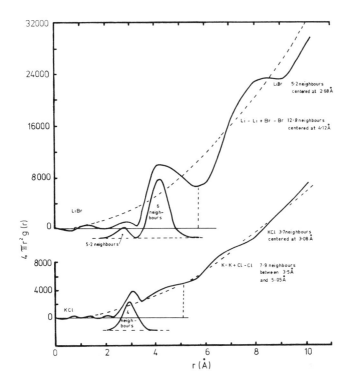

Figure 2. Radial distribution functions for molten LiBr and KCl by X-ray diffraction (from Levy et al.[6]).

A diffracted X-ray (or neutron) beam at an angle θ to the incident beam of wavelength λ has an intensity determined by a parameter s, where

$$s = (4\pi/\lambda)\sin(\theta/2) \tag{5}$$

Observed intensity curves (Levy et al.[6]) for molten KCl and LiBr are shown in Fig. 3.

This intensity at direction s is proportional to $H(s)$, which is the Fourier transform of $h(s)$, i.e., H(s) and h(s) are mutually related by the integrals

$$sH(s) = [2n/(2\pi)^{1/2}] \int_0^\infty rh(r)\sin(rs)\,dr \tag{6}$$

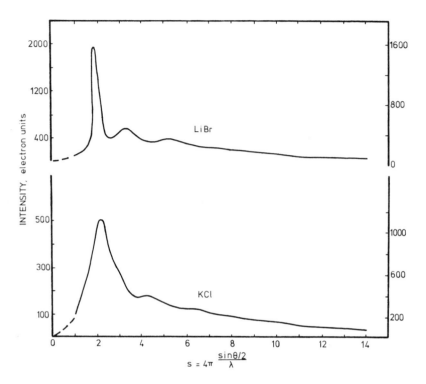

Figure 3. Observed intensity curves for molten LiBr and KCl by X-ray diffraction (from Levy et al.[6]).

and

$$rh(r) = [2/n(2\pi)^{1/2}] \int_0^\infty sH(s)\sin(rs)\,ds \qquad (7)$$

If the intensity $H(s)$ can be determined experimentally (e.g., by X-ray diffraction), then $h(r)$ and hence $g(r)$ [cf. equation (4)] can be found from (7).

On the other hand, if $g(r)$ is known from another method (see section on computer simulation), the X-ray diffraction pattern can be calculated from (6).

X-ray and neutron diffraction experiments show that each ion of one charge is surrounded by about 20% fewer nearest-neighbor ions of the other charge than in the solid, the interionic distance

Introduction

being slightly less in the liquid state than in the solid at the same temperature. The second shell of ions is assumed to be predominantly of the same charge as the reference ion, but the distance away is somewhat greater than in the solid and again the coordination number is less. One of the aspects relating to the interpretation of diffraction data is that only *average* values are given, so that it is difficult to use the data to determine the structure at any particular instant of time.

Another problem is that from X-ray diffraction measurements alone it is impossible to differentiate between the anions and cations making up the molten salt, the only simplification being that it is most likely that the nearest-neighbor shell will be composed predominantly of the oppositely charged ions (see, however, the discussion on computer simulation). Neutron diffraction studies, when combined with X-ray investigations, help to resolve this uncertainty by allowing comparison between diffraction experiments with different isotopes of the same ion and noting the changes produced in the neutron diffraction pattern.

The main structural features emerging from X-ray and neutron diffraction work, together with the knowledge that salts expand in volume by about 20% on melting,[7] can be interpreted using the following concepts.

(a) The average configurations of positive and negative ions will lead to microscopic local densities in the melt which are somewhat greater than the average density of the corresponding solid.

(b) There will be considerable microscopic volumes in the melt which have densities much lower than that of the solid. Many investigators have named such regions "holes," or "vacancies" if they are large enough on the average to accommodate an ion. In some theories regions of low densities are assumed to contain the vapor of the salt.[12]

Various workers, including Zarzycki,[8a,8b] Levy *et al.*,[6] and others, have obtained diffraction data by X-ray and neutron methods for a number of molten salts. The results have been interpreted by calculating radial distribution functions (RDF) for various combinations of ions, e.g., + and − nearest neighbors, + + and − − second nearest neighbors, etc. Although the method gives very useful and fundamental information, the results have the disadvantage of being time-averaged values and thus can be interpreted in different ways if individual instantaneous structural patterns are to be derived from them. Some results on molten alkali halides are shown in Table 1.

Table 1
Interionic Distance and Coordination Number for Some Alkali Halides at Their Melting Points (from Levy et al.[6])

Salt	Liquid at m.p.			Solid at m.p.		
	Cation–anion distance, Å	Coordination number (cation–anion)	Cation–cation or anion–anion distance, Å	Cation–anion distance, Å	Coordination number (cation–anion)	Cation–cation or anion–anion distance, Å
LiCl	2.47	4.0	3.85	2.66	6	3.76
CsCl	3.53	4.6	4.87	3.57	6	5.05
LiI	2.85	5.6	4.45	3.12	6	4.41
NaI	3.15	4.0	4.80	3.35	6	4.47

RDF data thus show that:

(i) Average coordination number for nearest neighbors is less in the melt than in the solid salt.

(ii) Positive and negative ions are, on the average, closer in the melt than in the crystal.

(iii) Like charged ions are further apart in the melt than in the solid.

These results can be combined with density measurements, which show that most alkali halides expand by about 20% on melting, to give various physical pictures of the molten salt. Figure 4 shows a possible interpretation of the melting of sodium chloride (using a two-dimensional representation). The smaller circles represent Na^+ and the larger circles, Cl^-. This picture is in agreement with the general deductions cited above.

Further reference will be made below to various models of molten salts. There also remains the question of representing two other classes of pure molten salts (apart from charge asymmetric salts):

(i) Salts containing ions that are not spherical, e.g., nitrates, carbonates, and sulfates, as well as the even more complicated class of molten silicates (in which polymeric entities are of considerable importance).

Introduction

 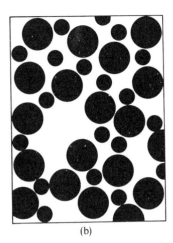

(a)   (b)

Figure 4. Two-dimensional representation of structure of (a) a 1–1 solid salt of the sodium chloride structure, (b) the same salt in its molten state. (In this figure the smaller circles represent one type of ion, e.g., the cation and the larger circles, the other.)

(ii) Salts that are not predominantly ionic over all temperature ranges, e.g., $ZnCl_2$.

This chapter will not concern itself with these classes.

In simple molten salts composed essentially of spherical ions the Coulombic interaction, with its long range of influence and its forces of attraction for some pairs of ions and of repulsion for other pairs, is difficult to fit into simple theories of liquids. Where charged particles in which one or both is of negligible dimensions are concerned plasma theory can be used, but when these particles have hard cores and long-range potentials of alternating sign the position is far more complicated. Questions to be answered, apart from electrostatic interactions, concern interaction between hard cores and long-range forces, the nature of the dielectric constant, and charge screening in these systems. More specific questions, such as, "What is the meaning and explanation of heats of activation of transport phenomena in molten salts?" and "How do we calculate activity coefficients for a component of a binary halide mixture?", require the setting up of special models for the particular purpose. The same model is not necessarily applicable to general considerations of molten salt structure on the one hand and to the specific properties listed above. On the other hand, the existence of models capable of rationalizing even small areas of the subject can be of tremendous practical use, and should such be available, would provide a considerable impetus to an attempt to find a more general theory.

Models of molten salts can be divided into two main categories:

(i) Operational models capable of yielding practical information using reasonably simple concepts which lead to equations containing as few as possible adjustable parameters. These include hole models (Bockris and Richards[9]), the liquid free volume model (Cohen and Turnbull[10]), the configurational entropy model (Adam and Gibbs[11]), and the significant structures model (Eyring and co-workers[12]).

(ii) Intermolecular force and related structural models derived from first principles and requiring such fundamental properties as ionic sizes and interionic forces (particularly the variation of these forces with distance between some centers). These include scaled particle theory (Stillinger,[1] Mayer[13]), the cell model (Lennard-Jones and Devonshire,[14] McQuarrie[15]), and, more recently, the computer simulation of molten alkali chlorides using Monte Carlo and molecular dynamics techniques (Woodcock and Singer,[16] Woodcock[76]).

These various approaches will be described in brief and an attempt will be made to evaluate the success of the various methods.

II. OPERATIONAL MODELS

1. Hole Models

A typical molar volume of an alkali halide melt can be taken as 50 cm^3. Of this volume ~ 10 cm^3 would be the additional volume gained on melting the solid (at its normal melting point). Clearly this additional volume need not be present in the form of holes big enough to accommodate an ion, but there are strong grounds for believing that these interstitial voids are arranged in such a manner as would allow a considerable proportion of them to be described as "holes." Even though RDF data apply to average configurations, the strong implication is that with decrease of average $+ -$ distance and increase of average $+ +$ and $- -$, the additional 10 cm^3 must be present in the form of quanta of space which, although not spherical, are large enough to accommodate ions which are able to use these spaces for transport phenomena, e.g., diffusion. The combination of the existence of the 20% volume expansion on melting together with a diminution of nearest-neighbor internuclear distance appears to make the existence of holes in the melt unambiguous.

Hole models generally are approximations to the implied situation in molten salt packing. The model has been developed into a very useful general guide for prediction of a number of properties by Bockris and co-workers.[9,17,18] This model originally, applied by Fürth[19] to molecular liquids, applies statistical fluctuation theory to the size of a hole generated in a liquid and obeying normal hydrodynamic relationships. The interparticle force relation is evaluated effectively by making use of the free energy of formation of a hole in the liquid calculated on the basis of the liquid–vapor surface tension. Bockris and Richards[9] were able to get good agreement between calculated and observed thermodynamic properties using this model, and later Bockris[17,18] was able to show that the model led to the deduction that the heats of activation for transport processes such as diffusion were given by $3.7RT_M$.

The most important equations of the Bockris et al. hole model are (i) the free energy of formation of a hole is given by

$$\Delta G_h = 4\pi r^2 \gamma N \tag{8}$$

where r is the most probable radius of the hole, γ is the surface tension, and N is Avogadro's number; (ii) the most probable volume of a hole is given by

$$V_h = \tfrac{4}{3}\pi r^3 = 0.68(kT/\gamma)^{3/2} \tag{9}$$

where r is the most probable radius, k is Boltzmann's constant, and T is the absolute temperature; and (iii) the enthalpy of formation of a hole is given by

$$\Delta H_h = 3.5RT_M \tag{10}$$

where T_M °K is the melting point of the salt.

The observed activation energy ($\Delta H^\ddagger$) of various transport processes, e.g., viscous flow and self-diffusion, can be expressed in terms of the equation

$$\Delta H^\ddagger = \Delta H_h + \Delta H_j \tag{11}$$

where ΔH_j is the enthalpy required for the particle to "jump" into the hole. Since it can be shown by measuring transport properties at constant volume as well as at constant pressure that

$$\Delta H_h \gg \Delta H_j \tag{12}$$

equation (11) can be expressed in a suitable form to be tested experimentally as

$$E_\eta \approx E_{D_+} \approx E_D \approx 3.7\, RT_M \tag{13}$$

where the E_η are the activation energies, i.e., the slope of the linear plot of $\ln \eta$ vs. $1/T$ multiplied by the gas constant R.

The derivation of equation (13) was first presented by Bockris and Richards,[17] but the details of the method were shown to lead to an inconsistency by Blander (cited by Yosim and Reiss[20]). Later Emi and Bockris[18] reexamined the derivation.

The expressions for work necessary to release an ion from the surface of a hole into a hole at temperature T is $A(T)/N_0$ where N_0 is the total number of ions and $A(T)$ the total work. Evaluation of A from first principles is difficult since the intermolecular forces involved must be known, but this work per particle was related by Emi and Bockris (following Fürth) to the work done against surface tension in forming the hole. If the average size of the hole is $\langle r_M \rangle$ at the melting temperature T_M, the work done to form the surface is $4\pi \langle r_M^2 \rangle$ at T_M. If it is assumed that the filling is done on the average by one ion at T_M, it follows that

$$A(T_M)/N_0 = 4\pi \langle r_M^2 \rangle \gamma_M \tag{14}$$

At a temperature $T\, (>T_M)$ the volume of the hole v_T will be $>v_M$ and a larger number of ions will be needed to fill the hole. Thus (14) becomes

$$A(T)/N_0 = 4\pi \langle r_T^2 \rangle \gamma_T / n \tag{15}$$

where n is ratio of number of ions per hole n_T' at T to the number n_M' at T_M. Since the difference between the volume of the liquid salt and that of the corresponding solid ΔV can be regarded as being due only to the holes, the number of holes per mole of salt becomes

$$N_h' = \Delta V_T / v_T \tag{16}$$

where v_T is now given by $(4/3)\pi \langle r_T^2 \rangle^{3/2}$. Thus for a 1:1 salt

$$n_T' = 2N_0/(\Delta V_T / v_T) \tag{17}$$

Operational Models

Hence

$$n_T'/n_M' = (\Delta V_M/\Delta V_T)(v_T/v_M) \tag{18}$$

where ΔV_M and v_M are the respective values of ΔV and v at T_M.

From Fürth's derivation of the distribution of hole sizes it can be shown that the average surface area of a hole is given by

$$4\pi \langle r^2 \rangle = 3.5kT/\gamma_T \tag{19}$$

and from equations (14), (15), (18), and (19) it follows that

$$A(T) = (T_M/T)^{1/2}(\gamma_T/\gamma_M)^{3/2}(\Delta V_T/\Delta V_M)A(T_M) \tag{20}$$

For simple molten salts at temperatures no more than 200° above the melting point we have

$$(T_M/T^{1/2})(\gamma_T/\gamma_M)^{3/2}(\Delta V_T/\Delta V_M) \approx 1 \tag{21}$$

Hence from equations (14), (20), (21), and (19)

$$A(T) \approx A(T_M) = 4\pi \langle r_M^2 \rangle \gamma_M N_0 = 3.5RT_M/\text{g-ion} \tag{22}$$

The evaluation of A (the work required to release one mole of ion pairs from the hole surfaces of the liquid into the holes) enables the equations for viscosity and (by using the Stokes–Einstein equation) the diffusion coefficient of the molten salt to be given in the forms

$$\eta = 0.6(\Delta V_T \gamma_T/V_T)(2\pi m/kT)^{1/2} \exp(3.5RT_M/RT) \tag{23}$$

where m is the mass of the ion pair and V_T is the molar volume of the molten salt, and

$$D = [0.17 V_T kT/V_T \gamma_T^{1/2}(2\pi m)^{1/2}] \exp(-3.5RT_M/RT) \tag{24}$$

From equations (23) and (24) the Arrhenius activation energies for viscosity and diffusion can readily be calculated. Although they contain temperature-dependent terms, these terms compensate each other over the temperature range considered (200°) and hence for practical purposes are temperature independent.

Equation (13) implies that the plot of E_η vs. T_M should be linear for all liquids and that the slope of the line should be $3.7R$. Bockris and Richards[17] collected activation energy data (for both viscous flow and diffusion) for a large number of liquids of different types,

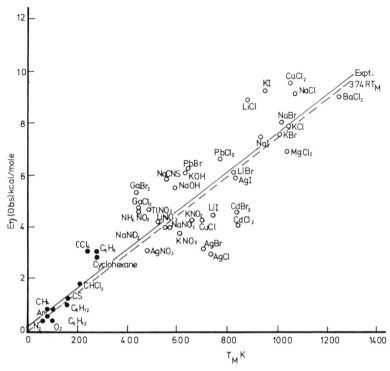

Figure 5. Energy of activation for viscous flow (E_η) for a number of liquids including molten salts as a function of melting temperature T_M °K. (Reprinted with permission from Bockris and Richards, *J. Phys. Chem.*, **69** (1965) 671. Copyright by the American Chemical Society.)

e.g., N_2, organic liquids, molten salts, and molten metals. The relevant data are shown in Fig. 5.

The scatter of points about the theoretical line is considerable but the trend of the relationship between activation energies and melting temperature is remarkably in agreement with equation (13). One of the most important aspects of this equation is that it does not contain any adjustable parameter. This is an important point when comparing the Bockris and Richards model with those of other investigators and will be referred to again later.

The values of E_η and E_D for a number of molten salts (both values being nearly identical) were found to give reasonable agreement with experimentally determined activation energies and as shown in Fig. 5, $E_D \approx E_\eta \approx 3.7RT_M$, for a wide range of liquids. For molten salts (mainly halides) the average values of $E_\eta(\text{calc})/RT_M$ and $E_D(\text{calc})/RT_M$ are about 3.9, which agrees reasonably with Fig. 5 (only E_η is shown in this diagram).

Emi and Bockris[18] were thus able, without adjustable parameters, to give an approximate numerical account of the heat of activation for transport at constant pressure for simple liquids. Rationalization of the empirical relations between the heats of activation of transport and the melting temperature has not yet been made by any other theory.

Assumptions made in derivation of the equations were:

(1) An ion in a hole behaves as an ideal gas. Since the size of the hole is, on the average, the same as that of the ion, this is equivalent to stating that an ion within a molten salt behaves as an ideal gas—there is little evidence to support this viewpoint—certainly the molar volumes, and hence the room available to the ions, are vastly different.

(2) The macroscopic surface tension between liquid surrounding a hole and vapor within the hole applies. Although this assumption does not appear reasonable at first sight, because the average hole is the same in size as the ion within it, it is probably acceptable if the surface tension is regarded as a time-averaged quantity. Equation (21) illustrates that it is in actual fact the ratio of surface tensions γ_T/γ_M which is important in this context. The hole theory will be shown later (see the summary) to compare favorably with other theories in its application to the calculation of molten salt properties.

2. Liquid Free–Volume Model

Free volume in a melt was first proposed by Zernicke and Prins[21] to explain the increase in volume of a salt on melting and further heating. The original concept was of an increase in the volume of each cell in the solid to allow for the volume requirements of the liquid. This volume was distributed equally among the various individual cells of the salt. Cohen and Turnbull,[10] on the other hand, considered that the free volume should be associated with the liquid as a whole, leading to the concept of "liquid free volume," v_f, or the excess volume that can be introduced into a liquid without change of energy. This "liquid free volume" can eventually lead to individual holes which allow transport through the liquid.

In order to evaluate their model, Cohen and Turnbull began with the potential ("cell potential" V) of a molecule within its coordination cell. At high density this potential was assumed to depend only on the average separation of particles and the coordination number. Thus the central volume ω will define a class of cell shapes for fixed immediate neighbor positions for which the cell potential remains constant within the usual thermal energy limit, kT. This becomes

$$\omega \approx K_c(R - R_c)^3 \qquad (25)$$

where R_c is the "hard-core" radius of the particle and is thus a little less than its equilibrium crystal radius R_0 in order to permit $V(R)$ to fluctuate up to kT, and K_c is the proportionality constant. R is the *average* spacing between the central particle and its nearest neighbors. The average "excess volume" $\Delta \bar{v}$ is thus

$$\Delta \bar{v} \approx K_c(R - R_0) \qquad (26)$$

at high relative densities.

The distribution of excess volume was shown qualitatively by Cohen and Turnbull to be nonuniform, and thus the potential varies from cell to cell. Below a certain temperature T_0, v_f is virtually zero, but at higher temperatures it approaches

$$v_f \approx K_c(\bar{R} - R_c) \qquad (27)$$

(where $\bar{R}$ is the average size of the coordination cell) and exceeds $\Delta \bar{v}$ due to the small decrease of the hard-core radius R_c as temperature increases.

The transition or reference temperature T_0 has been associated by Cohen and Turnbull with the transition from glass to liquid. With molten salts, values of T_0 often need to be taken to be around room temperature. The main uses of the Cohen and Turnbull model have been by Angell[5,22] to compare the transport properties of molten salts. Angell developed the theory of Cohen and Turnbull to apply not only to diffusion, but extended it to include electrical conductivity and viscosity. Thus the work of Angell would apply almost exclusively to glassy states (including viscous salts just above their melting points) rather than to ionic melts.

When the voids (holes) reach a certain critical size by random thermal redistribution of free volume through the liquid structure,

Operational Models

transport takes place by transfer of an adjacent particle into the void. The probability of such an occurrence is an exponential function of the ratio of the critical void volume to the total free volume, leading to the following expression for the diffusion coefficient D:

$$D = gau(-\gamma v^*/v_f) \qquad (28)$$

where g is a geometric factor, given the value of 1/6; a is the jump distance; u is the gas kinetic velocity of the particle; γ is an overlap factor for free volume; and v^* is the critical void volume. Using this equation with known molecular constants and making the assumptions outlined above for relating v_f with T_0, the expression reduces to

$$D = AT^{1/2}\exp[-k/(T-T_0)] \qquad (29)$$

where A and k are constants.

Upon differentiation, we find

$$d(\ln D)/d(1/T) = -\tfrac{1}{2}T - k[T/(T-T_0)]^2 \qquad (30)$$

Comparing the usual Arrhenius activation energy for diffusion, this is given by

$$D = A'\exp(-E_D/RT) \qquad (31)$$

where E_D is the activation energy for diffusion and A' is another constant.

Differentiation with respect to $1/T$ gives

$$d(\ln D)/d(1/T) = -E_D/R \qquad (32)$$

Combining (30) and (32), one obtains

$$d(\ln D)/d(1/T) = -E_D/R = -\tfrac{1}{2}T - k[T/(T-T_0)]^2 \qquad (33)$$

and this equation implies that if the Cohen and Turnbull model is valid, $(E_D - \tfrac{1}{2}RT)$ should be a linear function of $[T/(T-T_0)]^2$ which passes through the origin. To test this equation is very difficult in practice since the accuracy of diffusion data is not usually sufficient.

If it is assumed that the Nernst–Einstein equation holds, we have

$$\Lambda = (F^2/RT)(D_+ + D_-) \qquad (34)$$

where Λ is the equivalent conductance, F is Faraday's constant and D_+ and D_- are the diffusion coefficients for cation and anion, respectively, then

$$E_D = E_\Lambda + RT \tag{35}$$

provided there is near equivalence or marked dissimilarity between D_+ and D_- for a particular molten electrolyte. Thus according to the Cohen and Turnbull model, $(E_\Lambda + \frac{1}{2}RT)$ should be a linear function of $[T/(T - T_0)]^2$.

The usual method of checking the above equations is to put them into the form

$$D = AT^{1/2} \exp[-B/(T - T_0)] \tag{36}$$

$$\Lambda = A'T^{-1/2} \exp[-B'/(T - T_0)] \tag{37}$$

$$\eta = A''T^{1/2} \exp[B''/(T - T_0)] \tag{38}$$

Since the quantities D, Λ, and η are not very sensitive to the $T^{-1/2}$ or $T^{1/2}$ term, this is usually dropped.

Equations of the type (36)–(38) have long been used to describe the behavior of non-Arrhenius liquids empirically (Tamann and Hesse[23]). In some cases it is possible to identify T_0 with the experimentally determined glass transition temperature T_g (from heat capacity measurements or differential thermal analysis). T_g marks a second-order thermodynamic transition which is dependent on the time scale of the experiment due to the slowness of relaxation processes, whereas T_0 corresponds to the same change carried out infinitely slowly; hence $T_0 < T_g$.

The Cohen–Turnbull theory makes many arbitrary and unlikely assumptions. Its application to molten salt systems has been discussed by Dampier and Janz,[4] whose evaluation of this model will be discussed along with that of the Adam–Gibbs model below.

3. Relationship between Free Volume from Different Models and the Hole Volume

The hole volume V_h as used by Bockris and co-workers is (within a few percent) the additional volume introduced into the salt by its act of melting, and the volume of each hole is thus $\Delta V_M/N_h$, where N_h is the number of holes. Another volume term, known as the "free volume," has been used by a number of investigators. This has been discussed by Bockris and Richards, who measured this quantity by ultrasonic velocity determinations in molten salts, and found that the free volume, i.e., the volume in which an ion may move before striking its nearest-neighbor ions, is approximately 2% of the hole volume for a number of molten alkali halides. Thus the greater part of the additional volume introduced into the salt when it melts is due to holes, not additional "slackness" of structure. This is in keeping with the X-ray structure determination.

Cohen and Turnbull's "free volume" differs considerably from the "free volume" of Bockris and Richards. According to Cohen and Turnbull, the "free volume" is related to the "excess volume" introduced into the salt by its change of temperature on heating through the transition temperature T_0, and above the melting temperature their "free volume" is thus more closely related to the "hole volume" of Bockris and Richards rather than to their "free volume." The Cohen and Turnbull "free volume" becomes greater than their "excess volume" at higher temperatures due to the decrease in hard-core radii of the ions with increase of temperature above T_0.

4. The Adam and Gibbs Configurational–Entropy Theory

In spite of the work of Bockris et al.[9,17,18] an objection to the Cohen–Turnbull free volume theory in the minds of some critics is that diffusion occurs by molecular-size jumps. Adam and Gibbs[11] derived a theory in which diffusion was assumed to be due to cooperative rearrangements of groups of particles. It is the minimum size of the cooperatively rearranging groups which determines the temperature dependence of the transport process in this model. This minimum size can be obtained in terms of the macroscopic configurational entropy if the cooperatively rearranging groups are assumed not to interact significantly. Angell[24] confirmed that the electrical conductance is proportional to the transition probability $\overline{W}(T)$. According to Adam and Gibbs,

$$\overline{W}(T) = \overline{A} \exp[(-\Delta\mu)s_c^*/kTS_c] \qquad (39)$$

$$= \overline{A} \exp(-C/TS_c) \qquad (40)$$

where $\overline{A}$ and C are constants and $\Delta\mu$ is the potential energy hindering the cooperative rearrangement. s_c^* is the configurational entropy of the critical size cooperative region and S_c is the molar configurational entropy [i.e., $\Delta C_p \ln(T/T_0)$, where ΔC_p is the difference in heat capacities between liquid and glassy states].

If ΔC_p is constant,

$$\overline{W}(T) = \overline{A} \exp\{-C/[T \ln(T/T_0)]\} \tag{41}$$

Angell expanded the $\ln(T/T_0)$ term by using the series

$$\ln \frac{T}{T_0} = \frac{(T/T_0) - 1}{T/T_0} + \frac{1}{2}\left[\frac{(T/T_0) - 1}{T/T_0}\right]^2 + \cdots \tag{42}$$

for $T/T_0 > \frac{1}{2}$. Hence, neglecting all but the first term on the right-hand side,

$$\overline{W}(T) = \overline{A} \exp[-C/(T - T_0)] \tag{43}$$

where

$$C = (\Delta\mu)s_c^*/(R/N)$$

The transition probability is inversely proportional to the relaxation time, i.e., it is proportional to $1/\eta$. Hence

$$\eta = A' \exp[B'/(T - T_0)] \tag{44}$$

and

$$\Lambda = A'' \exp[-B''/(T - T_0)] \tag{45}$$

These equations differ from the Cohen–Turnbull theory by the absence of the $T^{-1/2}$ term. The frequency factor A has customarily been assumed to be temperature independent in any case. The usual method of testing the Adam–Gibbs theory is to test the linearity of a plot of $(E + \frac{1}{2}RT)$ as a function of $[T/(T - T_0)]^2$.

Equation (43) is usually used as the approximate expression of the Adam–Gibbs equation, but the neglect of the higher-order terms in the $\ln(T/T_0)$ expression is not valid for ratios of T/T_0 differing very much from unity. Since the values of T_0 are very commonly in the range 150–300°K, the use of equation (43) is clearly invalid for all but very low-melting salts (e.g., nitrates). For ionic salts, such as NaCl (MP 1074°K), the above approximation is clearly ridiculous.

5. The Significant Structures Model

Eyring and co-workers[12] based their significant structures model on the concept that a liquid consists of fragments of the crystal structure of the material separated by regions consisting of compressed gas of the same material. Thus the difficulties associated with the application of thermodynamics to the partially disordered structures associated with the liquid state were (in theory) bypassed by substituting the two states of matter for which precise thermodynamic relations can be established.

In detail, the method consists in evaluating the thermodynamic partition functions of the solidlike regions and the gaslike regions separately and combining them to give the complete partition function of the liquid. This is achieved by using two adjustable parameters which may be evaluated by means of a comparison liquid. In a later paper Eyring and co-workers[25] introduced a third adjustable parameter by considering the equilibrium between assumed monomer and dimer species in the "gaslike" regions of the liquid.

The final expression for the molten salt partition function f_l thus becomes

$$f_l = \left\{ \exp \frac{(E_s/2RT)(V/V_s)^{1/3}}{[1 - \exp(-\theta/T)]^3} \right.$$
$$\times \left[1 + n\frac{V - V_s}{V_s} \exp \frac{-aE_s(V/V_s)^{1/3}}{2RT(V - V_s)/V_s} \right] \right\}^{2NV_s/V}$$
$$\times \left\{ \left(\frac{F_1 eV}{N} \right)^{N(V-V_s)/V} \left(1 + 2\frac{n_2}{n_1} \right)^{n_1} \left[\frac{K_2 N}{eV} \left(\frac{n_1}{n_2} + 2 \right) \right]^{n_2} \right\} \quad (46)$$

where

$$F_1 = \left(\frac{2\pi mkT}{h^2} \right)^{3/2} \frac{8\pi^2 I_1 kT}{\sigma' h^2} \left[1 - \exp \frac{-hv}{kT} \right]^{-1}$$

N is Avogadro's number, n is the number of molecules, and the subscripts 1 and 2 denote monomer and dimer gas molecules, respectively. V is the molar volume and V_s is that of the solid species at the melting point, K is the equilibrium constant for the monomer–

dimer reaction,

$$2MX \rightleftharpoons (MX)_2 \tag{47}$$

F is the partition function per unit volume of gaslike molecules, i.e., for the monomer, $F_1 = f_1/(V - V_s)$, etc., m is the molecular mass, k is Boltzmann's constant, h is Planck's constant, I is the moment of inertia, σ' is the symmetry number, and v is the ground-state vibrational frequency. E_s is the sublimation energy per mole of solid, θ is the Einstein characteristic temperature of the solid, and n and a are adjustable parameters which Ree et al.[12] state can be calculated from the number of nearest neighbors and from the sublimation energy.

By making various assumptions, Eyring and co-workers have calculated equations relating viscosity and conductivity of liquids to the assumed solidlike and gaslike structures. In these calculations at least one other adjustable parameter, the "proportionality constant," is introduced.

Lu et al.[25] have compared the calculated and measured molar volume, vapor pressure, entropy, heat capacity, thermal coefficient of expansion, compressibility, viscosity, and specific conductivity of molten salts by means of the above method of calculation. Some comparisons are shown in Tables 2 and 3.

Although there is reasonable agreement between calculated and measured molar volume and entropy for molten NaCl and KCl (Table 2), the viscosity and conductance calculations leave much to

Table 2

Molar Volume and Entropy[25]

Salt	T, °K	V, cm³		S, e.u.	
		Calc.	Obs.	Calc.	Obs.
NaCl	1074 (m.p.)	37.42	37.56	36.37	40.77
	1200	39.17	39.29	37.92	42.57
	1500	43.12	44.12	41.26	46.14
KCl	1043 (m.p.)	48.77	48.80	39.00	42.21
	1100	49.93	49.88	39.80	43.06
	1200	51.96	51.91	41.14	44.45
	1500	58.35	59.11	44.93	48.02

Table 3
Specific Conductance of Molten Salts[25]

Salt	T, °K	κ, $\Omega^{-1}\text{cm}^{-1}$ Calc.	Obs.
NaCl	1080	2.400	3.596
	1200	2.832	3.954
	1290	3.052	4.179
NaBr	1030	1.516	2.922
	1200	1.934	3.432
KCl	1050	1.502	2.174
	1200	1.926	2.522
KI	1000	0.832	1.369
	1180	1.086	1.656

be desired. The incorporation of so many adjustable parameters into the significant structures model makes any attempt to compare calculated and experimental values meaningless (cf. the case also of water).

Regardless of the deficiencies of the theory of significant structures, several workers have expanded its application; e.g., Jhon et al.[26] introduced the additional concept of complex ions when discussing its application to the properties of the molten mercuric halides. Assuming an autocomplex formation equilibrium,

$$2HgX_2 \rightleftharpoons HgX^+ + HgX_3^- \tag{48}$$

these workers introduced a further constant into the treatment; the equilibrium constant for the reaction (48). This constant was evaluated from conductivity data by assuming the validity of Walden's rule, a rule which has been regarded as not applicable to molten salts.[27] The calculations yield results which are in good quantitative agreement with experiment for such properties as viscosity of the molten mercuric halides.

Regardless of the intuitively pleasing nature of the Eyring significant structures theories, the use of several adjustable parameters and the doubtful validity of many of the assumptions made in evaluating various properties make the application of these theories to molten salts of little value.

Regardless of the amount of effort spent in such calculations, it can hardly be claimed that the significant structures theory adds much to our knowledge of the structure of molten salts.

III. MODELS INVOLVING INTERMOLECULAR FORCES

An attempt will now be made to describe the calculation of the properties of a molten salt from first principles, as opposed to the use of operational models where in most cases extensive use of experimental data is made.

There are essentially two distinct steps to this procedure; the first is to obtain the intermolecular potential energy for the system, and the second is to calculate values of the bulk properties from this microscopic information. The first step may be accomplished by means of quantum mechanical calculations, by finding the parameters of an assumed potential energy function from some measured bulk property, or, more usually, by a combination of both approaches.[28,29] The second step may be tackled by the methods of statistical mechanics,[40,46] and since these two steps are distinct, they will be discussed in separate sections.

1. Intermolecular Potentials in Molten Salts. Basic Theory

In this section we will outline the basic theory of intermolecular potentials. The fundamental equation which describes the behavior of electrons and nuclei is the Schrödinger equation,[30] which for the stationary state of a system of N electrons and P nuclei is

$$\mathscr{H}\Psi = E\Psi \qquad (49)$$

where E is the total energy of the system, Ψ is the wave function, the Hamiltonian operator $\mathscr{H}$ is

$$\mathscr{H} = -\sum_{k=1}^{P} \frac{\hbar^2}{2M_k}\nabla_k^2 - \sum_{i=1}^{N} \frac{\hbar^2}{2m}\nabla_i^2 + V_{ee}(\mathbf{r}) + V_{en}(\mathbf{r}, \mathbf{R}) + V_{nn}(\mathbf{R})$$

where M_k is mass of nucleus k; m is the mass of an electron; $\mathbf{r}$ labels the electron coordinates and $\mathbf{R}$ the nuclear coordinates; and the electron–electron, electron–nuclei, and nuclei–nuclei potential terms are

$$V_{nn}(\mathbf{R}) = \tfrac{1}{2} \sum_{k \neq l = 1}^{P} Z_k Z_l e^2 |\mathbf{R}_k - \mathbf{R}_l|^{-1}$$

$$V_{en}(\mathbf{r}, \mathbf{R}) = \sum_{k=1}^{P} \sum_{i=1}^{N} Z_k e^2 |\mathbf{R}_k - \mathbf{r}_i|^{-1}$$

$$V_{ee}(\mathbf{r}) = \tfrac{1}{2} \sum_{i \neq j = 1}^{N} e^2 |\mathbf{r}_i - \mathbf{r}_j|^{-1}$$

respectively, with $Z_k e$ the charge on nucleus k and e the charge on an electron. We have neglected all relativistic and spin effects in $\mathscr{H}$ since they contribute little to the intermolecular potential.

A great simplification can be made by realizing that the nuclei are far more massive than the electrons and hence that their motions can be separated.[28–30] Therefore, we can write, approximately,

$$\Psi = \Psi_e(\mathbf{r}, \mathbf{R}) \Psi_n(\mathbf{R}) \qquad (50)$$

where $\Psi_e(\mathbf{r}, \mathbf{R})$ depends on the electronic coordinates and only parametrically on the nuclear coordinates $\mathbf{R}$, and $\Psi_n(\mathbf{R})$ depends only on R, which upon substitution in (49) leads to two equations,

$$\mathscr{H}_e(\mathbf{r}, \mathbf{R}) \Psi_e(\mathbf{r}, \mathbf{R}) = E_e(\mathbf{R}) \Psi_e(\mathbf{r}, \mathbf{R}) \qquad (51)$$

$$\mathscr{H}_n(\mathbf{R}) \Psi_n(\mathbf{R}) = E \Psi_n(\mathbf{R}) \qquad (52)$$

where

$$\mathscr{H}_e = - \sum_{i=1}^{N} (\hbar^2/2m) \nabla_i^2 + V_{ee}(\mathbf{r}) + V_{en}(\mathbf{r}, \mathbf{R})$$

$$\mathscr{H}_n = - \sum_{k=1}^{P} (\hbar^2/2M_k) \nabla_k^2 + V_{nn}(\mathbf{R}) + E_e(\mathbf{R})$$

Thus this separation of nuclear and electronic motion (the Born–Oppenheimer approximation) enables us to solve an equation (51) for the electronic energy for a fixed set of nuclear coordinates. This energy $E_e(\mathbf{R})$ then provides part of the potential in which the nuclei move, i.e., the intermolecular potential $U'(\mathbf{R})$,

$$U'(\mathbf{R}) = V_{nn}(\mathbf{R}) + E_e(\mathbf{R}) \qquad (53)$$

Equation (53) is usually modified slightly by adjusting the energy of the system so that the potential energy is zero when the molecules are infinitely separated,

$$U(\mathbf{R}) = U'(\mathbf{R}) - E_\infty \qquad (54)$$

where E_∞ is the energy of separated molecules.

Two different wave mechanical approaches have been used to find practical solutions: the approach based on the molecular orbital method and its extensions, and the Heitler–London method and its extensions.

Exact solutions of (51) may be written[30]

$$\Psi_e = \Psi_e^0 + \chi_e \qquad (55)$$

where Ψ_e^0 is a Hartree–Fock or molecular orbital function, which is a single determinant of doubly occupied one-electron functions (spin orbitals) $\phi_i(i)$ for a closed-shell system, i.e.,

$$\Psi_e^0 = (N!)^{-1/2} \det|\phi_1(1) \cdots \phi_N(N)| \qquad (56)$$

and χ_e is a correlation function which allows for the instantaneous correlations of electronic motion not allowed for by the independent particle wave function Ψ_e^0. The electronic energy may then be written[30]

$$E = E_{HF} + E_{correl} \qquad (57)$$

where E_{HF} is the Hartree–Fock energy and E_{correl} is the correlation correction.

In practice Ψ_e^0 is obtained to a high degree of accuracy but χ_e is approximated in some manner—usually by means of a superimposition of configurations technique, i.e., expanding χ_e in a finite set of determinants.[35,36] χ_e may also be thought of as a way of introducing dispersion energy into the MO method.[36]

The alternative approach has been to use wave functions expressed in terms of separated-molecule wave functions,[28,31] which is equivalent to the Heitler–London method with corrections for dispersion and higher-order exchange effects. Mathematically, we have

$$\Psi_e = \sum_{i=1}^{\infty} C_i Q \chi_i \qquad (58)$$

where C_i are coefficients, Q is a symmetry projection operator which gives a function of correct symmetry ("allows for electron exchange between the molecules"), and the χ_i are products of ground- and excited-state wave functions of the separated molecules.

The exact energy corresponding to (58) is

$$E_e = E_\infty + \frac{\langle \chi_0 | VQ\chi_0 \rangle}{\langle \chi_0 | Q\chi_0 \rangle} + \frac{\langle \chi_0 | VT_1 V | \chi_0 \rangle}{\langle \chi_0 | Q\chi_0 \rangle} \qquad (59)$$

where $H_e = H_0 + V$, V is the intermolecular part of H_e, T_1 is an operator which is equivalent to a certain sum over products of excited-state wave functions of separated molecules (correctly symmetrized).[31,32]

The exact energy cannot of course be obtained, and the usual method is to approximate T_1 by a power series which is equivalent to perturbation theory,[31,32] the result to second order being

$$E_e = E_\infty + E^1 + E^2 \qquad (60)$$

where

$$E^1 = \frac{\langle \chi_0 | VQ\chi_0 \rangle}{\langle \chi_0 | Q\chi_0 \rangle}, \qquad E^2 = \frac{\langle \chi_0 | VT_1^1 V | \chi_0 \rangle}{\langle \chi_0 | Q\chi_0 \rangle}$$

where $T_1^{\,1}$ is a first-order approximation to T_1.

It may be shown that E^1 is in fact the Heitler–London energy of the system. The second-order term E^2 is not unique and many different second-order energies may be obtained; however, in all of them we may write

$$E^2 = E^2_{\text{pol}} + E^2_{\text{exch}} \qquad (61)$$

where E^2_{pol} is the second-order polarization energy of dispersion and induction, which does not depend on electron exchange, and E^2_{exch} does depend on exchange. If approximations to (55) and (58) are taken to infinite order, both would give the same energy.

In practice for He–He interactions the results[32] of (58) are in good agreement with the results of an approximation[33,34] to (55); thus we will think of the two schemes as largely equivalent.

Since exact solutions or even very accurate ones using the above fundamental ideas are very difficult for systems of interest, e.g., KCl, the problem of intermolecular forces has usually been circumvented by the use of models. The known long-range forces are usually used, but the other parameters of the models (usually two-body additive potentials) are obtained from experiment, mostly from solid-state properties.[65,66,69] The pair potentials obtained may be thought of as "effective" pair potentials, i.e., pair potentials which give the same results for bulk properties as would be given by the true, many-body potential.

2. Statistical Mechanics of Molten Salts

A link between the microscopic properties of molecules (e.g., the intermolecular potential) and the bulk, observable properties of a macroscopic system is by means of statistical mechanics.[40–43] One usually divides statistical mechanics into equilibrium and non-equilibrium branches since these two sections require different techniques and relate to equilibrium and transport phenomena, respectively. We shall follow this normal division and use only classical theory, since this is adequate for the systems discussed.

(i) Equilibrium Statistical Mechanics

This branch of statistical mechanics considers systems with no gradients of temperature, composition, or pressure. If we consider a system, e.g., a molten salt of N ions, then this system has fN degrees of freedom if each ion has f degrees of freedom. Thus the dynamic state or microstate of the system may be defined by specifying the values of $2fN$ variables, e.g., $3N$ positions and $3N$ momenta for N monatomic ions.

This dynamic state of a system Γ_t changes with time and the change is described by Newton's laws of motion, i.e., for a system of N molecules of mass m

$$m\, d^2\mathbf{r}_i/dt^2 = -(d/d\mathbf{r}_i)U_N(\mathbf{r}_i, \ldots, \mathbf{r}_N) \tag{62}$$

where $\mathbf{r}_i$ denotes the position coordinate of molecules and $i = 1, 2, \ldots, 3N$.

Equilibrium properties X_{eq} only depend on the time average of a mechanical variable X over an infinite time interval, i.e.,

$$X_{eq} = \lim_{t \to \infty} (1/t) \int_0^t X(\Gamma_t)\, dt \tag{63}$$

where X depends on the microstate.

On the other hand, the observable state of a macroscopic system or the macrostate of the system depends on a very few variables, e.g., pressure p, volume V, and temperature T. Hence the observable state or macrostate of a system of many molecules depends only on the microstate of its constituent molecules in some averaged way.

The method of equilibrium statistical mechanics replaces the above-mentioned time average of a mechanical variable by a probability weighted average over all possible microstates of a system, consistent with the macrostate of interest. This collection of microstates is called an ensemble and the probability weighted average, an ensemble average.

One should note that there are other ways of stating the basis of equilibrium statistical mechanics than that indicated in the above paragraphs, i.e., other than the replacement of time averages by ensemble averages (the ergodic approach). However, the above method is better suited to the discussion of the Monte Carlo and molecular dynamics methods.

Thus we have different ensembles and ensemble averages corresponding to different ways of defining the macrostate. However, in the thermodynamic limit, i.e., as $N \to \infty$, $V \to \infty$, $N/V = \rho$, with N the number of molecules, V the volume, and ρ the density, all ensembles give the same result for a macroscopic property and we shall consider only the canonical ensemble since this is the one most commonly used.

The probability density in the canonical ensemble is given by[44]

$$P_N(\mathbf{r}_i, \mathbf{p}_i) = e^{-\beta H} \bigg/ \int e^{-\beta H} \, d\mathbf{r}_1 \cdots d\mathbf{p}_N \qquad (64)$$

where P_N is the probability density, $\beta = 1/kT$, k is Boltzmann's constant, $\mathbf{p}_i$ are the momenta of the molecules, and

$$H = \sum_i (\mathbf{p}_i^2/2m_i) + U_N \qquad (65)$$

It is usual to write

$$P_N(\mathbf{r}_i, \mathbf{p}_i) = \left\{ \prod_j h^{-3N_j}(N_j!)^{-1} \right\} (e^{-\beta H}/Z) \qquad (66)$$

where Z is the canonical partition function defined by

$$Z = [1/\prod_j h^{-3N_j}(N_j!)^{-1}] \int e^{-\beta H} d\mathbf{r}_1 \cdots d\mathbf{p}_N$$

and all the thermodynamic information about the system is in fact contained in Z since it may be shown that[43]

$$\ln Z = -\beta A \tag{67}$$

where A is the Helmholtz free energy of the system.

All other thermodynamic properties of the system can then be obtained from the normal thermodynamic laws, e.g.,

$$pV/NkT = \rho(d/d\rho)(A/NkT)$$

where p is the pressure, V the volume, etc.

For a system of particles with no internal degrees of freedom (e.g., KCl) the integrations over momenta in Z may be carried out and we have[44]

$$Z = \left\{\prod_j \lambda_j^{-3N_j}(N_j!)^{-1}\right\} \int \exp(-\beta U_N) d\mathbf{r}_1 \cdots d\mathbf{r}_N \tag{68}$$

where j runs over all the different species (e.g., for KCl, $j = 1, 2$) of number N_j, $N = \sum N_j$; and $\lambda_j = h/(2\pi m_j kT)^{1/2}$ (h is Planck's constant) arises from integration over the momenta.

Obviously, directly evaluating (68) is not much more inviting than solving (62); however, much progress has been made in recent years of finding very good approximations to Z for liquids[44–49] and in subsequent sections we shall discuss some of these approximations for molten salts and see what has been learnt of molten salt structure from them.

(*a*) *Simple models.* The operational models of Section II.1 are all essentially attempts to approximate Z for a molten salt without recourse to intermolecular potentials, or may be thought of as such. Thus they attempt to obtain an approximate, parameterized form of Z. This of course is useful to correlate known experimental data and to predict other properties.

However, these models are very valuable for treating systems too complex for more exact models and do give insight into the

Models Involving Intermolecular Forces

structure of complex melts and allow properties to be correlated, as already pointed out.

(b) *Radial distribution function.* The energy of a one-component system interacting according to an additive potential $U_N = \frac{1}{2}\sum_{i \neq j} u(r_{ij})$ is given by[44]

$$E = $$

$$\left(\int H e^{-\beta H}\, d\mathbf{r}_1 \cdots d\mathbf{r}_N\, d\mathbf{p}_1 \cdots d\mathbf{p}_N\right) \Big/ \left(\int e^{-\beta H}\, d\mathbf{r}_1 \cdots d\mathbf{r}_N\, d\mathbf{p}_1 \cdots d\mathbf{p}_N\right)$$

$$= \tfrac{3}{2}NkT + \tfrac{1}{2}\left[\int \sum_{i \neq j} u(r_{ij}) e^{-\beta U_N}\, d\mathbf{r}_1 \cdots d\mathbf{r}_N\right] \Big/ \left(\int e^{-\beta U_N}\, d\mathbf{r}_1 \cdots d\mathbf{r}_N\right)$$

$$= \tfrac{3}{2}NkT + \tfrac{1}{2}\int n^{(2)}(r_{12})\, d\mathbf{r}_1\, d\mathbf{r}_2 \qquad (69)$$

where the h-particle distribution function is defined by

$$n^{(h)}(\mathbf{r}_1,\ldots,\mathbf{r}_h) = \frac{N!}{(N-h)!}\int P_N(\mathbf{r}_i)\, d\mathbf{r}_{h+1}\cdots d\mathbf{r}_N$$

$$= \frac{N!}{(N-h)!}\frac{\{\int e^{-\beta U_N}\, d\mathbf{r}_{h+1}\cdots d\mathbf{r}_N\}}{\{\int e^{-\beta U_N}\, d\mathbf{r}_1 \cdots d\mathbf{r}_N\}} \qquad (70)$$

We may further define the radial distribution function $g^{(2)}(r_{12})$ by

$$n^{(2)}(r_{12}) = [N(N-1)/V]g^{(2)}(r_{12}) \qquad (71)$$

and in the thermodynamic limit we have

$$n^{(2)}(r_{12}) = \rho^2 g^{(2)}(r_{12})$$

as used earlier [cf. equation (3)].

It can be shown that

$$E = \tfrac{3}{2}NkT + 2\pi N\rho \int_0^\infty g(r)u(r)r^2\, dr \qquad (72)$$

where $g(r) = g^{(2)}(r_{12})$,

$$\frac{pV}{NkT} = 1 - \frac{2\pi}{3}\frac{\rho}{kT}\int_0^\infty g(r)\frac{\partial u}{\partial r}r^3\,dr \qquad (73)$$

and

$$kT(\partial\rho/\partial\rho)_T = 1 + 4\pi\rho\int_0^\infty [g(r) - 1]r^2\,dr \qquad (74)$$

Hence all the thermodynamic properties of a system whose intermolecular potential is pairwise additive are obtainable from $g(r)$. In fact, three-body effects may also be included by use of $g^{(3)}(r_{12}, r_{13}, r_{23})$.[38]

However, $g(r)$ is more important than even this would suggest, since it gives information about the structure, albeit a time-averaged structure, of a liquid. It must be emphasized, however, that $g(r)$ is determined experimentally by measuring the scattering of X rays or neutrons from a fluid,[42,46] as pointed out earlier in this chapter. Thus $g(r)$ provides further experimental information about the structure and thermodynamic properties of a fluid and a test for theories of the fluid state, provided it can be determined initially using X-ray or neutron methods.

Since molten salts are mixtures of at least two distinct species, then at least three radial distribution functions can be defined, e.g., for a simple binary salt melt we have $g_{++}(r)$, $g_{+-}(r)$, and $g_{--}(r)$ as the cation–cation, cation–anion, and anion–anion radial distribution functions, respectively.

Equations (72) and (73) must be generalized to[51]

$$E = C^{-1}\left\{\sum_{i=1}^v \tfrac{3}{2}C_i kT + \sum_{i=1}^v \sum_{j=1}^v \tfrac{1}{2}C_i C_j \int u_{ij}(r)g_{ij}(r)\,d\mathbf{r}\right\} \qquad (75)$$

and

$$p = \sum_{i=1}^v C_i kT - \sum_{i=1}^v \sum_{j=1}^v \tfrac{1}{3}C_i C_j \int (du_{ij}/dr_{ij})g_{ij}(r)r\,d\mathbf{r} \qquad (76)$$

where there are v distinct species of concentration C_i and $C = \sum_{i=1}^v C_i$.

The relevant equation relating radiation scattering and $g(r)$ is[1]

$$\frac{I_{sc}(s)}{I_0} = \frac{1}{4\pi R^2} \left\{ \sum_{i=1}^{v} N_i f_i^2 + \frac{1}{V} \sum_{i,j=1}^{v} N_i N_j f_i f_j \right.$$

$$\left. \times \int_0^\infty 4\pi r^2 \, dr \frac{\sin sr}{sr} [g_{ij}(r) - 1] \, dr \right\} \quad (77)$$

where $s = (4\pi/\lambda)\sin(\theta/2)$, λ is the wavelength of the radiation, I_{sc} is the scattered intensity of a beam of radiation of incident intensity I_0 scattered through an angle θ to the incident direction at a distance R from the sample, and the f_i are the ionic scattering factors.

A further reason for introducing $g(r)$ is that we may learn something of its properties from statistical mechanics and also obtain exact equations for it, which may be solved approximately.[49,52,97]

At low densities

$$g(r) = e^{-\beta u(r)}[1 + g_1(r)\rho + g_2(r)\rho^2 + \cdots] \quad (78)$$

where $g_i(r)$ is independent of ρ but is related to $u(r)$.[49,50,53] This expansion in powers of the density ρ (the virial expansion) has proven of great value in the study of normal fluids.[49,50] Most importantly, by summing the virial series one can obtain exact integral equations for $g(r)$ which can be solved approximately in density regions where (78) is a very poor approximation to $g(r)$.

However, a difficulty arises in the case of systems with a long-range Coulomb potential; all the individual integrals in (78) diverge.[51,52] These integrals must first be resummed to obtain a convergent virial series with the potential being a shielded Coulomb or Debye–Hückel potential rather than the direct Coulomb potential. Then this series may be summed and an integral equation for $g(r)$ obtained.[52]

Allnatt[52] has carried out this procedure starting from Meeron's[53] series expansion for g_{ij}, i.e.,

$$g_{ij} = \exp(k_{ij} - U_{ij}^*\beta + \alpha_{ij}) \quad (79)$$

where

$$k_{ij} = -Z_i Z_j e^2 \beta \exp(-\kappa r_{ij})/Dr_{ij} \quad (80)$$

$$\kappa^2 = 4\pi e^2 \beta \left(\sum_{k=1}^{\nu} C_k Z_k^2 \right) \Big/ D \tag{81}$$

$$U_{ij} = U_{ij}^* + (Z_i Z_j e^2 / D r_{ij}) \tag{82}$$

and the α_{ij} are integrals over sums of products of k functions and ψ functions, where

$$\psi_{ij} = \exp(k_{ij} - U_{ij}^*\beta) - 1 - k_{ij}$$

and D is the dielectric constant of the medium.

The terms in this expansion may be represented by diagrams and these diagrams are divided into sets by topological arguments,[52] and

$$\alpha_{ij} = \tau_{ij} + \zeta_{ij} \tag{83}$$

where τ_{ij} is sum of all contributions in which the corresponding diagrams having cutting points and ζ_{ij} is the remainder. Allnatt[52] shows by summation that

$$\log g_{ij} = \tau_{ij} + \zeta_{ij} + k_{ij} - U_{ij}^*\beta$$

or

$$g_{ij} = (1 + \tau_{ij} + \gamma_{ij}) \exp(k_{ij} - U_{ij}^*\beta) \tag{84}$$

where γ_{ij} are the sum of all α_{ij} diagrams with certain restrictions on their topology. Thus we have an equation analogous to the usual integral equation for $g(r)$ for normal fluids but containing the shielded Coulomb potential k_{ij}. Approximate solutions of this exact equation may be obtained by approximating, or by neglecting, some of the diagrams, as has been done extensively for normal fluids.[44,54,97]

Allnatt's approach makes use of the Debye–Hückel potential, which is only valid for infinitely dilute solutions of electrolytes in a nonconducting solvent such as water. As such, its application to molten electrolytes cannot be justified and any correlation between calculations based on this method and the actual properties of molten salts must be taken as due largely to the use of calibration terms.

Another approach to radial distribution functions of ionic systems was investigated by Stillinger et al.[1,55] They considered only simple charge-symmetric salts with identical core potentials between all ion pairs, i.e., ions of equal "size." However, they did show how to generalize this result to other systems.

The mean radial distribution function g_m for this system is

$$g_m(r, \lambda) = \tfrac{1}{2}[g_{++}(r, \lambda) + g_{+-}(r,\lambda)] \tag{85}$$

λ being a "charging" parameter. They then constructed g_m so that it is the pair correlation function acting between a partially coupled particle and another particle in a fluid of $N - 2$ particles all identical with the second particle and all interacting only through short-range potentials.

Thus the molten salt is replaced by a single-component fluid with a single, modified particle. However, the equations of this model have never been solved exactly and few conclusions can be drawn from this work.

(c) *Lattice theories.* These theories are an attempt to approximate Z for a molten salt by generalizing the often used lattice theories of normal liquids.[40,41,56] As with all theories of molten salts, modifications must be made due to the long-range nature of the Coulomb potential, e.g., including more than just nearest-neighbor interaction. Since Stillinger[1] describes lattice theories very completely, we will only give a brief outline of the method and its results.

Lattice theories imagine the space occupied by the molecules of a liquid to be divided up into small volume elements or cells. The cells are usually chosen so as to form a regular lattice, i.e., all the cells are identical. The lack of conformity of these requirements with the known details of molten salt structure makes the application of such theories of very doubtful value.

We may write down the partition function for such a system as

$$Z_{\text{cell}} = \left(\prod_k N_k!\, \lambda_k^{3N_k}\right)^{-1} \sum_{j_1,\ldots,j_N = 1}^{\Omega} \int \exp\left[-\beta \sum_{i<j=1}^{N} u_{ij}(r_{ij})\right]$$
$$\times dr_1 \cdots dr_N \tag{86}$$

where the indices $j_1, \ldots, j_N$ run over all cells, Ω is the number of cells, (i.e., $1 \leq j \leq \Omega$), and the cell volume is $\omega = V/\Omega$.

The assumption of single occupancy of cells is usually made and thus we may assign an occupation number ζ_j to each cell ω_j. This parameter is $+1$ if there is a cation in cell ω_j, zero if the cell is empty, and -1 if it contains an anion. Then by using an effective lattice potential, $\tilde{U}_\Omega$,

$$Z_{cell} = \prod_k \left(\frac{\omega}{\lambda_k^3}\right)^{N_k} \sum_{\{\zeta_j\}}' \exp[-\beta \tilde{U}_\Omega(\zeta_1 \cdots \zeta_\Omega)] \qquad (87)$$

where the sum is over the set of all acceptable sets of $\{\zeta_j\}$. It is usually further assumed that we may write

$$\tilde{U}_\Omega(\zeta_1 \cdots \zeta_\Omega) = \sum_{i<j=1}^{\Omega} \tilde{u}_{ij}(\zeta_i, \zeta_j) \qquad (88)$$

i.e., the effective lattice potential is pairwise additive.

Stillinger[1] argues that

$$\tilde{u}_{ij}(\zeta_i, \zeta_j) = [\zeta_i \zeta_j (Ze)^2 / Dr_{ij}] + \eta(r_{ij}, \zeta_i, \zeta_j) \qquad (89)$$

where r_{ij} is the distance between the cell centers; $\eta(r_{ij}, \zeta_i, \zeta_j)$ is short ranged and is zero if ζ_i or ζ_j is zero.

This cell potential can then be used to define a normalized probability P_{cell} and lattice correlation functions $\gamma^{(n)}$ by analogy with P_n and $g^{(n)}$, i.e.,

$$P_{cell}(\zeta_1 \cdots \zeta_\Omega) = \frac{1}{Z_{cell}} \prod_k \left(\frac{\omega}{\lambda_k^3}\right)^{N_k} \exp[-\beta \tilde{U}_\Omega(\zeta_1 \cdots \zeta_\Omega)] \qquad (90)$$

and

$$\left(\prod_{i=1}^n x_{\zeta_i}\right) \gamma_{\zeta_1 \cdots \zeta_n}^{(n)}(\mathbf{r}_1 \cdots \mathbf{r}_n) = \sum_{\{\zeta_j\}}'' P_{cell}(\zeta_1 \cdots \zeta_\Omega) \qquad (91)$$

These are essentially coarse-grained versions of the corresponding exact quantities and thus these lattice correlation functions give us all the thermodynamic information about a lattice system. No exact solution has been achieved for this lattice model.

McQuarrie[15] has applied a simpler lattice theory to molten salts; the Lennard-Jones–Devonshire free volume theory.[14,40,56] This model uses a "smeared out" cell potential, i.e., a central molecule is surrounded by its neighbors "smeared out" over the surface of a spherical shell whose radius is equal to the distance between the central molecule and its neighbors. McQuarrie makes no provision for holes or vacancies in the lattice; thus the model predicts a somewhat too ordered picture of the melt and cannot be regarded as physically meaningful.

Models Involving Intermolecular Forces

Using this Lennard-Jones–Devonshire (LJD) model for the cell potential, Z becomes (for a binary salt)

$$Z = 2(N_1!N_2!)^{-1}(v_f\lambda_1^{-3})^{N_1}(v_f\lambda_2^{-3})^{N_2}\exp[-\tfrac{1}{2}\beta\phi(0,v)] \quad (92)$$

where N_1 and N_2 are the numbers of cations and anions, $\phi(0,v)$ is the LJD cell potential, and v_f is the "free volume" in which a central ion may move.

A number of approximations are used by McQuarrie to enable equation (92) to be applied to molten salts and enable an evaluation of all the thermodynamic data relating to the melt. McQuarrie thus calculated the equation of state, critical constants, and entropies of fusion and vaporization of some alkali halides,[15] with the results shown in Table 4. As expected from this rather too ordered model of

Table 4
Entropies of Vaporization and Fusion from the Lennard-Jones–Devonshire Lattice Theory[a]

Salt	ΔS_v Calc.	ΔS_v Obs.	Calc./Obs.	ΔS_f Calc.	ΔS_f Obs.	Calc/Obs.
LiF	55.9	26.1	0.47	—	—	—
LiCl	46.2	21.8	0.47	—	—	—
LiBr	42.8	22.4	0.52	—	—	—
LiI	40.0	28.2	0.70	—	—	—
NaF	51.2	25.0	0.49	—	—	—
NaCl	44.4	23.5	0.53	5.7	6.23–6.3	0.90–0.91
NaBr	42.9	23.2	0.54	—	—	—
NaI	39.9	24.2	0.61	—	—	—
KF	46.3	23.3	0.50	—	—	—
KCl	41.9	23.1	0.55	4.8	5.8–6.01	0.80–0.83
KBr	39.5	22.4	0.56	5.2	6.06–7	0.74–0.86
KI	38.2	21.7	0.57	4.2	6.02	0.70
RbF	45.0	23.5	0.52	—	—	—
RbCl	40.5	22.32	0.55	—	—	—
RbBr	39.6	22.84	0.58	4.6	3.9–5.77	0.80–1.18
RbI	37.8	22.80	0.60	4.7	3.27–5.73	0.82–1.44

[a] Taken from Tables III and IV, Ref. 15. ΔS in e.u. (Reprinted with permission. Copyright by the American Chemical Society.)

the melt, entropies of vaporization are poor (about half the experimental values), but the entropies of fusion are quite reasonable.

Since lattice theories have so far not proven of great value for the liquid state,[56] rigorous solutions of the cell partition function as given by Stillinger[1] will not be expected to give accurate results. However, more liquidlike cell theories may be better.[98]

(*d*) *Dimensional methods.* These dimensional techniques are essentially a statistical mechanical principle of corresponding states analysis.[40,57]

Pitzer[57] showed by a dimensional analysis of the partition function Z for simple substances that one may deduce the principle of corresponding states from statistical mechanics. The principle holds if there are certain restrictions imposed on the potential function, e.g., a two-body additive central potential with only two parameters (one a distance parameter, the other a depth parameter) immediately leads to the principle.

Subsequent work has been done to broaden the scope of Pitzer's work, e.g., including three-body[58] and quantum effects,[40] and the method has also been applied to molten salts by Reiss *et al.*[59] They applied a similar dimensional analysis to systems with a long-range Coulomb potential and tried to find out what useful results could be obtained, although the approximations they made in the treatment of Z introduce doubts regarding their conclusions.

Following Blander,[60] we have

$$A = -kT \ln Z \tag{93}$$

where Z is the partition function, and we may further write

$$Z = KQ \tag{94}$$

where K is the kinetic energy integral, which may be ignored in a dimensional analysis, and Q is the configurational integral given by

$$Q = (N!)^{-2} \int \cdots \int e^{-\beta U_N} (d\mathbf{r}_c)^N (d\mathbf{r}_a)^N \tag{95}$$

$$U_N = \sum_c \sum_a u_{ca} + \sum_{c<c} u_{cc'} + \sum_{a<a'} u_{aa'} \tag{96}$$

where c labels cations and a the anions. It is then further assumed that[60]

$$u_{ij} = (1/d) f(r/d) \tag{97}$$

e.g.,

$$u_{ij} = A r_{ij}^{-n} - e^2 r_{ij}^{-1}$$

Models Involving Intermolecular Forces

where d is a dimensional length parameter. To make the dimensional analysis of Q possible, since we have three pair potentials and thus three d's, the further assumption is made that Coulomb repulsions between like ions lead to a strong tendency toward local electroneutrality.

Thus configurations in which anions or cations are very close give a relatively high potential and thus contribute negligibly to Q. Hence in most significant configurations the contribution of like-ion repulsions is very small and may be neglected. This assumption is used in some other theories of molten salts and is confirmed by both X-ray scattering and neutron diffraction work[42] and Monte Carlo[16,70] and molecular dynamics calculations[76,77]. Exceptions to this rule will occur when the anion and cation ratios are very different and anion–anion "contacts" are important, e.g., in LiI.

On the above assumption[60]

$$U_N = (1/d)F(r/d) \tag{98}$$

since all the individual pair potentials in significant configurations have the same form and are characterized by a single parameter. It then follows that[60]

$$Q = d^{6N}I[(dT), (V/d^3), (A/d^2)] \tag{99}$$

where A is the interfacial area, i.e., the area of the hole concerned, and I is Q with all distances measured in units of d and volumes in units of d^3.

Now the pressure is given by[60]

$$p(T, V) = kT\left(\frac{\partial \ln Z}{\partial V}\right) = \frac{k}{d^3}\frac{Td}{d}\left[\frac{\partial \ln Z(Td, V/d^3)}{\partial(V/d^3)}\right]_{T,N}$$

$$= \frac{1}{d^4}p\left(Td, \frac{V}{d^3}\right)_{T,N} \tag{100}$$

and thus we may write a "molecular theory" reduced equation of state

$$\pi = d^4p = \pi(\tau, \theta) \tag{101}$$

where $\tau = Td$ and $\theta = V/d^3$.

One may also define a "thermodynamic" reduced equation of state, i.e., an equation of state reduced with respect to thermodynamic as opposed to molecular variables, by defining $p/p_M = \pi/\pi_M = \pi'$, $T/T_M = \tau/\tau_M = \tau'$, and $V/V_M = \theta/\theta_M = \theta'$; then

$$\pi' = \pi'(\tau', \theta') \tag{102}$$

Now although (102) is less informative than (101), it is more general and it will follow from a more general pair potential than (97). Thus we may write $d = \tau_M/T_M$, $\tau'' = \tau/\tau_M, = T/T_M = \tau'$, $\theta'' = \theta\tau_M^3 = T_M^3 V$ and we have

$$\pi'' = \pi/\tau_M^4 = p/T_M^4 = \pi''(\tau'', \theta'') \tag{103}$$

which is again a "molecular theory" reduced equation of state.

The above results may be generalized to salts with more general pair potentials than (97).[59] For example, for a system of anions and cations of charge z then[59]

$$\pi = \pi(\tau, \theta) = (d^4/z^2)p, \quad \tau = dT/z^2, \quad \text{and} \quad \theta = V/d^3 \tag{104}$$

Another example is a system with a dielectric constant D appearing in the pair potential; in this case we have[59]

$$\pi = \pi(\tau, \theta) = (Dd^4/z^2)p, \quad \tau = DdT/z^2, \quad \text{and} \quad \theta = V/d^3 \tag{105}$$

Finally for a charge-asymmetric salt with cation of charge z_1 and anion of charge z_2 then[91]

$$\pi = \pi(\tau, \theta, z_1, z_2) = Dd^4 p, \quad \tau = dDT, \quad \text{and} \quad \theta = V/d^3 \tag{106}$$

The previous equations have been applied to correlate the properties of a variety of salts,[59,60] examples being the melting temperature, vapor pressures, and surface tensions for simple salts (see Tables 5 and 6).

It may be seen that the results are good except for Li salts, where the anion-to-cation ratio is large and anion–anion interactions would be expected to be of importance.

Also, the results for more complex salts, e.g., BeF_2, $ZnCl_2$, and $HgCl_2$, are not as good, but this would be expected since all these salts are at least partly covalent, i.e., a simple potential such as

(97) with or without a dielectric term included does not describe the potential energy in these systems.

Thus we may conclude that dimensional analysis techniques are very useful to correlate the properties of simple ionic systems with a minimum number of parameters.

However, since the analysis rests on the assumption that only cation–anion interactions are important, the method may fail if the anion and cation "sizes" are vastly different, e.g., for Li salts. Also, it may fail for partly covalent systems, i.e., systems whose pair potentials cannot be described by an equation of the form of (97) with or without a dielectric term included.

Table 5
Melting Points and Interatomic Distances for Symmetric Compounds[59]

Compound	Melting point T_M, °K	Interatomic distance in solid, cm × 10^8	$dT_M/z^2 = \tau_M$ cm deg × 10^5
Mgo	3073	2.10	1.61
CaO	2873	2.40	1.73
SrO	2733	2.54	1.74
BaO	2198	2.75	1.51
NaF	1265	2.31	2.92
NaCl	1074	2.81	3.02
NaBr	1023	2.98	3.04
NaI	933	3.23	3.01
KF	1129	2.67	3.02
KCl	1045	3.14	3.28
KBr	1013	3.29	3.34
KI	958	3.53	3.39
RbF	1048	2.82	2.96
RbCl	998	3.29	3.26
RbBr	953	3.43	3.27
RbI	913	3.66	3.34
CsF	955	3.01	2.88
CsCl	918	3.47	3.18
CsBr	909	3.62	3.29
CsI	894	3.83	3.42
LiF	1121	2.01	2.25
LiCl	887	2.57	2.27
LiBr	823	2.75	2.27
LiI	718	3.02	2.21

Table 6
Corresponding-States Entropies of Vaporization of the Monomer from the Liquid at the Melting Temperature, $\Delta H_{vl}/T_M$, Vapor Pressures π'', and Surface Tensions Γ'' for Molten Alkali Halides[60]

Salt	T_M, °K	$\Delta H_{vl}/T_M$, e.u.	$\pi''_{1.30}$, mm deg$^{-4} \times 10^{12}$	$\Gamma''_{1.10}[=(\gamma/T_M^3)_{1.10}]$,[a] dyn cm$^{-1} \times 10^9$
NaF	1265	43.1	33.0	93
NaCl	1074	42.2	27.8	86
NaBr	1023	41.8	31.9	87
NaI	933	41.9	23.5	98
KF	1129	41.4	44.2	93
KCl	1045	41.9	35.5	80
KBr	1013	41.4	36.2	78
KI	958	41.8	33.1	81
RbF	1048	41.7	37.4	102
RbCl	988	42.5	27.9	94
RbBr	953	42.3	22.9	97
RbI	913	42.2	20.6	99
CsF	955	41.1	59.6	114
CsCl	918	44.3	29.7	109
CsBr	909	42.1	24.0	102
CsI	894	42.7	26.8	96
LiF	1121	49.3	4.5	171
LiCl	887	50.1	7.4	188
LiBr	823	51.2	5.3	—
LiI	718	51.6	0.8	—

[a] γ (dyn cm^{-1}) = surface tension.

(e) *Scaled Particle Theory.* This theory of fluids was developed by Reiss, Frisch, Helfand, and Lebowitz (see Ref. 61) from considerations of a fluid of hard spheres of diameter d. For such a fluid the equation of state may be written in terms of the radial distribution function at contact $g(d)$, i.e.,

$$p/\rho kT = 1 + \tfrac{2}{3}\pi d^3 g(d) \tag{107}$$

Thus one need only determine $g(d)$ to obtain all the thermodynamic properties of this system.

The theory by which $g(d)$ is obtained is largely a geometric one which concerns itself with the packing of hard spheres in a fluid of hard spheres. The value of $g(d)$ is obtained by considering a "solute" molecule in a cavity in the "solvent" of other hard-sphere molecules.

If $W_0(r)$ is the probability of finding a spherical cavity of at least radius r centered about a specific point in the fluid, then the probability of finding a cavity of radius lying between r and $r + dr$ is

$$-dW_0(r)/dr = W_0(r)4\pi r^2 \rho G(r) \qquad (108)$$

where $\rho G(r)$ is the average density of hard spheres in contact with the boundary of the spherical cavity. It may be shown that

$$G(d) = g(d) \quad \text{and} \quad G(r) = (1/\rho kT)[p + (2\gamma/r)]$$

where γ is the surface tension. For r not too small

$$G(r) = \frac{1}{\rho kT}\left(p + \frac{2\gamma_0}{r} + \frac{4\gamma_0 \delta d}{r^2}\right) \qquad (109)$$

where γ_0 and δ are parameters, and further

$$G(r) = 1/(1 - \tfrac{4}{3}\pi r^3 \rho) \quad \text{for} \quad r < \tfrac{1}{2}d$$

In scaled particle theory the assumption is made that (109) is valid down to $r = d/2$, and since $G(r)$ and its first derivative are continuous at $r = d/2$, γ_0, and δ, then $G(d)$ may be constructed from the above information.

After this is done substitution of $G(d)$ in (107) gives

$$p/\rho kT = (1 + x + x^2)/(1 - x)^3 \qquad (110)$$

where $x = (\pi/6)\rho d^3$. This equation of state is identical with that obtained from the compressibility equation of state and the Percus–Yevick theory (see Reiss[61]) and is a very good approximation to the exact equation for hard spheres. The same approach has, however, not so far been successful for real liquids.

Thus in the absence of an exact calculation of the analog of $G(r)$ for real liquids another approach has been developed. For real liquids the "soft," i.e., attractive part of the intermolecular potential is assumed to act primarily to establish the overall density of the fluid, while the internal structure is determined by the packing of the hard cores, i.e., the repulsive part of the potential. This idea is in accord with the very successful perturbation and variation theories of liquids.[46] Thus no equation of state is obtainable since the soft part of the potential is introduced only implicitly through the measured density ρ at temperature T. However, other quantities, e.g., compressibilities, heats of fusion, and expansivities, may be obtained by this method in terms of a single parameter d, which is usually obtained from some independent source.

The application to fused salts is based on the dimensional analysis or principle of corresponding states idea, i.e., that the properties of a melt depend only on one parameter d, the sum of the ionic radii of the cation and anion and not on their individual radii, and this parameter d has been determined in a variety of ways for molten salts.

However, all this work may be divided into two categories, that which uses fixed values of d taken from some other source, and that which uses the experimental bulk property to calculate d (sometimes as a function of temperature and density). Obviously the second approach is only useful to correlate known data unless, of course, the values of d obtained from one property are found to be of use to calculate other properties.

Reiss[61] quotes unpublished work of Yosim and Owens, who use gas-phase bond lengths to calculate compressibilities κ_T, expansivities α_p, and heats of fusion ΔH_f for the molten alkali halides. The results for κ_T and α_p are in reasonable agreement with experiment, but good results for ΔH_f are only obtained if it is assumed that the entire communal entropy $2R$ appears upon fusion (see Tables 7 and 8). This last idea has also been suggested by Singer and Woodcock[16] on the basis of Monte Carlo calculations.

Reiss and Mayer[62] used the same approach to calculate the surface tension for molten halides and once again found reasonable agreement with experiment, except for "partially covalent" salts (see Tables 7 and 8).

However, in subsequent work Yosim and Owens[63] used mean ionic diameters taken from X-ray diffraction data on the melts rather than gas-phase bond lengths to approximate d. They obtained good agreement with experimentally measured values of entropies at four temperatures, and with the heat capacity and entropy of fusion at the melting point (see Tables 9 and 10). In fact the results for these properties are better than those obtained by use of the gas-phase bond lengths. However, agreement with experiment is not good for the compressibilities and is only good for the entropies of fusion if the entire communal entropy is not assumed to appear upon fusion (a rather doubtful assumption).

Table 7
Scaled Particle Results for the Alkali Halides[a]

Salt	T[b] °C	γ Calc.	γ Obs.	α_p Calc.	α_p Obs.	κ_T Calc.	κ_T Obs.	ΔH_f Calc.	ΔH_f Obs.
LiF	—	—	—	0.48	0.27	—	—	7.07	6.47
LiCl	—	—	—	0.55	0.30	45	22	5.46	4.72
LiBr	—	—	—	0.59	0.26	56	23	4.97	4.22
LiI	—	—	—	0.65	0.30	—	—	4.18	3.50
NaF	—	—	—	0.39	0.30	—	—	7.98	8.03
NaCl	1000	111	98	0.41	0.36	38	34	6.98	6.69
NaBr	900	95	91	0.43	0.36	45	35	6.50	6.24
NaI	700	85	84	0.46	0.36	58	45	5.91	5.64
KF	—	—	—	0.42	0.36	—	—	6.59	6.75
KCl	900	96	90	0.40	0.40	42	44	6.41	6.27
KBr	800	87	85	0.40	0.41	47	47	6.17	6.10
KI	800	76	69	0.41	0.41	55	57	5.78	5.74
RbF	—	—	—	0.45	0.36	—	—	6.02	5.82
RbCl	828	81	89	0.41	0.41	—	—	5.53	5.67
RbBr	831	83	81	0.41	0.41	—	—	4.47	5.57
RbI	772	72	72	0.42	0.41	—	—	5.08	5.27
CsF	826	85	96	0.50	0.36	—	—	5.07	5.19
CsCl	830	79	78	0.44	0.39	53	45	4.83	4.84
CsBr	808	76	72	0.43	0.40	55	59	6.84	5.64
CsI	821	70	63	0.43	0.38	—	—	7.62	5.64

[a] These values are taken from Reiss,[61] and unpublished results of Yosim and Owens; d was taken to be the gas-phase bond length of the molecules; surface tensions are from Reiss and Mayer.[62] Surface tension γ in dyn cm^{-1}. Expansitivity α_p at $T = 1.1 T_M$ in °K^{-1} × 10^3. Compressibilities κ_T at $T = 1.1 T_M$ in cm^2 dyn^{-1} × 10^{12}. Heat of fusion ΔH_f in kcal mol^{-1}. Note: Reiss[61] uses the symbol β for isothermal compressibilities, whereas Stillinger[64] and Woodcock and Singer[16] use the symbol κ_T.

[b] Temperature at which the surface tension was measured.

Table 8
Scaled Particle Theory Surface Tensions

Salt	Temp., °C	Density, g cm^{-3}	Gas equilibrium distance, Å	γ (Obs.)	γ (Calc.)
BaCl$_2$	1000	2.982	2.82	157	152
CdCl$_2$	600	3.366	2.21	83	79
HgCl$_2$	293	4.670	2.29	56	36
HgBr$_2$	241	5.113	2.41	65	37
BiCl$_3$	271	3.554	2.48	66	52
BiBr$_3$	281	4.099	2.63	64	47

Surface tensions in dyn cm^{-1}. Taken from Reiss and Mayer.[62]

Stillinger[64] used the alternative method of determining d from experimental data and used experimental κ_T values to find d as a function of temperature. He found values which are smaller than the sum of the Pauling ionic radii in the crystal but usually larger than the gas-phase bond distances.

Stillinger further found that the values of d decrease with increasing temperature, as they do for normal liquids. He also found that d in a homologous series, e.g., the chlorides of Li, Na, K, and Cs, increase down the periodic table, as one would expect. Reiss[61] also reports values of d for NaCl, NaBr, NaI, and KI as a function of temperature derived from surface tension and compressibility data.

Finally, Mayer[13] has used experimental surface tensions, compressibilities, and expansivities at a single temperature to calculate d and has in fact found values close to the gas-phase bond lengths, except for Li salts. He also used the d values from one property to calculate the values of the other properties and once again found good results, except for the Li case (see Table 11).

We may conclude that the scaled particle theory is very useful for obtaining reasonable values of a variety of properties of ionic melts and for correlating properties using only one parameter. This parameter may be obtained independently of the property being calculated and use of gas-phase molecule bond lengths usually leads to reasonable results.[13,61,62] However, some properties appear to be better described if the values of d are obtained from X-ray data on the melt, e.g., entropies.[63]

Table 9
Entropies of Molten Alkali Halides at Reduced Temperatures T/T_M of 1.0, 1.1, 1.2, and 1.3[a]

Salt	$T/T_M = 1.0$		$T/T_M = 1.1$		$T/T_M = 1.2$		$T/T_M = 1.3$	
	Calc.	Expt.	Calc.	Expt.	Calc.	Expt.	Calc.	Expt.
LiF	34.2	30.6	35.3	32.1	36.3	33.4	37.3	34.6
LiCl	37.8	33.5	38.8	34.9	39.8	36.3	40.8	37.5
LiBr	41.6	34.4	42.7	35.9	43.6	37.2	44.5	38.5
LiI	45.2	34.6	46.2	36.0	47.1	37.3	48.0	38.5
NaF	37.9	37.4	39.3	38.9	40.6	40.4	41.8	41.7
NaCl	40.4	40.8	41.9	42.4	43.2	43.8	44.5	45.3
NaBr	43.5	42.4	44.9	44.0	46.3	45.4	47.6	46.7
NaI	47.1	43.2	48.4	44.8	49.6	46.2	50.8	47.5
CsF	47.4	40.3	48.7	42.0	49.8	43.5	51.0	44.9
CsCl	41.6	45[101,102]	43.6	47	45.4	49	47.1	50
CsBr	46.7	47.8	48.3	49.5	49.9	51.1	51.4	52.6
CsI	46.8	50.7	48.6	52.4	50.2	53.9	51.8	55.4

[a] From Ref. 63. The entropies (in cal per mole deg) of all the liquids but CsCl were obtained from tabulations of $S°_{298}$,[102,103] solid heat capacities,[104] entropies of fusion,[105] and heat capacities of the liquids.[106]

Table 10
Heat Capacity, Entropy of Fusion, and Compressibilities of Alkali Halides at the Melting Point[63]

Salt	Heat capacity, cal mole^{-1} deg^{-1}		Entropy of fusion, cal mole^{-1} deg^{-1}		Compressibilities, × 10^{12} cm^2 dyn^{-1}	
	Calc.	Obs.	Calc.	Obs.[a]	Calc.	Obs.[9]
LiF	11.4	15.5[101]	6.27	5.77	—	—
LiCl	11.0	15.3[106]	5.96	5.39	9.8	19.2
LiBr	10.6	15.6[106]	6.12	5.13	9.7	21.4
LiI	10.3	15.1[106]	4.30	4.72	—	—
NaF	14.0	16.4[101]	7.2	6.33	—	—
NaCl	14.9	17.0[106]	7.07	6.23	13.2	28.7
NaBr	15.0	16.7[106]	6.72	6.12	14.4	31.5
NaI	13.3	16.5[106]	4.74	6.04	19.9	37.3
CsF	12.8	17.7[106]	3.05	5.32	—	—
CsCl	20.7	18.5[106]	4.98	5.27	13.0	38.9
CsBr	17.2	18.5[106]	10.95	6.20	20.0	49.5
CsI	18.4	18.0[106]	14.63	6.27	—	—

[a] Table 1, Ref. f of Yosim and Owens.[63]

Table 11
Comparison of Measured Thermal Expansivities, Surface Tensions, and Compressibilities of Molten Salts at 1073°K with Those Calculated by a Scaled Particle Method[a]

Compound	d_γ	d_κ	d_α	d_{gas}	γ (meas.)	γ_κ	γ_α	κ (meas.)	κ_γ	κ_α	α (meas.)	α_γ	α_κ
CsCl	2.92	2.93	2.95	2.91	80	81	82	0.51	0.53	0.50	0.41	0.41	0.41
CsBr	3.02	2.98	3.03	3.07	72	69	73	0.67	0.62	0.61	0.42	0.42	0.43
CsI	3.22	—	3.28	3.32	64	—	69	—	0.74	0.65	0.40	0.42	—
NaCl	2.46	2.47	2.63	2.36	116	119	149	0.29	0.30	0.19	0.35	0.41	0.40
NaBr	2.55	2.63	2.77	2.50	96	107	129	0.34	0.41	0.24	0.36	0.43	0.40
NaI	2.70	2.81	2.98	2.71	75	86	105	0.47	0.59	0.33	0.36	0.45	0.41
LiCl	2.22	2.26	2.56	2.03	124	133	216	0.25	0.28	0.10	0.30	0.43	0.41
LiBr	—	2.43	2.83	2.17	—	118	223	0.29	—	0.09	0.28	—	0.41

[a] From Ref. 13. No compressibility data were found for CsI, RbCl, RbBr, and RbI. Units for parameter d are angstroms; d_γ, d_κ, and d_α were computed from the measured γ, κ, and α, respectively, and Equations (2), (4), and (3) of Ref. 13. d_{gas} is the measured interatomic distance for the salt vapor, taken from Ref. 107. Surface tension γ(dyn cm^{-1}) from Ref. 108, except for Cs, Rb, and Li salts, from Ref. 109. Isothermal compressibility κ (10^{-10} cm^2 dyn^{-1}) from Ref. 108. Thermal expansivity α (10^{-3} deg^{-1}) from Ref. 108. Note that Mayer[13] uses the symbol β for isothermal compressibilities, as does Reiss[61] (note, however, that Refs. 16 and 64 use the symbol κ_T, as is used in this chapter).

The scaled particle theory shares with dimensional methods the weaknesses of considering only cation–anion interactions and not allowing for covalency effects.

(*f*) *Monte Carlo Method.* This method essentially uses a high-speed computer to evaluate ensemble averages.[45–48] However, even with modern, high-capacity computers this cannot be done for a system containing of the order of 10^{23} molecules. Thus the liquid or solid is modeled by a system of N particles (N usually being of the order of 10^2–10^3) interacting according to a given potential function and enclosed in a box or cell of volume V (usually a cube).

The particles are assigned an initial position in the box (usually corresponding to a regular lattice) and are then given random displacements from this original configuration to a new one. The total (potential) energy changes from U_N to $U_N + \Delta U_N$. If ΔU_N is negative, the move is accepted and the new configuration replaces the old one, but if it is positive, it is only accepted with a probability (the Boltzmann probability) proportional to $\exp(-\beta \Delta U_N)$. Repetition of this process of generating random configurations produces a chain of configurations (a Markov chain[45]) with probability density proportional to $\exp(-\beta U_N)$. Thus the overall chain average of a mechanical variable X (e.g., U_N) tends to the canonical ensemble average of X, written as $\langle X \rangle$. Hence the above procedure models the canonical (i.e., N, V, T) ensemble.

Examples of the relationships between the quantities generated and bulk properties are[45]

$$U/NkT = \tfrac{3}{2} + \langle U_N \rangle / NkT \tag{111}$$

$$pV/NkT = 1 - (\rho/3kT) \left\langle \sum_{i<j} \mathbf{r}_{ij} \cdot \nabla_{\mathbf{r}_{ij}} U \right\rangle \tag{112}$$

$$C_v/Nk = \tfrac{3}{2} + [(\langle U_N^2 \rangle - \langle U_N \rangle^2)/N(kT)^2] \tag{113}$$

and

$$g(r) = \langle n(r, r + \Delta r) \rangle / (2\pi \rho r^2 \, \Delta r) \tag{114}$$

where U is the internal energy, C_v is the heat capacity at constant volume, and $n(r, r + \Delta r)$ is the average number of particles within the distance r and $r + \Delta r$ of a given particle. $\mathbf{r}_i$ denotes the position vector of the *i*th particle and $\mathbf{r}_{ij}$ denotes $\mathbf{r}_j - \mathbf{r}_i$.

In practice the averages are carried out over the order of a few hundred thousand configurations and in fact the first 100,000 or so are rejected to allow the system to reach equilibrium.

Furthermore, since N is a rather small number (10^2–10^3), the system of N molecules in one cell of volume V (determined by the required density) would show rather drastic surface effects. Thus to minimize this effect, the basic box is surrounded by replicas of itself, i.e., periodic boundary conditions are imposed on the system and when a particle leaves the basic cell another one moves in through the opposite edge. Except for the critical state region, these periodic boundary conditions have little effect on the accuracy of the method.[45-48] As with all other theories of molten salts, Monte Carlo (MC) calculations are complicated by the presence of the long-range Coulomb potential, which makes the evaluation of U_N difficult. Thus although the MC method gives excellent agreement with experiment for normal liquids,[38,39,46] it was not obvious that it could be implemented for ionic systems at high density.

There are two methods normally used to evaluate the potential energy of a large system of ions, namely those of Evjen and Ewald.[16,43,65,66] The Evjen method truncates the Coulomb sum outside a cube of side L, centered on the ion in question and containing an equal number of positive and negative ions so as to maintain overall electroneutrality. This method has been applied to molten salts by Forland et al.[67,68] and Woodcock and Singer.[16] However, it appears to be unsatisfactory for systems of high effective charge density, although suitable for perfect lattices and dilute electrolytes.[16] Its use in an MC process, starting from a lattice configuration, leads to a gradual decrease in Coulomb energy to a value about 25% too low, and Woodcock and Singer[16] also noticed marked spurious changes in $g(r)$. The situation was not improved by using initial configurations more characteristic of a liquid, or by trebling L.

Thus Woodcock and Singer[16] chose the less rapid but more stable Ewald method. This uses a mathematical transformation which essentially expresses the Coulomb sum as the sum of two mutually cancelling charge distributions, i.e., for a given distribution of point charges $\rho(\mathbf{r})$

$$\rho(\mathbf{r}) = \sum_k Z_k \delta(\mathbf{r} - \mathbf{r}_k) = \rho_1(\mathbf{r}) + \rho_2(\mathbf{r}) \tag{115}$$

where Z_k is the charge of the kth ion, δ denotes the Dirac delta function,

$$\rho_1(\mathbf{r}) = \sum_k Z_k \delta(\mathbf{r} - \mathbf{r}_k) - \alpha^3 \pi^{-3/2} \exp(-\alpha^2 |\mathbf{r} - \mathbf{r}_k|^2)$$

and

$$\rho_2(\mathbf{r}) = \sum_k Z_k [\alpha^3 \pi^{-3/2} \exp(-\alpha^2 |\mathbf{r} - \mathbf{r}_k|^2) - L^{-3}]$$

In $\rho_1(\mathbf{r})$ the second term is a normalized Gaussian charge distribution of half-width α^{-1} and of opposite sign, centered on each ion; $\rho_2(\mathbf{r})$ consists of an identical Gaussian charge distribution of the same sign as the ion, and of a uniform normalized charge distribution of the opposite sign. The Gaussian distribution in ρ_1 and ρ_2 cancel each other, as do uniform distributions corresponding to equal numbers of positive and negative ions as used in the Evjen method.

The total Coulomb potential energy U_N^e may then be written[16]

$$U_N^e = U_1 + U_2 \tag{116}$$

where

$$U_1 = \frac{1}{2} \left\{ \sum_k Z_k \left[\sum_{k'}{}' Z_{k'} \sum_n \frac{1 - \mathrm{erf}(\alpha r_{kk'n})}{r_{kk'n}} - \frac{Z_k \alpha}{\sqrt{\pi}} \right] \right\}$$

$$U_2 = \frac{1}{2\pi L} \sum_k \sum_{k'}{}' Z_k Z_{k'} \sum_n \frac{\exp(-\pi^2 |\mathbf{n}|^2 \alpha^2 L^2) \cos(2\pi \mathbf{n} \cdot \mathbf{r}_{kk'}/L)}{|\mathbf{n}|^2}$$

$$\mathrm{erf}(x) = 2\pi^{-1/2} \int_0^x \exp(-u^2)\, du$$

$r_{kk'}$ is the distance between ions k and k' in a cube of side L centered on $\mathbf{r}_k$; $\mathbf{n}$ is a vector with integer components; $r_{kk'n} = |\mathbf{r}_k - \mathbf{r}_{k'} + L\mathbf{n}|$; and the prime in the summation of U_1 indicates that $k = k'$ is omitted when $\mathbf{n} = (0, 0, 0)$ and in U_2 the prime indicates $\mathbf{n} = (0, 0, 0)$ is omitted.

This procedure is slower than the Evjen method but is more stable and, if tabulation is used, it is reported that only an additional 30% is added to the computing time over that required for the Evjen method.[16]

There have been three studies of molten salt structure by the MC technique. The first was made by Forland et al.,[67,68] who used a very simple pairwise additive potential,

$$U_N = \sum_{i<j}^{N} u_{ij} \qquad (117)$$

where

$$u_{ij} = (Z_i Z_j / r_{ij}) + (b_{ij}/r_{ij}^n) \qquad (118)$$

Forland et al. used a two-dimensional cell for NaCl[67] and a three-dimensional cell for LiCl[68] (with small numbers of ions). These results are in qualitative agreement with experiment; however, both the simplicity of their pair potential and their use of the Evjen method limit the usefulness of the conclusions that may be drawn from their results.

Woodcock and Singer[16] and Singer and Lewis[72] used a much more realistic pair potential developed for crystalline salts,

$$u_{ij}(r) = Z_i Z_j r^{-1} + b_{ij} \exp[B(\sigma_{ij} - r)] + c_{ij} r^{-6} + d_{ij} r^{-8} \qquad (119)$$

where the constant b_{ij} has the same value for all alkali halides except Li salts, B has a common value for the three ion pairs in the salt, σ_{ij} is the sum of the ionic radii, and c_{ij} and d_{ij} are the van der Waals coefficients C_6 and C_8 ($|Z_i Z_j| = 23.067 \times 10^{-20}$ erg cm for alkali halides). The Ewald method was used to sum the Coulomb potential.

The parameters of this pair potential were taken from the work of Tosi and Fumi,[69] who obtained them from solid-state data and hence these pair potentials may be thought of as *effective* pair potentials. A value of $\alpha = 5.714/L$ in equation (115) was found to be optimum for the convergence of the Coulomb sum by truncating U_1 at $|\mathbf{r}_k - \mathbf{r}_{k'}| = L/2$ and the Fourier series of U_2 at $|\mathbf{n}| = 1$.

The summation of the non-Coulombic terms in (119) was truncated at $|\mathbf{r}_k - \mathbf{r}_{k'}| = L/2$ and a long-range correction made for the effect of distant particles by replacing them by a medium of uniform density, as is the usual practice for short-range potentials.[48] Furthermore, the contribution of the r^{-8} term was only evaluated every 2000th step but since this term only contributes about 0.5% to U_N, this is of little consequence.

Most of the calculations were performed on a 216-particle system, but a check with a 64-particle system showed the same values of U and p within the limits of statistical error. The computations were started from an NaCl-type lattice, the first 10^5 MC steps rejected, and a further $2-4 \times 10^5$ MC steps used to evaluate U, p, etc., and the usual statistical tests of the errors involved were also made.[16]

We will first discuss the thermodynamic properties obtained, to indicate the accuracy of the method, then investigate the picture of the melt which emerges and what light these results throw on operational theories. Since the most extensive study has been made on KCl, we will discuss this salt first and in great detail. The KCl calculations were made for 24 V, T points with one isotherm in the solid range at the melting point T_M and four isotherms in the liquid range at T_M, $1.25 T_M$, $2.0 T_M$ and $2.75 T_M$.[16] Some results are shown in Table 12.

Calculated internal energies U agree within 0.5% of experiment and the differences between liquid and solid energies at T_M are also very good, as can be seen from Table 13. Computed molar heat capacities agree quite well with experiment and $C_p (p = 0)$ is approximately constant over a wide range of T, in accord with the dimensional analysis for ionic systems.[59]

The pressure calculated from (112) agrees, within combined theoretical and experimental errors, with recent experimental data. Woodcock and Singer[16] also calculated the following: the thermal pressure coefficient $\beta_v = (\partial P/\partial T)_v$; the thermal expansivities $\alpha_p = (1/V)(\partial V/\partial T)_p = [1 + (1/V)(\partial H/\partial P)_T]/T$, and the isothermal compressibilities $-(1/V)(\partial V/\partial P)_T = \kappa_T = \alpha_p/\beta_v$. Once again satisfactory agreement with experiment is obtained (see Table 12).

Entropy changes calculated from the C_v, T and P, V data also agree with experiment. This can be seen from Table 12. Similarly the calculated melting temperature agrees with experiment.

The following interesting points about the molten salt arise directly from the above calculated quantities. First, $(\partial U/\partial V)_T$ is approximately 50% less in the liquid than in the solid at T_M, in contrast with a change of only about 5% for nonpolar liquids.[39] This discrepancy must thus reflect the effect of the long-range potential.

Even more interesting data are obtained from an analysis of the potential energy into contributions from Coulomb, short-range repulsion, dipole–dipole, and dipole–quadrupole dispersion energies,

Table 12
Comparison of MC Results for KCl with Experimental Data[a]

	$T = 1045°K$		$T = 1306°K$	
	Expt.	MC	Expt.	MC
U, 10^3 J mole^{-1}	−626.5	−628.0	−608.3	−611.8
V, cm^3 mole^{-1}	49.9	50.20	54.7	53.91
κ_T, 10^{-6} bar^{-1}	36.5	36.1	58.5	52.7
α_p, 10^{-4} K^{-1}	3.58	3.33	3.92	3.37
β_v, bar °K^{-1}	9.81	9.23	6.70	6.40
C_v, J °K^{-1} mole^{-1}	46.9	50.5	45.2	48.0
C_p, J °K^{-1} mole^{-1}	66.9	66.8	66.9	63.4
S, J °K^{-1} mole^{-1}	177.8	175.1	192.7	190.0

[a]Taken from Woodcock and Singer.[16]

$\langle U_N^E \rangle$, $\langle U_N^R \rangle$, $\langle U_N^{DD} \rangle$, and $\langle U_N^{DQ} \rangle$, respectively. The relative contributions are approximately 100:12.5:3.0:0.5 and the short-range repulsion may be further analyzed into 96–97% between unlike and only 3–4% (or 0.6% to U_N) between like ions (see Tables 13 and 14).

These results thus confirm the hypotheses upon which other models, e.g., scaled particle and dimensional analysis theories, are based. It will be very interesting to have similar results for LiI, where the anion is much larger than the cation and anion–anion interactions are supposedly the cause of poor results in both scaled particle and dimensional analysis theories.

A point which emerges from the pressure calculation is that although the dipole–quadrupole dispersion term contributes only about 0.5% to U, it contributes about 5% to p. Thus care is necessary before dismissing the effect of small terms in the potential energy expression, since different properties depend in different ways on U_N (see also Ref. 38).

From the good agreement between theory and experiment we may conclude that the Tosi–Fumi pair potential is a good effective pair potential for KCl and that further, more detailed deductions about the structure of the melt may be obtained by a closer analysis of the MC results.

Besides thermodynamic data, Woodcock and Singer calculate and discuss the structure via $g(r)$. They discuss $g(r)$ corresponding to

Table 13
Changes on Fusion for KCl[a]

	T = 1045°K				Percent
	Solid	Liquid	Δ_{MC}	$\Delta_{expt.}$	change (MC)
U, 10^3 J mole^{-1}	−654.3	−628.0	26.3	25.5	−4.0
$\langle U_N^E \rangle$, 10^3 J mole^{-1}	−747.9	−722.0	25.9	—	−3.5
$\langle U_N^R \rangle$, 10^3 J mole^{-1}	97.9	94.8	−3.1	—	−3.2
$\langle U_N^{DD} + U_N^{DQ} \rangle$, 10^3 J mole^{-1}	−30.4	−25.9	4.5	—	−15.8
V, cm^3 mol^{-1}	41.75	50.20	8.45	8.3, 9.09, 7.23	20.2
κ_T, 10^{-6} bar^{-1}	10.1	36.1	26.0	—	257
α_p, 10^{-4} bar^{-1}	1.97	3.33	1.96	—	99.5
β_v, bar °K^{-1}	19.4	9.23	−10.2	—	−52.4
C_v, J °K^{-1} mole^{-1}	48.4	50.5	2.1	—	4.3
C_p, J °K^{-1} mole^{-1}	65.3	66.8	1.5	9.2, 3.5	2.4

[a]Taken from Woodcock and Singer.[16]

Table 14
Contribution of the Different Types of Potential Energy at $T = T_M$ [a]

		$\langle U_N^E \rangle$	$\langle U_N^R \rangle$	$\langle U_N^{DD} \rangle$	$\langle U_N^{DQ} \rangle$	$100\dfrac{\langle U_N^{R++} \rangle + \langle U_N^{R--} \rangle}{\langle U_N^R \rangle}$	$100\dfrac{\langle U_N^{R++} \rangle + \langle U_N^{R--} \rangle}{\langle U_N \rangle}$
LiCl	(s)	100	15	2	0.3	16	3
LiCl	(l)	100	18	2	0.3	12	2
NaCl	(s)	100	12.5	2	0.3	7	1
NaCl	(l)	100	12.5	2	0.3	6.5	1
KF	(s)	100	16	3	0.4	9	2
KF	(l)	100	17	3	0.5	7.4	1.5
KCl	(s)	100	13	4	0.5	4	0.6
KCl	(l)	100	13	4	0.5	4.5	0.7
KBr	(s)	100	13	3.5	0.4	4	0.6
KBr	(l)	100	13	3	0.3	6	0.9
KI	(s)	100	12.5	4	0.5	5	0.7
KI	(l)	100	13.0	4	0.5	6	0.9
RbCl	(s)	100	12.5	4	0.6	3	0.4
RbCl	(l)	100	12.5	4	0.5	4	0.6
CsF	(s)	100	12	4.5	0.6	9	1.1
CsF	(l)	100	12.5	4	0.6	9.5	1.3

[a] Unpublished results of Singer and Lewis.[72] $\langle U_N^E \rangle$ = electrostatic energy; $\langle U_N^R \rangle$ = exponential repulsive energy; $\langle U_N^{DD} \rangle$ = dipole dipole energy; $\langle U_N^{DQ} \rangle$ = dipole quadrupole energy.

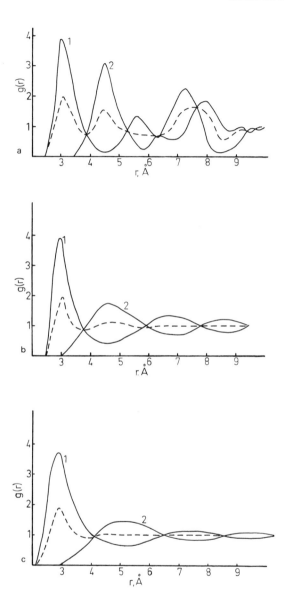

Figure 6. Radial distribution functions for KCl[16]: (a) solid at 1045°K, (b) liquid at 1045°K, (c) liquid at 2874°K. Solid lines 1 are $g_u(r)$; solid lines 2 are $g_l(r)$; broken lines are $g_m(r)$.

Models Involving Intermolecular Forces

all pairs $g_m(r)$, unlike-ion pairs $g_u(r)$, and like-ion pairs $g_l(r)$, and since K^+ and Cl^- are practically identical in size, then $g_{++}(r)$ is indistinguishable from $g_{--}(r)$ (see Fig. 6). They further define the following quantities to further characterize $g(r)$: The distance of closest approach d, the position r^{max} and height h of the main peak, and the position r^{min} of the minimum following the main peak. They also discuss apparent coordination numbers n, defined by

$$n_m = (N-1)V^{-1}4\pi \int_0^{r_m^{min}} r^2 g_m(r)\, dr \qquad (120)$$

$$n_u = \tfrac{1}{2}NV^{-1}4\pi \int_0^{r_u^{min}} r^2 g_u(r)\, dr \qquad (121)$$

and

$$n_l = (\tfrac{1}{2}N-1)V^{-1}4\pi \int_0^{r_l^{min}} r^2 g_l(r)\, dr \qquad (122)$$

An analysis of this information shows that even the solid near T_M is not a perfect lattice since r_m^{max} is 0.18 Å smaller than the perfect lattice value and n_m is 0.5 Å smaller.

When n_m is resolved into like-ion, $n_{m,l}$, and unlike-ion, $n_{m,u}$, contributions an even more interesting result occurs. This is that there is approximately a 4% penetration of like ions into the apparent first coordination shell, the number of ions in this shell being 5.3 and not 6 as for a perfect lattice. Upon fusion n_m and r_m^{max} decrease, whereas $n_{m,l}$ increases, showing even more disorder in the structure. The percentage of like ions in the first coordination shell increases from 7 to 15% from 1045 to 2874°K. This casts some doubt on the experimental resolution of $g_m(r)$ since it is usually assumed that $n_{m,l} = 0$. However, it must be noted that this may be an artifact of the actual equation used for the pair potential, or of the fact that only an effective pair potential is used; not a potential which takes two-, three-, four-, etc., body terms into account. If this penetration is correct, then the identification of the first peaks of g_m and g_u leads also to estimates of apparent coordination numbers which are too low. Because of the rapid rise of g_l at $r < r_m^{min}$, $r_m^{min} < r_u^{min}$, n_m is found to be less than n_u.

If one uses g_u rather than g_m to define apparent coordination numbers, then $n_u = 6$ for solid and 5.5 for liquid at T_M and $n_m = 5.3$ and 4.0.

The behavior of r_u^{max} and h_u on expansion (volume increase) indicates that ion pairs are formed with separations smaller than the minimum in U_{+-}. The behavior of r^{max} and h for nonpolar liquids contrasts with that of ionic melts in that r^{max} is very close to the minimum of $u(r)$ and the height of the first peak decreases strongly on fusion and with increasing T in the liquid, while r^{max} remains constant or increases slightly. Otherwise, the radial distribution function data are very similar to those of nonpolar liquids except that both g_u and g_l show oscillations up to about 10 Å, in contrast to g_m, which shows little structure past 5 Å (see Fig. 6). This indicates, in accord with the radial distribution function theory of Stillinger et al.[55] and the lattice theory of Stillinger,[1] that we have alternating spherical shells of predominantly positive and negative charges [see Section III.2(iia)].

A further confirmation of the accuracy of the MC results is obtained by comparing r_m^{max} with experiment, an error of only ± 0.1 Å being apparent (see Table 15).

Table 15
Radial Distribution Function Data for Molten KCl

	r_m^{max},[a] Å	n_m
Lark–Horowitz and Miller,[110] X-ray diffraction	3.14	5.8
Zarzycki,[8b] X-ray diffraction	3.14	5.2
Levy et al.,[6] X-ray diffraction	3.08	3.7
Levy et al.,[6] neutron diffraction	3.10	3.5
Zarzycki/Levy[b]	3.20	4.5
MC/MD[c]	3.05 ± 0.1	4.3

[a] Maximum of $r^2 g(r)$, not of $g(r)$ as stated in Ref. 16.
[b] Recalculated by Levy from Ref. 8b.
[c] Calculated by data given by Woodcock.[77]

We shall now show that the MC results may also be used to investigate the validity of other theories.

First, the calculations show that the root mean square fluctuations of the electrostatic potential acting on an ion are about 9.2%

of U_N for the solid and 11.2% for the liquid at T_M. This is contrary to the LJD cell theory of McQuarrie,[15] who assumes that the Madelung constant implies that the ions may be regarded as moving in a uniformly charged sphere, i.e., are not affected by the electrostatic potential. More detailed insight into the often used cell model and the significance of free volume may be gained from the following analysis, due to Gosling and Singer.[71]

The range of attempted MC displacement defines a volume v_d around each particle. For a liquid near T_M it is possible to choose $v_d < L^3/N$ but so large that the chance of acceptance of moves to positions outside v_d is negligible. The ratio of accepted to attempted MC moves is then an MC estimate of

$$\left\langle \int_{v_d} \omega(\mathbf{r}_1)\, d\mathbf{r}_1 \right\rangle_{N-1} / v_d = v_f/v_d \tag{123}$$

where

$$\omega(\mathbf{r}_k) = \exp[-\tfrac{1}{2}\beta(U_k - U_k^{\min})] \quad \text{and} \quad U_k = \sum_{k' \neq k} u(|\mathbf{r}_{k'} - \mathbf{r}_k|)$$

and

$$U_k^{\min} = \sum_{k'} u(|\mathbf{r}_{k'} - \mathbf{r}_k^{\min}|)$$

is the minimum of the potential well in the volume v_d, the "cell" of particle k; $\langle \ \rangle_{N-1}$ indicates the average over all positions of particles $k' \neq k$; and v_f is the mean free volume (per particle).

Then

$$\begin{aligned}
Q' &= \int \cdots \int \exp(-\beta U_N)\, d\mathbf{r}_1 \cdots d\mathbf{r}_N \\
&= \int_{v_d} \cdots \int_{v_d} \prod_{k=1}^{N} \omega(\mathbf{r}_k) \exp(-\tfrac{1}{2}\beta U_k^{\min})\, d\mathbf{r}_k \\
&\approx \left(\left\langle \int_{v_d} \omega(\mathbf{r}_1) \exp(-\tfrac{1}{2}\beta U_1^{\min})\, d\mathbf{r}_1 \right\rangle_{N-1} \right)^N
\end{aligned} \tag{124}$$

the last step following from the normal assumption of cell theory and is not rigorous but is implied by the fact that the sequence of acceptances of MC moves and follows the statistics of independent events.[71]

Now $Q = CQ'$, where $C = 1$ for the solid and $N^N/N!$ for the liquid, i.e., the entire communal entropy $2R$ is assumed to appear immediately upon fusion (as in scaled particle theory, if the hard-sphere diameters are taken to be the "gas-phase" values[61]). This idea is supported by calculations on Ar and by the fact that the entropies calculated for KCl differ from experiment by only about $0.2R$[16]. Thus the MC data imply that a cell theory will be accurate for the liquid only if the correct ensemble averages are used for the volume and for the depth of the cell potential.

This casts doubt on the fundamental validity of the Murgulescu and Vasu[100] *cellular* hole theory. It also shows that the gaslike regions of significant structures theory are not physically meaningful (as others have suggested[1]). The deductions from MC data are, however, in excellent agreement with a more general hole theory (see Section II.1).

The free volumes obtained from the MC method[16] are also in agreement with the experimental values[9] (within the large experimental error involved), adding further evidence for the validity of using the Tosi–Fumi effective pair potential for the molten salts under investigation.

Another interesting quantity calculated by the MC method is the mean Madelung constant M, defined by

$$M = \tfrac{1}{2} N^{-4} V^{1/3} \sum_{k \neq l = 1}^{N} Z_k Z_l \langle 1/r_{kl} \rangle \qquad (125)$$

which for an NaCl lattice is 1.74756 and for an infinitely dilute electrolyte is zero.

The MC results show that M increases approximately linearly with V in the solid and liquid isotherms at T_M but increases by only about 3% upon fusion. Furthermore, it decreases as T increases and as V decreases. This indicates that in the liquid there is an increasing cancellation of positive and negative potentials as a result of increased thermal motion but a less complete cancellation of potentials due to ion pair formation, thus confirming the normal ideas of the structure of melts.

Another point which arises from the MC results is that the latent heat of fusion is largely due to the increase in Coulomb energy rather than because the Madelung constant decreases.[15] This is because the small increase of M on fusion does not compensate the volume increase effect. Also, the fact that expansion of the solid to the volume of the liquid at T_M causes it to be electrostatically unstable compared with a liquid at this temperature gives us some insight into the reason for fusion.

We may conclude that the recent MC calculations on KCl[16] have shown that using an effective pair potential obtained from solid-state data, one may calculate properties of a melt from first principles to a high degree of accuracy. Furthermore, this has led to considerable insight into the structure of the melt and the validity or otherwise of other theories. Finally, this has enabled a resolution of quantities into contributions from various effects, e.g., $g(r)$ into $g_m(r)$, $g_u(r)$, and $g_l(r)$, and to the evaluation of quantities very difficult or impossible to measure experimentally, e.g., M and v_f.

Preliminary work on molten LiCl, LiI, NaCl, KF, KBr, KI, RbCl, and CsF has been carried out by Lewis and Singer.[72] Their results for these molten alkali halides are in line with those for molten KCl.

(ii) Nonequilibrium Statistical Mechanics

Nonequilibrium statistical mechanics essentially concerns itself with the solution of Newton's equation of motion for a system of molecules, i.e., solving equation (62) rather than evaluating the partition function, which is the main concern of equilibrium theory.

In such theories one studies[46,49,73] the time evolution of the dynamic variables required to characterize the state of the molecules which make up the system, i.e., the microstate. If X is a mechanical property of the system, then $X(\Gamma_t)$ is the information needed, Γ_t being the state of the system at time t.

Equilibrium properties are obtained from this information by direct time averaging, using equation (63), rather than replacing these time averages by ensemble averages as in equilibrium theory. Of course, time-dependent properties are also obtainable by these nonequilibrium techniques, e.g., the time-dependent correlation function[73]

$$G_d(r, t) = (V/N)n(r, t)/(4\pi r^2 \Delta r) \qquad (126)$$

where $n(r, t)$ is the time-dependent analogue of $n(r)$,[73] and $n(r)$ is $\langle n(r, r + \Delta r) \rangle$ in equation (114).

Transport properties can also be calculated from the basic nonequilibrium data, e.g., the coefficients of diffusion, viscosity, and thermal and electrical conductivity. These may be obtained from the time evolution of the mechanical variables of the system in two ways, either through the Einstein relations or via the time correlation function approach.[73-75,93]

An example of these relationships is given by the diffusion coefficient D[75],

$$D = \lim_{t \to \infty} (1/6t)\langle[x(t) - x(0)]^2\rangle \tag{127}$$

$$= \tfrac{1}{3}\int_0^\infty \langle \dot{x}(0) \cdot \dot{x}(t)\rangle \, dt \tag{128}$$

where $\langle \; \rangle$ indicates time averaging in (127) and averaging over an equilibrium ensemble in (128). Other relations of this type may be found in the literature[73-75,93] for all the normal transport coefficients.

We will now discuss two approaches to nonequilibrium theory which have been applied to molten salt systems, the molecular dynamics method[45-48,75,76] and the distribution function method.[81]

(a) *Molecular Dynamics Method.* This method seeks to solve numerically, Newton's equations of motion [i.e., equation (62)] for a system of N molecules in a box of volume V with periodic boundary conditions.[45-48,73,75,76]

Thus the molecular dynamics (MD) method is an attempt to solve the classical kinetic theory problems, just as the MC method is an attempt to evaluate the classical ensemble averages. Both methods use the same basic model, i.e., N molecules in a box of volume V, with periodic boundary conditions imposed to eliminate surface effects.

In the MD approach the particles are generally initially placed in a regular lattice arrangement with velocities chosen at random, usually with a Gaussian distribution law. Then the equations of motion of the particles are solved by a finite differences method about every 10^{-14} sec for a total time of about 10^{-11} sec. Thus the dynamic behavior of the system of particles is followed as a function of time, i.e., the time evolution of the microstate of the system is obtained.

The equilibrium properties of the system are then calculated from equations (111)–(114) as direct time averages rather than ensemble averages (as they would be in the MC method). However, whereas the MC method (as normally used[45–48]) models the canonical (fixed N, V, T) ensemble, the usual MD method models an isolated thermodynamic system, i.e., a microcanonical ensemble (fixed N, V, U).

Thus the temperature of the system is not fixed but varies with time as the kinetic energy varies, i.e.,

$$T = \frac{1}{(t - t')3Nk} \int_t^{t'} \sum_{i=1}^{N} m_i \mathbf{v}_i(t) \cdot \mathbf{v}_i(t) \, dt \qquad (129)$$

where $\mathbf{v}_i(t)$ is the velocity of the ith particle at time t. In its application to molten salts, however, Woodcock[76] modified the basic MD method by adjusting the velocities and positions of the particles in such a way as to keep T approximately constant. This procedure enables this method to model the canonical, NVT ensemble as does the MC approach.

An outline of the procedure used in the isothermal MD[76] method is as follows. One records the positions of all the particles in the basic cell, i.e., by the vector $\mathbf{r}_i(t)$ for the ith ion at an initial time t, and the velocities are specified by allocating positions at a previous time $t - \Delta t$, i.e., $\mathbf{r}_i(t - \Delta t)$.

For small Δt we may use a finite difference approximation to (62), i.e.,

$$\mathbf{r}_i(t + \Delta t) = 2\mathbf{r}_i(t) - \mathbf{r}_i(t - \Delta t) + (\Delta t)^2 \left\{ (d/d\mathbf{r}_i) \sum_{j \neq i}^{N} u_{ij}(\mathbf{r}) \right\} / m_i \qquad (130)$$

and the velocities in the interval t to $t + \Delta t$ are

$$\mathbf{v}_i(t) = [\mathbf{r}_i(t + \Delta t) - \mathbf{r}_i(t)]/(\Delta t) \qquad (131)$$

The instantaneous temperature is, from (129),

$$T(t) = (3Nk)^{-1} \sum_{i=1}^{N} m_i \mathbf{v}_i^2(t) \qquad (132)$$

The state of the system at time $t' = t + \Delta t$ is given by the updating procedure

$$\mathbf{r}_i(t' - \Delta t) = \mathbf{r}_i(t) \qquad (133)$$

$$\mathbf{r}_i(t') = \mathbf{r}_i(t + \Delta t) \qquad (134)$$

If the required mean temperature is T^0, then,

$$T^0 = (m_\alpha/3k)\overline{(\mathbf{v}_i^2)}_\alpha^0 \qquad (135)$$

where $\overline{(\mathbf{v}_i^2)}_\alpha^0$ is the required mean-squared velocity of species of component α of mass m_α.

Thus we may define a function $f_\alpha(t)$ such that

$$[f_\alpha(t)]^2 = \overline{(\mathbf{v}_i^2)}_\alpha^0 \bigg/ \left[N_\alpha^{-1} \sum_{i=1}^{N_\alpha} \mathbf{v}_i^2(t) \right] \qquad (136)$$

which may be used to correct the velocities and positions of the particles so as to maintain T approximately equal to T^0. This is done by replacing equation (133) by

$$\mathbf{r}_i(t' - \Delta t) = \mathbf{r}_i(t + \Delta t) - \{(\mathbf{r}_i(t + \Delta t) - \mathbf{r}_i(t)\} f_\alpha(t) \qquad (137)$$

The use of $\mathbf{r}_i(t - \Delta t)$ in (130) to calculate $\mathbf{r}_i(t + \Delta t)$ retards or accelerates the particles uniformly so as to maintain approximately constant T.[76]

Except for this isothermal step, this is the normal MD procedure[73,75] and the additional labor required to maintain T constant is reported to be minimal.[76]

Woodcock has applied this isothermal MD method to KCl, NaCl, and LiCl,[76] once again using the Tosi–Fumi pair potential.[69] The computations were started from an NaCl lattice and after 10×10^{-14} sec the temperature is reported[76] to approach T^0 very closely. Within 50×10^{-14} sec the pressure had increased to approximately the correct value and the initial crystalline order had disappeared. Thus the isothermal MD method appears to be quite practical to implement for ionic systems. As is usual, the total time span used was 10^{-11} sec. It is of interest to note that the ratio of root-mean-squared velocity of the ions to the average speed corresponded to the correct Maxwellian distribution within the error involved in the method.

The equilibrium properties obtained agree well with both experiment and MC calculations, within combined uncertainties (see Table 16). Like ions were found to contribute 3%, 6%, and 13% of the total short-range repulsive energy in KCl, NaCl, and LiCl, respectively.[76] This indicates that as the anion-to-cation radius ratio increases, the dimensional analysis and scaled particle methods

Table 16
Some Calculated and Experimental Properties of Molten Alkali Chlorides at 1273°K[76a]

	KCl			NaCl		LiCl	
	MD[a]	MC	Expt.[76]	MD	Expt.[76]	MD	Expt.[76]
V, cm^3 mole^{-1}	—	54.0	—	—	40.2	—	31.8
p, 10^3 bar	−0.29	−0.35	0	0.94	0	1.18	0
U, 10^3 J mole^{-1}	−611.5	−613.4	−610.3	−678.8	−681.9	−788.7	−766.5
E_{++}^r, 10^3 J mole^{-1}	1.6	1.6	—	0.6	—	0.3	—
E_{+-}^r, 10^3 J mole^{-1}	86.2	86.5	—	93.2	—	134.9	—
E_{--}^r, 10^3 J mole^{-1}	2.4	2.3	—	6.6	—	16.8	—
$E^{dd,c}$, 10^3 J mole^{-1}	−20.9	−21.4	—	−14.3	—	−14.6	—
$E^{dq,c}$, 10^3 J mole^{-1}	−3.2	−3.2	—	−2.6	—	−3.1	—
M_+	1.812	1.817	—	1.829	—	2.028	—
M_-	1.815	1.819	—	1.845	—	2.067	—
C_v, J °K^{-1} mole^{-1}	45.1	45.9	45.2	46.8	47.3	47.4	49.4
S, J °K^{-1} mole^{-1}	194.3	189.5	192.2	182.9	181.3	162.7	163.2
V_+^f, 10^{-24} cm^3	2.57	—	—	1.86	—	1.45	—
V_-^f, 10^{-24} cm^3	2.47	—	—	1.58	—	1.04	—
$V_+^f + V_-^f$	—	—	2.66	—	2.06	—	3.28
D_+, 10^{-4} cm^2 sec^{-1}	1.35	—	1.48	1.54	1.58	2.55	—
D_-, 10^{-4} cm^2 sec^{-1}	1.25	—	1.34	1.10	1.26	0.78	—

[a] $E = \langle U_N \rangle$.

(which both consider only anion–cation interactions) would be expected to become less accurate. This is in agreement with experiment.[61,62]

Reported radial distribution function data confirm the comments made in Section III 2(if) as to the structure of the melt.

The MD method applied to molten salts appears to have two advantages over the MC method[76]: (1) equilibration is more rapid, and (2) statistical errors show that the MD method is perhaps a factor of two more efficient in terms of computer requirements. However, the MD method has the inherent advantage over the MC approach that time-dependent quantities and transport properties may be calculated from the MD data.

As mentioned in Section III 2(ii), transport coefficients may be obtained from the basic MD information by two methods, either as a long-time limit of a mechanical variable[74,75] or in terms of an autocorrelation function.[93]

So far only D_+ and D_- have been reported for molten salts using the MD method. These were calculated as the limiting slopes of $\overline{\Delta r^2}$ versus time, i.e., using (127). The values obtained for KCl and NaCl are in quite good agreement with experiment (see Table 16), and interestingly $\overline{\Delta r^2}$ is approximately the same for both types of ions in a salt at small t (to about 10^{-12} sec). For times greater than this, around 1.5×10^{-12} sec the curves show ripples, after which they settle down to about their limiting values. This has been interpreted[76] as indicating a tendency of neutral pairs or small clusters of ions to exist for reasonable lengths of time.

The behavior of other dynamic variables with time and more importantly the behavior of time-dependent correlation functions, which have led to so much detailed information about the structure of normal liquids,[94] would help clear up this point. A study of the above-mentioned quantities would also give more information on the structure of the melt and mechanisms of transport, in particular the mechanism of electrical conduction in molten salts.

Although this has not yet been undertaken, much detailed information from the MD approach about the time-averaged structure of molten KCl has recently been published[77] and we will now briefly discuss this.

Now $g_{ij}(r)$ may be written in terms of the average number $\bar{n}_j(r)$ of particles of type j in a sphere of radius r centered on a particle of type i as

$$g_{ij}(r) = [d\bar{n}_j(r)/dr]/\rho_j^0(r) \tag{138}$$

where $\rho_j^0(r) = (N_j/v)4\pi r^2$. This suggests that something may be learnt about the structure of a liquid by defining and calculating higher moments of the distribution of particles, e.g., the second moments of the radial distribution function or the radial fluctuation function,[77]

$$W_{jk}^i(r) = [\overline{n_j n_k(r)} - \bar{n}_j(r)\bar{n}_k(r)]/\tfrac{4}{3}\pi r^3 \rho \tag{139}$$

For a general pure fused salt system there will be six components of the indiscriminate radial fluctuation function, but as with g(r) for KCl, we need consider only like and unlike components, i.e., $W_u(r) = W_{--}^+(r) + W_{++}^-(r)$, $W_1(r) = W_{++}^+(r) + W_{--}^-(r)$, $W_{u1}(r) = W_{+-}^+(r) + W_{+-}^-(r)$, and $W_l(r) = W_u(r) + W_1(r) + W_{u1}(r)$.

One could go on and define higher moments, but they would require large amounts of computer time to generate.

Woodcock[77] has used the MD method to determine g(r) and W(r) for KCl at one temperature, the pair potential being the same as that for the MC calculations of Woodcock and Singer.[16] The results obtained confirm those obtained by the MC procedure (as do the values of the bulk properties[76]), e.g., the height and width of the first peak of g(r) is indicative of partial ion pairing or ion clustering. The narrow half-width in particular implies that the partial ion-association model is a more accurate description of the melt than a quasilattice theory.

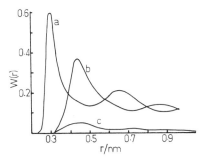

Figure 7. Radial fluctuation functions for KCl[77]: (a) $W_u(r)$, (b) $W_l(r)$, (c) $W_{ul}(r)$.

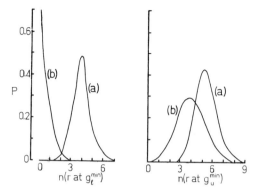

Figure 8. Normalized probabilities of radial occupation numbers for (a) unlike ions and (b) like ions in KCl.[77]

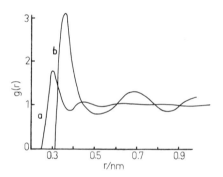

Figure 9. The indiscriminate $g_i(r)$ for (a) KCl and (b) Ar.[77]

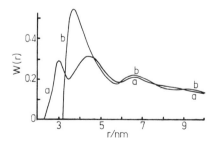

Figure 10. The indiscriminate $W_i(r)$ for (a) KCl and (b) Ar.[77]

The picture of alternating shells of positive and negative charge around a given ion which emerged from the MC calculations also comes out rather clearly from the graphs of $g(r)$ and $W(r)$ (see Figs. 7, 9, and 10).

Interestingly, the probability distribution $P(n_j, r)$ for radial occupation numbers n_j at a particular r show that as many as three like ions may be present in what is normally considered to be the first coordination limit for the unlike-ion shell, a result indicated by the MC results.[16] The probability of the occupancy of these ions at these distances is, however, very small (see Fig. 8). It would be very interesting to see if this happens with other salts and with other pair potentials.

Comparison with the results of an MD calculation on Ar,[77] using a Lennard-Jones 12-6 potential is very instructive in that liquid Ar appears to be a much more closely packed structure than KCl (see Figs. 9 and 10). The KCl structure has much more open structure with large voids (a term Woodcock uses in preference to the term "holes"), a picture which is confirmed by making models of typical MC configurations.[79] The conclusions arrived at by Stillinger,[1] i.e., that the overall spatial distribution of ion cores is little affected by the presence of Coulombic potential and that $g_m(r)$ for molten KCl should be practically identical to that of liquid Ar at a similar number density, are thus shown to be quite incorrect.

The idea that transport processes, i.e., the motion of ions in the melt, are little affected by the presence of the Coulomb potential[81–84,87] also needs reappraisal now that this detailed picture of the structure of molten salts has been obtained.

The very recent studies of K_2MgCl_4 by the MD method using a simple central, modified Tosi–Fumi pair potential (Sundheim and Woodcock[80]) shows that the MD procedure is capable of being applied to even relatively complex systems while still giving meaningful results. This is backed up by other recent MD calculations on nonsimple salts.[99] Thus we may look forward to more detailed insight into the structure of molten salts by means of the MD technique.

However, there is much more work necessary on even simple systems relating to different pair potentials, many-body effects, time-dependent and transport properties, and the effects of temperature and pressure on these properties.

(b) *Distribution function and friction coefficient methods.* In equilibrium statistical mechanics one may proceed via the partition function or via the equilibrium distribution function; similarly in nonequilibrium theory one may use either an autocorrelation function[93] or a time-dependent distribution function.[81,88–90]

Unlike the equilibrium function, the nonequilibrium distribution functions are time dependent as well as dependent on particle distribution. The time dependence of the N-body distribution function is governed by Newton's equations of motion (or equivalently the Liouville equation) and the MD method described in the last section essentially solves these equations for a system of N molecules in a cubic box with periodic boundary conditions.

Other approaches to nonequilibrium phenomena have been attempted in order to calculate the lower-order distribution functions,[81,88–90] usually the one- and two-body reduced distribution functions. These are sufficient to calculate normal transport properties. Such methods usually derive approximate equations for the lower-order distribution functions by making approximations to the hierarchy of equations linking them to the higher-order functions.

The most widely used of such theories is that due to Rice and Allnatt,[81,95] in which the pair potential is separated into a repulsive or "hard" part $u^h(r)$ and a slowly varying "soft" part $u^s(r)$. This division of the potential means that the motions of the molecules in the liquid can be treated on two time scales, one referring to the short-duration, large energy and momentum exchange events due to u^h, and one for the more numerous weaker collision events due to u^s. The short-range, "hard" encounters are then treated by the Enskog, hard-sphere method[40] and the "soft" part by a quasi-Brownian motion approach based on the Fokker–Planck treatment of Kirkwood.[81]

The method used to relate the transport properties to this information is via the friction coefficient ζ, defined by[81]

$$\zeta = (1/3kT\tau) \int_0^\tau \int_0^s \langle F(t) \cdot F(t + s') \rangle_1 \, ds' \, ds \qquad (140)$$

τ is a time interval chosen so that the basic dynamic events which occur during this interval are independent of both prior events and future events.

In (140) the force autocorrelation function is

$$\langle F(0) \cdot F(s)\rangle = \langle F^h(0) \cdot F^h(s)\rangle + \langle [F^h(0) + F^s(0)] \cdot F^s(s)\rangle + \langle F^s(0) \cdot F^s(s)\rangle \quad (141)$$

where $F(s)$ is the total force acting on a representative ion at time $t = s$. Thus

$$\zeta = \zeta^h + \zeta^s + \zeta^{hs} \quad (142)$$

The Einstein relation can be written

$$D = kT\zeta^{-1} \quad (143)$$

and this may be used to calculate the friction coefficient from experimentally determined D values or may be used to obtain D, if ζ is obtained theoretically.

In the Rice–Allnatt theory[81–86]

$$\zeta \approx \zeta^h + \zeta^s \quad (144)$$

and the "hard" friction coefficient ζ^h is obtained by the Enskog hard-sphere theory,

$$\zeta^h = (8/3)(N\sigma^3/V)(\pi mkT/\sigma^2)^{1/2}g(\sigma) \quad (145)$$

where σ is the effective core diameter of an ion pair, $g(\sigma)$ is the equilibrium radial distribution function at "contact," and the "soft" friction coefficient is usually obtained by the linear trajectory approximation[85,86,92] which enables one to relate ζ^s to integrals over the equilibrium radial distribution function.

The transport coefficients may then be related to ζ^h, ζ^s, $g(\sigma)$, $g(r)$, and the low-density collision integrals; the full expressions may be found in Refs. 81 and 85 for pure substances and in Ref. 82 for mixtures and we will not reproduce them here. However, for viscosity η and thermal conductivity λ we may write

$$\eta = \eta_K + \eta_V(r = \sigma) + \eta_V(r > \sigma) \quad (146)$$

and

$$\lambda = \lambda_K + \lambda_V(r = \sigma) + \lambda_V(r > \sigma) \quad (147)$$

where η_K and λ_K are the kinetic contributions to η and λ, and η_V and λ_V are the potential contributions.

This approach gives reasonable results for liquid noble gases,[81] but recent work[85,86] has shown that if all the quantities are obtained from theory, calculated properties do not agree with their experimental values.

To apply the method to molten salts, the equations must be generalized to a mixture of species; however, Rice[81,83,84] has introduced a simplification in that he considers a melt of an equal number of positive and negative spherically symmetric ions, with pair potentials consisting of a hard sphere, a short-range van der Waals part, and a long-range Coulomb part. The hard-core encounters for like ions are assumed to contribute nothing to transport and the friction coefficient is found not to be affected by the Coulomb potential.

Thus it is assumed that the primary role of the Coulomb potential is to determine the local geometry of the melt and not to influence the transport mechanisms. This reduces the problem to that of a pseudo-one-component fluid, as is done in some equilibrium theories. Similar equations to (145)–(147) are then obtained. For a molten salt electrical conductivity can also be calculated.

For ions in a normal fluid the force on an ion due to an external field E must vanish at least as E^2, but this is not so if the fluid surrounding the ion is also charged. In this case the external field exerts a force of opposite sign on the ion in question and on its first neighbors. This leads to a deviation of the local environment from spherical symmetry and in turn produces an internal field on the ion opposed to the external field. Since the net field on an ion is always less than the applied field, the diffusion mobility would be expected to be greater than the conductance mobility, i.e., there will be a deviation Δ from the Nernst–Einstein equation.

The extent of the deviation Δ may be obtained from the Rice–Allnatt theory[81] and is given by

$$\Delta = (4\rho_j^*/3kT) \int_{\sigma_{ij}}^{\infty} (du_{ij}/dr_{ij}) g_0^{(2)}(r_{ij}) r_{ij}^3 \, dr_{ij} \qquad (148)$$

where we are considering the motion of species j with respect to species i.

A simple analysis[81] for KCl taking $g_0(r)$ from liquid Ar at the same number density gave reasonable agreement with the experiment. However, caution should be exercised in using (148) since it involves the derivative of the pair potential, which makes Δ sensitive to small errors in $u(r)$. Furthermore, MD calculations[77] have shown that the use of $g_0(r)$ from Ar at a similar number density has led to very inaccurate results for molten KCl.

Ichikawa and Shimoji[82] have recently used the basic Rice–Allnatt model but without the assumption that the system is a pseudo-one-component system. They took $g(r)$ from diffraction data, ζ^s was calculated by means of the small step diffusion model for mixtures,[95] and for the pair potential they chose

$$u_{ij}(r) = u_{ij}^h(r) + u_{ij}^s(r)$$

$$u_{ij}^h(r) = \infty, \quad r < \sigma_{ij}$$

$$= 0, \quad r > \sigma_{ij}$$

$$u_{ij}^s(r) = 0, \quad r < \sigma_{ij}$$

$$= A_{ij}e^{-(B_{ij}-r_{ij})/c_{ij}} - (C_{ij}/r_{ij}^6) - (D_{ij}/r_{ij}^8) + (Z_i Z_j e^2/Dr_{ij}),$$

$$r > \sigma_{ij} \quad (149)$$

D is the local dielectric constant and σ_{ij} is the effective core diameter of the ion pair ij. They used the full multicomponent equations for the transport properties.[82]

This method was applied to calculate the diffusion coefficients D_+ and D_- and the viscosities η of molten NaCl, KCl, RbCl, CsCl, and NaI.[82] The results show that Rice's assumption of a pseudo-one-component fluid is reasonable only for RbCl and CsCl, but that for the other salts ζ_{++} and ζ_{--} are too large to be neglected. The actual values of D_+, D_-, and η obtained are in good agreement with experiment (see Table 17 and 18). This seems to indicate that, used in this manner, the Rice–Allnatt model gives a reasonable description of transport processes in ionic melts.

However, in appraising the theories of Rice[81] and Ichikawa and Shimoji,[82] it is necessary to bear in mind that published tests of their validity are of doubtful value. This is because of the heavy reliance on calibration by using experimental results to supply some of the quantities appearing in the theories.

Morrison and Lind[87] have also performed some calculations using a method similar to the Rice–Allnatt model, i.e., by using a

Table 17
Comparison between Theoretical and Experimental Shear Viscosities η[82]

Molten salt	T, °C	η, mP Theor.	η, mP Expt.
NaCl	800	12.5	14.6,[111] 15.9[113]
KCl	800	12.5	11.1,[111] 11.3[114]
RbCl	750	18.2	12.4,[111]
CsCl	700	11.5	11.5,[111]
NaI	700	19.9	13.6,[112] 15.9[115]

friction coefficient formalism. They used three different methods, the Brownian motion model [i.e., relating ζ to $g(r)$], the acoustical model of Rice, and the hard-sphere model, to calculate the friction coefficient for KCl and NaI and the coefficient of viscosity for KCl. They estimated $g(r)$ from X-ray data, and the pair potential (both of which are needed to calculate from the Brownian motion model) was taken to be

$$u = \frac{e^2}{r} + \frac{c}{r^6} + \frac{d}{r^8} + C \exp\left(\frac{D-r}{F}\right) \tag{150}$$

with parameters taken from solid-state data.

The results of calculations using friction coefficient formalisms leave much to be desired. At present this lack of agreement with

Table 18
Comparison between Theoretical and Experimental Diffusion Coefficients[a]

Molten salt	T, °C	D_+ Theor.	D_+ Expt.	D_- Theor.	D_- Expt.
NaCl	800	8.01	8.5[116,118]	6.02	5.9[116,118]
KCl	800	6.47	7.2[118]	6.30	6.5[118]
RbCl	750	4.95	4.9[116]	4.95	4.4[116]
CcCl	700	4.52	3.9[116]	4.66	4.4[116]
NaI	700	9.46	7.9[116,117]	4.02	4.2[116,117]

[a] From Ref. 82. D values in m^2 sec^{-1} × 10^9.

experiment and the calibration necessary to obtain reasonable numerical results make this avenue of investigation of rather doubtful value. Recently reported additional work on friction coefficient methods[85,86] and on distribution function methods[88-90] could, however, lead to other models being developed along these lines and to some clarification of the approximations used by former workers.

The most promising avenues for future investigation in the investigation of models of molten salts would appear to be weighted heavily in favor of MC and MD calculations.

It is interesting to note that the physical picture of an ionic melt given by MC and MD methods is similar to that arising from the hole model of Bockris and co-workers.[9,17,18] The hole model calculations of physical properties such as compressibility also agree closely with the MC and MD methods and with experiment.

REFERENCES

[1] F. H. Stillinger, in *Molten Salt Chemistry*, Ed. by M. Blander, Interscience, New York, 1964, pp. 1–108.
[2] H. Bloom and J. O'M Bockris, in *Fused Salts*, Ed. by B. R. Sundheim, McGraw-Hill, New York, 1964, pp. 1–62.
[3] M. Blander, in *Molten Salts*, Ed. by G. Mamantov, Marcel Dekker, New York, 1969, pp. 1–54.
[4] F. W. Dampier and G. J. Janz, *J. Electrochem Soc.* **118** (1971) 1900.
[5] C. A. Angell, *Ann. Rev. Phys. Chem.* **22** (1971) 429–464.
[6] H. A. Levy, P. A. Agron, M. A. Bredig, and M. D. Danford, *Ann. N.Y. Acad. Sci.* **79** (1960) 762.
[7] H. Bloom, *The Chemistry of Molten Salts*, Benjamin, New York, 1967.
[8] J. Zarzycki, *J. Phys. Radium* (Suppl. Phys. Appl.) **18** (1957) 65A.
[8a] J. Zarzycki, *J. Phys. Radium* **19** (1958) 13A.
[9] J. O'M. Bockris and N. E. Richards, *Proc. Roy. Soc.* **A241** (1957) 44.
[10] M. H. Cohen and D. Turnbull, *J. Chem. Phys.* **31** (1959) 1164.
[11] G. Adam and J. H. Gibbs, *J. Chem. Phys.* **43** (1965) 139.
[12] T. S. Ree, T. Ree, H. Eyring, and R. Perkins, *J. Phys. Chem.* **69** (1965) 3322.
[13] S. W. Mayer, *J. Chem. Phys.* **40** (1964) 2429.
[14] J. E. Lennard-Jones and A. F. Devonshire, *Proc. Roy. Soc.* **A163** (1937) 53; **A169** (1939) 317; **A170** (1939) 464.
[15] D. A. McQuarrie, *J. Phys. Chem.* **66** (1962) 1508.
[16] L. V. Woodcock and K. Singer, *Trans. Faraday Soc.* **67** (1971) 12.
[17] J. O'M. Bockris and S. R. Richards, *J. Phys. Chem.* **69** (1965) 671.
[18] T. Emi and J. O'M. Bockris, *J. Phys. Chem.* **74** (1970) 159.
[19] R. Fürth, *Proc. Cambridge Phil. Soc.* **37** (1941) 252.
[20] S. J. Yosim and H. Reiss, *Ann. Rev. Phys. Chem.* **19** (1968), 59.
[21] F. Zernicke and J. A. Prins, *Z. Physik.* **41** (1927) 184.
[22] C. A. Angell, *J. Phys. Chem.* **68** (1964) 1917.

[23] G. Tamann and W. Hesse, Z. Anorg. Allgem. Chem. **156** (1926) 245.
[24] C. A. Angell, J. Phys. Chem. **70** (1966) 2793.
[25] Wei-Chen Lu, T. Ree, V. G. Gerrard, and H. Eyring, J. Chem. Phys. **49** (1968) 797.
[26] M. S. Jhon, G. Clemena, and E. R. Van Artsdalen, J. Phys. Chem. **72** (1968) 4155.
[27] H. Bloom and E. Heymann, Proc. Roy. Soc. **A188** (1947) 392.
[28] H. Margenau and N. R. Kestner, *Theory of Intermolecular Forces*, Pergamon, New York, 1969.
[29] J. O. Hirschfelder (Ed.), *Advances in Chemical Physics*, Vol. 12, Wiley—Interscience, New York, 1967.
[30] R. G. Parr, *Quantum Theory of Molecular Electronic Structure*, Benjamin, New York, 1964.
[31] P.-O. Löwdin, Int. J. Quantum Chem. **2** (1968) 867.
[32] I. K. Snook and T. H. Spurling, to be published.
[33] P. Bertoncini and A. C. Wahl, Phys. Rev. Letters **25** (1970) 991.
[34] D. R. McLaughlin and H. F. Schaeffer, III, Chem. Phys. Letters, **12** (1971) 244.
[35] T. L. Gilbert and A. C. Wahl, J. Chem. Phys. **47** (1967) 3425.
[36] G. Das and A. C. Wahl, Phys. Rev. **A4** (1971) 825.
[37] P. T. Wedepohl, Proc. Phys. Soc. (London) **92** (1967) 79.
[38] J. A. Barker, R. A. Fisher, and R. O. Watts, Mol. Phys. **21** (1971) 657.
[39] I. R. McDonald and K. Singer, J. Chem. Phys. **50** (1969) 2308; Disc. Faraday Soc. **43** (1967) 40.
[40] J. O. Hirschfelder, C. F. Curtiss, and R. B. Bird, *Molecular Theory of Liquids and Gases*, Wiley, New York, 1954.
[41] T. L. Hill, *Introduction to Statistical Mechanics*, Addison-Wesley, Massachusetts, 1960.
[42] H. A. Levy and M. D. Danford, "Diffraction studies of the structure of molten salts," in *Molten Salt Chemistry*, Ed. by M. Blander, Interscience, New York, 1964, p. 109.
[43] *Physical Chemistry, an Advanced Treatise*, Vol. II, *Statistical Mechanics*, Academic Press, New York, 1967.
[44] D. Henderson and S. G. Davison, Ref. 43, Chapter 7, p. 339.
[45] R. O. Watts, Rev. Pure and Appl. Chem. **21** (1971) 167.
[46] *Physical Chemistry, an Advanced Treatise*, Vols. 8A, 8B, Academic Press, New York, 1971.
[47] G. A. Neece and B. Widom, Ann. Rev. Phys. Chem. **20** (1969) 167.
[48] I. R. McDonald and K. Singer, Quart. Rev. **24** (1970) 238.
[49] F. H. Ree, Ref. 46, Vol. 8A, p. 233.
[50] E. A. Mason and T. H. Spurling, *The Virial Equation of State*, Pergamon, New York, 1971.
[51] H. T. Davis, Ref. 43, Chapter 8, p. 452.
[52] A. R. Allnatt, Mol. Phys. **8** (1964) 533.
[53] E. Meeron, J. Chem. Phys. **28** (1958) 630.
[54] D. J. Henderson, J. A. Barker, and R. O. Watts, IBM J. Res. Devel. **14** (1970) 668.
[55] F. H. Stillinger, J. G. Kirkwood, and P. J. Wojtowicz, J. Chem. Phys. **32** (1960) 1837.
[56] J. A. Barker, *Lattice Theories of the Liquid State*, Pergamon, London, 1963.
[57] K. S. Pitzer, J. Chem. Phys. **7** (1939) 583.
[58] J. A. Barker, D. Henderson, and W. R. Smith, Phys. Rev. Letters, **21** (1968) 134.
[59] H. Reiss, S. W. Mayer, and J. L. Katz, J. Chem. Phys. **35** (1961) 820.
[60] M. Blander, Adv. Chem. Phys. **11** (1967) 83.
[61] H. L. Frisch, Adv. Chem. Phys. **6** (1964) 253; H. Reiss, Adv. Chem. Phys. **9** (1965) 1.
[62] H. Reiss and S. W. Mayer, J. Chem. Phys. **34** (1961) 2001.

[63] S. J. Yosim and B. B. Owens, *J. Chem. Phys.* **41** (1964) 2032.
[64] F. H. Stillinger, *J. Chem. Phys.* **35** (1961) 1581.
[65] M. P. Tosi, *Solid State Phys.* **16** (1964) 1.
[66] T. C. Waddington, *Adv. Inorg. Radiochem.* **1** (1959) 157.
[67] T. Forland, T. Østvold, and J. Krogh-Moe, *Acta Chem. Scand.* **22** (1968) 2415.
[68] J. Krogh-Moe, T. Forland, and T. Østvold, *Acta. Chem. Scand.* **23** (1969) 2421.
[69] M. P. Tosi and F. G. Fumi, *J. Phys. Chem. Solids* **25** (1964) 31, 45.
[70] K. Singer and L. V. Woodcock, in *Proc. Culham Conf. Computational Phys.*, HMSO, 1969, paper No. 25.
[71] E. M. Gosling and K. Singer, *Proc. Int. Conf. Thermodynamics*, IUPAP and IUPAC, Cardiff, 1970.
[72] K. Singer and J. W. E. Lewis, unpublished work.
[73] B. J. Berne, Ref. 46, Vol. 8B, p. 539; S-H. Chen, Ref. 46, Vol. 8A, p. 108.
[74] E. Helfand, *Phys. Rev.* **119** (1960) 1; D. M. Gass, *J. Chem. Phys.* **51** (1969) 4560.
[75] B. J. Alder, D. M. Gass, and T. E. Wainwright, *J. Chem. Phys.* **53** (1970) 3813.
[76] L. V. Woodcock, *Chem. Phys. Letters* **10** (1971) 257.
[77] L. V. Woodcock, *Proc. Roy. Soc. (London)* **A328** (1972) 83.
[78] J. H. R. Clarke and L. V. Woodcock, *J. Chem. Phys.* **57** (1972) 1006.
[79] L. V. Woodcock, *Nature Phys. Sci.* **232** (1971) 63.
[80] B. R. Sundheim and L. V. Woodcock, *Chem. Phys. Letters* **15** (1972) 191.
[81] S. A. Rice and P. Gray, *The Statistical Mechanics of Simple Liquids*, Interscience, New York, 1965.
[82] K. Ichikawa and M. Shimoji, *Trans. Faraday Soc.* **66** (1970) 843.
[83] S. A. Rice, *Trans. Faraday Soc.* **58** (1962) 499.
[84] B. Berne and S. A. Rice, *J. Chem. Phys.* **40** (1964) 1347.
[85] A. F. Collings and L. A. Woolf, *Aust. J. Chem.* **24** (1971) 225.
[86] A. F. Collings, R. O. Watts and L. A. Collings, *Mol. Phys.* **20** (1971) 1121.
[87] G. Morrison and J. E. Lind, Jr., *J. Phys. Chem.* **72** (1968) 3001.
[88] M. J. Foster and G. H. A. Cole, *Mol. Phys.* **20** (1971) 417.
[89] P. M. Allen, *Physica* **52** (1971) 237.
[90] I. Prigogine, G. Nicholis, and J. Misguich, *J. Chem. Phys.* **43** (1965) 4516.
[91] E. Rhodes, in *Fused Salts as Liquids, Water and Aqueous Solutions*, Ed. by Horne, New York, Wiley–Interscience, 1972.
[92] E. Helfand, *Phys. Fluids* **4** (1961) 681.
[93] R. Zwanzig, *Ann. Rev. Phys. Chem.* **16** (1965) 67.
[94] A. Rahman, *J. Chem. Phys.* **45** (1966) 2585; *Phys. Rev.* **136** (1964) A405; B. J. Berne and D. Forster, *Ann. Rev. Phys. Chem.* **22** (1971) 563; B. J. Berne and G. D. Harp, *Advan. Chem. Phys.* **17** (1970), 63.
[95] S. A. Rice and A. R. Allnatt, *J. Chem. Phys.* **34** (1961) 409.
[96] S. A. Rice, *Mol. Phys.* **4** (1961) 305.
[97] R. O. Watts, "Integral Equation Approximations," in *The Theory of Liquids*, Specialist Reports of the Chemical Society (London), Statistical Mechanics, Vol. 1, 1973.
[98] D. J. Adams and A. J. Matheson, *JCS Faraday Trans. II* **68** (1972) 1536.
[99] A. Rahman, R. H. Fowler, and A. H. Narten, *J. Chem. Phys.* **57** (1972), 3010.
[100] I. G. Murgulescu and G. H. Vasu, *Revue Roumaine Chem.* **11** (1966) 681.
[101] D. Smith, E. Kaylor, E. Walden, Jr., and J. Gayle, U.S. Bur. Mines, Rept. Invest. 5832 (1961).
[102] O. Kubeschewski and E. Evans, *Metallurgical Thermochemistry*, 3rd ed., Pergamon Press, New York, 1958.
[103] K. K. Kelley, and E. G. King, U.S. Bur. Mines, Bull. 592 (1961).

[104] K. K. Kelley, U.S. Bur. Mines, Bull. 584 (1960).
[105] A. S. Dworkin and M. A. Bredig, *J. Phys. Chem.* **64** (1960) 269.
[106] A. S. Dworkin and M. A. Bredig (private communication).
[107] A. Honig, M. Mandel, M. L. Stitch, and C. M. Townes, *Phys. Rev.* **96** (1954) 629.
[108] H. Bloom and J. O'M. Bockris, in *Modern Aspects of Electrochemistry*, No. 2, Ed. by J. O'M. Bockris, Butterworths, London, 1959, Chapter 3.
[109] F. M. Jaeger, *Z. Anorg. Chem.* **101** (1917) 1.
[110] K. Lark-Horowitz and E. P. Miller, *Phys. Rev.* **51** (1937) 61.
[111] I. G. Murgulescu and G. H. Zuca, *Z. Phys. Chem. (Leipzig)* **222** (1963) 300.
[112] I. G. Murgulescu and G. H. Zuca, *Rev. Roumanie Chim.* **10** (1965) 123.
[113] R. S. Dantuma, *Z. anorg. Chem.* **175** (1928) 1.
[114] J. D. Edwards, C. S. Taylor, A. S. Russell and L. F. Maranville, *J. Electrochem. Soc.* **99** (1952) 527.
[115] S. Karpachev and A. Stromberg, *Zh. Fiz. Khim.*, **11** (1938) 852.
[116] J. O'M. Bockris and G. W. Hooper, *Disc. Faraday Soc.* **32** (1961) 218.
[117] S. B. Tricklebank, L. Nanis and J. O'M. Bockris, *J. Phys. Chem.* **68** (1964) 58.
[118] J. O'M. Bockris, S. R. Richards and L. Nanis, *J. Phys. Chem.* **69** (1965) 1627.

4

The Electrical Double Layer: The Current Status of Data and Models, with Particular Emphasis on the Solvent

R. M. Reeves

Department of Chemistry, University of Bristol, Bristol, England

I. INTRODUCTION

The intention of this discussion is not to present a complete review of recent developments in the entire double-layer field, but to attempt a discussion and intercomparison of the current theoretical descriptions of the double layer, with particular emphasis being laid on the theoretical molecular approach. Although there now exists a body of "molecular theory," in order to assess the validity and applicability and to understand the consequences of these theories, it is necessary to describe and discuss theories not having an obvious molecular basis. The general background of experimental data is now distributed throughout the literature in a number of reviews. The principal trends in these data must be outlined since it is from these data that the theoretical approaches are derived. In general the discussion will be limited to the mercury–solution interface, although the theories in many cases are apparently independent of the metal type. The characteristics of the gallium–solution interface are also included, principally because this can be seen as a limiting case of solvent behavior at the interface.

In the first section the basic double-layer model employed by most workers in the field is described and the diffuse layer theory outlined. The second section is concerned with the interrelationship between the adsorbent and adsorbate. The approach assumed is not conventional but is intended to illustrate the range of interactions that may be involved in a realistic model. The third section is concerned solely with the mercury–solution interface. The thermodynamic description is of course applicable to any ideally polarizable interface, but the remainder of the section describes various aspects of the results of studies of the mercury–solution double layer.

The fourth section describes the models proposed to represent the systems discussed in Section III. A brief description of the earlier models in the field is included. The remainder of the section is concerned with a description and discussion of the theories that are currently of interest. The division of the theories emphasizes the current polarization of views in the field. Little work has been done at a theoretical molecular level on organic systems and this deficiency is illustrated in this section. The final part of this section is concerned with the gallium–solution interface, at which the contribution of the solvent predominates.

In the final section an attempt is made to show the extent to which the theories proposed in Section IV are applicable to experimental data. As a result, a number of suggestions are made as to the areas of interest in which critical observations are necessary before further theoretical models can be formulated.

1. Basic Double-Layer Model

The macroscopic model on which most of the theoretical work is based is shown in Fig. 1. The terminology used in this diagram is standard and has been fully described elsewhere. The Gouy[1]–Chapman[2]–Stern (GCS) model as modified by Grahame[3] will be the basis of the following discussion. This model essentially consists in dividing the region between the metal surface and the bulk of the solution into a diffuse layer whose properties are described by the GCS theory and a layer adjacent to the electrode. The latter inner region will be seen to have properties which are markedly different from the bulk or diffuse regions. Grahame[3] found it convenient to divide the inner region into two zones, one bounded by the inner Helmholtz plane (IHP), the plane of adsorption of the specifically adsorbed ions, and a second region between the IHP and the outer Helmholtz plane (OHP), which is the average plane bounding the inner region of the diffuse layer. This latter plane is characterized by the plane of closest approach of the nonspecifically

Introduction

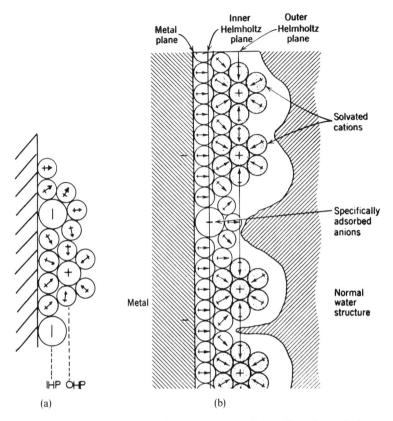

Figure 1. (a) Hypothetical model of double layer according to Gouy–Stern–Grahame. (b) Hypothetical model of double layer according to Bockris–Devanathan–Muller.[107]

adsorbed ions. The thickness and structure of the inner layer will comprise most of this discussion, but since nearly all the models assume a linear separation of the inner region from the diffuse layer, the properties of the inner region depend on how precisely the diffuse region can be described.

2. The Diffuse Layer

The GCS theory has been frequently reviewed (e.g. Refs. 1–5), and will not be described in detail in this article. The theory is based on the Poisson–Boltzmann relation and hence depends on the applicability of the Poisson and Boltzmann equations in this region. This has been criticized by Delahay, who points out that an explicit formulation for ion polarization and other interaction terms with the electric field must be taken into account. Prigogine et al.[6] have recently formulated the problem using the theory of local thermodynamic balance and have concluded that the classical diffuse layer model is materially correct. Such effects as dielectric saturation in the inner part of the diffuse layer, where a field of $\sim 10^9$ V m^{-1} may exist, and ion pair formation are also considered, but they have little influence on the precision with which the theory describes this region. The influence of saturation effects and dielectric inhomogeneity in the inner part of the diffuse layer has also been treated by Buff and Goel[7,8] with particular reference to the problems of ion imaging at the OHP. These calculations are discussed in a later section.

Very few proofs of the applicability of the GCS theory have been published. The first attempt by Joshi and Parsons[9] showed that there were significant differences between the observed and calculated ionic concentrations, but in a later study Parsons and Trasatti[10] showed that there were errors in the earlier calculations on HCl and BaCl$_2$ mixtures and that on correction the results agreed closely with the GCS theory. However, some small deviations were noted and were attributed to ion pair formation by the Ba^{2+} ions. In order to avoid these difficulties, the system Mg^{2+} and K$^+$ in the presence of chloride ions was adopted for study.

In the diffuse layer the charge due to the ith ionic charge is given by[11]

$$q^i = Gz_i \int_{u_2}^{u^i} \frac{n_i(u^{z_i} - 1)\,du}{u[\sum n_i(u^{z_i} - 1)]^{1/2}} \tag{1}$$

where $G = (kT\varepsilon/2\pi)^{1/2}$, $u = \exp(-e\phi/kT)$, ϕ is the potential at a distance x from the electrode, n_i is the concentration of ion species i, and z_i is the valence of the ith ion species. The electrode charge is given by

$$q = \pm 2G[\sum n_i(u_i^{z_i} - 1)]^{1/2} \tag{2}$$

The values of q^1 can be derived from equation (1) as a function of ϕ_2 for given compositions. The results of such calculations are shown in Figure 2. The experimental surface excess seems always

Introduction

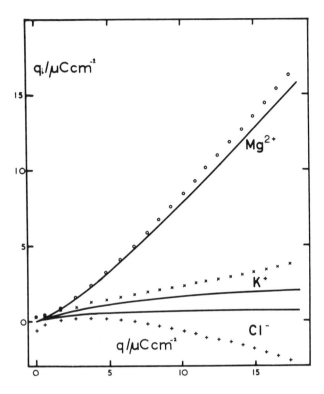

Figure 2. Experimental (points) and calculated (lines) surface concentrations of K^+ and Mg^{2+} in 20.8 mM KCl + 5.55 mM $MgCl_2$. Lines are calculated assuming no specific adsorption of Cl^-, using diffuse layer theory.[10]

to be slightly greater than that predicted by the simple diffuse layer theory. Similar deviations seem to be observed in the single salt systems KCl and $MgCl_2$ and it seems that the deviations are not a result of the GCS theory. This theory is probably adequate under most conditions, a conclusion which is confirmed by Anson et al.[12] as a result of charge injection experiments.

The deviations from the theory at present remain unresolved satisfactorily. Their obvious cause, experimental artefacts, seems unlikely, but further work is necessary to identify unequivocally the source of the deviations.

II. SOME CONSIDERATIONS OF THE PROPERTIES OF A SOLVENT IN THE REGION ADJACENT TO A SURFACE

1. Introduction

The purpose of this section is to draw together a wide range of evidence for the unusual properties of the solvent in the region adjacent to the interface. This involves consideration not only of the mercury–solution interface, but of a wider range of interfaces, both metallic and nonmetallic. From the experimental data it will become clear that the structuring of the solvent in the interfacial region is a feature of many systems and that in the presence of mercury, as an ideal surface system, the derivation of a specific solvent contribution must constitute a major part of any complete theory of the double layer at mercury. In comparing these data with the data at the metal–solution interface, the electrical variable must be identified and it is evident that sensible comparisons can only be made at $q = 0$.

2. General Properties of a Solvent Near Interfaces

The properties of interfacial water have been discussed for many decades and much of the older literature was reviewed by Henniker.[13] Arguments in favor of long-range structuring were reviewed more recently by Derjaguin.[14] By studying water in narrow capillaries constructed of quartz, anomalous water (called polywater) was suggested as an explanation of the results. This may be evidence that water is capable of existing in states other than those classically recognized, the structure being induced by the proximity of the interface. Much evidence for anomalous water may be derived from the thermal properties of interfacial functions. Although discontinuities are observed in the thermal properties of the bulk solvent, in the presence of an interface extra discontinuities frequently occur.[15] These anomalies have been discussed by Drost-Hansen.[16] A number of demonstrations of these phenomena are available. Peschel and Aldfinger[17a] have shown that a series of maxima can be obtained in the measurement of the disjoining pressure of two highly polished quartz plates at 100 Å separation which, it is suggested, is the result of surface structuring of the solvent. In a system more closely related to the electrochemical system, Lyklema[18] has observed a definite inflection in the surface charge versus temperature function at the AgI sol. This sudden "melting" of the structure near the interface is fully supported by differential thermal analysis data, which, obtained in the absence of an excess of bulk solution, displays similar trends. This evidence is not as conclusive as it seems, since there are other difficulties with these systems. In all probability, the surface structure of the particle also changes with temperature, with consequent changes in solvent structuring. Nucleation studies also provide evidence for unusual solution properties.[19] The number of crystallites found shows anomalies at ~ 15 and 70°C. In these systems the discrete structure of the solvent will depend on the ion types involved. These studies do not help in the elucidation of problems in pure or dilute ionic solutions, but may be useful in studies involving high ionic concentrations. These studies do suggest that any structuring that is present will be highly temperature dependent.

Theoretical studies of the interfacial region are frequently beset by the problem of estimating the dielectric constant of the interfacial region.[107,188,233] In the diffuse region it seems that the approximation to the bulk value of the dielectric constant is reasonable. In the inner region direct measurement of the inner region dielectric constant alone presents formidable problems. The usual method is to introduce a model, and the values obtained consequently depend on the model assumed. Some studies of the dielectric properties of interfaces have been made. Palmer et al.[20] studied the dielectric properties of water between sheets of mica and have reported very low values of the dielectric constant. The measurements, performed at 2–3 MHz, indicate that the dielectric constant decreases with thickness from ~ 64 with the thickest films employed to ~ 20 for films of $\sim 5\,\mu m$, to <10 for films $\sim 2\,\mu m$. The authors concluded that "the water film tends to act not as solid water but as 'liquid ice'." Other studies have also provided similar evidence.[21] Again, although the evidence seems superficially conclusive, there are extreme difficulties in the interpretation of the data, in particular, the models involved to separate the influence of bulk and interfacial phases.

NMR is yet another technique which may be used to detect the molecular configuration of the adsorbed solvent. The fundamental nature of this technique has attracted a number of workers.[22-24] Studies of the adsorption of water onto mica and alumina have been made. Pickett and Rogers in particular have studied the effects of building up layers of water on surfaces. Their results suggest that at low coverages the water is disorganized and mobile while at higher coverages the water has a significantly oriented structure near the surface which tends to extend its influence to the next layers as they are added, to produce a long-range ordering.

This evidence has been mainly in support of long-range ordering phenomena but there is equally convincing evidence against such effects. Water vapor adsorption isotherms on clays indicate that the adsorption energy decreases rapidly over a distance of about 10 Å. Similarly, density measurements of the water adsorbed by expanding clays suggest that water is of less than normal density only for the first two monolayers, beyond which it rapidly attains its bulk value.[25] These data are not undisputed and Anderson and Low[26] have suggested that the anomalous density region extends to about 60 Å. There are many difficulties in interpreting the experimental data and disagreement on the magnitude of the effective deviations from normal density are not uncommon. The basic problem in these studies is to define precisely the structure of the clay system under study.

This evidence, although of a semiquantitative nature and involving difficulties in interpretation, does clearly show that the presence of an interface significantly perturbs the solvent structure for a considerable distance into the bulk solvent. It serves to emphasize that any electrochemical model for the interface must contain a quantitative solvent model. The electrochemist has a distinct advantage for the types of interfaces discussed above in that control can be exercised over the electrical variable at the interface. In addition, at many metals the surface can be rather precisely defined and many of the indeterminate parameters in the above systems excluded.

3. The Aqueous–Air Interface

The air–solution interface should provide a simple example of interfacial solution properties. Although variation of the field across the interface is not readily achieved, the model of a solvent in contact with a dielectric presumably saturated with solvent vapor represents one of the simplest possible situations. The asymmetric field across the surface makes it probable that there will be some degree of preferential orientation of the solvent. Most of the evidence indicates that the potential drop due to dipole orientation is ~ 0.2 V.[27] A highly ordered surface as suggested by Fletcher[28] would give a surface potential of about 5 V. Stillinger and Ben-Naim[29] used a model of water consisting of a point dipole and a point quadrupole encased in a spherical exclusion envelope. The water molecule at the surface will be oriented in such a way that it will place the maximum amount of its electron field in the region of high dielectric constant (solution) rather than in the low-dielectric-constant vapor; in this way the water molecule will minimize its free energy. The field symmetry of a centrally placed dipole moment would not give any overall orientation and hence no surface potential. The permanent quadrupole moment of the molecule leads to an asymmetric field for the molecule which gives the observed surface potential. By minimizing the free energy of the model, the surface potential at 25°C was shown to be ~ 30 mV, the oxygen being oriented toward the vapor.

The temperature coefficient of the surface tension of water is negative and this indicates a positive surface excess entropy $S°$,[30]

$$S° = -(\partial \gamma / \partial T) = 0.157 \quad \text{erg cm}^1 \, °\text{K}^{-1} \qquad (3)$$

which, if due to a monolayer of water, would be equivalent to 2.18 cal °K mole^{-1}, the entropy change in bulk water on raising the temperature by 40°C. This effect is localized in the surface layer, as is confirmed by optical studies.[31]

If inorganic salt systems are now studied, the surface tension is generally found to be greater than that of pure water. This suggests that there is a deficiency of water in the surface layer (cf. Section II.5), which Langmuir estimated to be ~ 4 Å thick.[32] This layer thickness depends on the salt concentration and decreases for the alkali chlorides from 5 Å for an 0.1 M solution to 3 Å for a 1–2 M solution.[33,70] The theory of repulsion of ions from the surface of aqueous electrolytes by their electrical images agrees approximately with experimental data for 1:1 electrolytes.[34,35] The ions ClO_3^- and CNS^- seem exceptional in their behavior. These ions lower the surface tension of water with increasing ionic concentration, which suggests that they are not subject to the same repulsion forces as are the simple alkali metal cations or Cl^-, OH^-, or

SO_4^{2-}. Frumkin measured the surface potential of many aqueous electrolyte solutions and has shown that solutions of most simple salts have surface potentials more positive than that of pure water. This is interpreted in terms of anions being repelled less strongly from the surface than cations.[36] These effectively attractive forces may be in part due to ionic hydration effects. Randles[37a] has used a multilayer model to interpret these data. If the cations and anions have different approach distances to the surface, and the chemical potential of the ions in the vicinity of the surface is suitably modified, the change in surface potential can be estimated. Good agreement with the experimental data was obtained at low concentrations. At higher ionic concentrations the smeared-out charge model employed in the first approximation is clearly inadequate. Another factor which might well be considered is the extent to which the ion is able to be accommodated in the bulk phase and some contribution from the degree of "structure breaking" of the ion should be included.[38] The relatively disordered interfacial region would seem ideal for accommodating such structure-breaking ions as ClO_4^- and PF_6^-.

Since the differences in surface potential $\chi_{soln} - \chi_{water}$ are about -10 to $+50$ mV compared with the value of water itself, which is <0.2 V,[37b] the observed changes in this system are probably principally due to the solute and not the solvent.

4. The Role of the Metal

The most commonly studied metal–solution interface is the "smooth, reproducible" mercury interface with its large range of ideal polarizability. It is possible to study with almost equal facility the liquid gallium–solution interface at 30°C. The ideally polarizable range is smaller than for mercury and the pH of the solution must be controlled. Studies at other solid metal–solution interfaces have generally been of lower precision and techniques have only recently been developed which can give sufficient reliability in the data to apply critical analysis to the results.[39–41] Even at the gallium–solution interface studies have not been extensive,[42–48] although from these data important trends in the adsorption of ions and information on the possible role of the metal in the adsorption process can be ascertained. The general characteristics of the

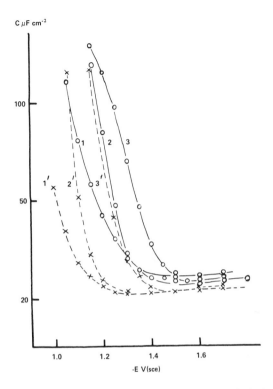

Figure 3. Differential capacity curves on liquid (dashed lines) and solid (continuous lines) gallium at 10 kHz in 1 N solutions of (1,1') $NaClO_4$; (2,2') KCl; (3,3') KI.[44]

capacity versus potential functions in Fig. 3 resemble those of aqueous iodide rather than chloride systems at mercury. A cathodic capacity rise is observed but there is no sign of the pronounced maximum observed on mercury even in the oxyanion systems. The surface activity of ions, $SO_4^{2-} < Cl^- < Br^- < I^-$, is in the same order as on mercury, but the surface excesses are smaller and no specific adsorption of ClO_4^- can be detected. The slopes of E_{pzc} (at zero charge q on the electrode) versus $\log a_\pm$ also differ from the behavior of the mercury system. In the latter system the slopes are in the order $I^- < Cl^-$, while the gallium $I^- > Cl^-$.[50] The ClO_4^- system at gallium poses certain problems since one of the usual tests of specific adsorption, or rather the absence of it, is to assume the series capacitor model for the diffuse and inner regions and to use a plot of $1/C$ versus $1/C_d$ which, in the absence of specific adsorption, is of unit slope since the inner layer capacity is supposedly independent of concentration in the absence of specific adsorption of ions. The slope in this system is not unity but the plots are linear and C_i is $\sim 78\ \mu\text{F cm}^{-2}$.

In order to obtain a unit slope, it is suggested that the diffuse layer dielectric constant must be increased to 120, which, it is proposed, may be a consequence of long-range ordering of the solvent by the gallium electrode.[50] Frumkin et al.[47–50] suggest that the dipoles of water are oriented at negative potentials with their negative ends toward the electrode. The gain in free energy when an uncharged gallium interface is wetted with water is 18.55 μJ cm^{-2} compared with $\sim 12.5\ \mu\text{J cm}^{-2}$ for mercury and if the water is firmly bound, the electrostatic interaction of the dipoles with the adsorbing ions will tend to reduce the specific adsorption.[51] In addition, the interfacial tension of gallium is $\sim 64.6\text{–}7\ \mu\text{J cm}^{-2}$ as compared with $\sim 42.6\text{–}7\ \mu\text{J cm}^{-2}$ for mercury. The changes in interfacial tension are generally smaller at gallium than at mercury. This evidence indicates that the metal plays an important role in determining the characteristics of the interfacial solvent. In the next section one approach which attempts to relate these changes systematically is discussed.

5. The Surface and the Work Function

The search for relationships between the nature of the metal and characteristic properties of the metal in contact with a working solution originated with the observation that the difference in the E_{pzc} for two metals is approximately equal to the difference in work function (or contact potential) Φ of the metals.[52] Analysis of the available experimental data suggests that a linear relationship may exist between these two quantities[53]

$$E_{\text{pzc}} = A\Phi - K \tag{4}$$

where K is a constant, usually derived from results on mercury, for which the data are most reliable. The slope A of the plot was suggested to be unity by Frumkin[53] and Argade and Gileadi[54] but 0.86 by Vasenin.[55] The difference between the slope of unity and the observed slope was attributed by Vasenin to the variation of orientation of the water molecules with Φ. In molten salts, for example, ΔE_{pzc} for various metals seems to agree closely with the difference in their contact potentials,[53] although to compare these two solvent systems may be dangerous.[56]

The biggest probable errors in such comparisons are in the values of the work function. A priori, the work of extraction of electrons from metal to solution and from metal to vacuum should be similar. The differences will be due to (1) the probable modifications of the double layer in the surface region of the metal due to the asymmetric distribution of electrons across the surface, and (2) the additional double layer in the solution arising from the layer of partially or totally oriented solvent dipoles.[11] Two surface potentials are associated with these two double layers, χ^m, the surface potential of a metal exposed to vacuum and g^m_{dipole}, the surface potential of the metal in contact with a solution; these are related in principle, through

$$g^m_{\text{dipole}} = \chi^m + \delta\chi^m \tag{5}$$

where $\delta\chi^m$ is due to modifications in the surface layer upon contact with water. On the solution side, if χ^s is the surface potential of water exposed to air and g^s_{dipole} that developed after contact with the metal, we have

$$g^s_{\text{dipole}} = \chi^s + \delta\chi^s \tag{6}$$

The potential at the point of zero charge is now related to Φ through

$$E_{pzc} = \Phi + \delta\chi^m - g^s_{\text{dipole}} + K' \tag{7}$$

where K' is a constant including all other potential contributions to the system.[56] After an extensive review of the precision and methods of determining the pertinent parameters, Trasatti suggests that two E_{pzc} versus Φ relationships exist, one for the transition metals,

$$E_{pzc} = \Phi - 5.01 \tag{8a}$$

and another for sp metals

$$E_{pzc} = \Phi - 4.69 \tag{8b}$$

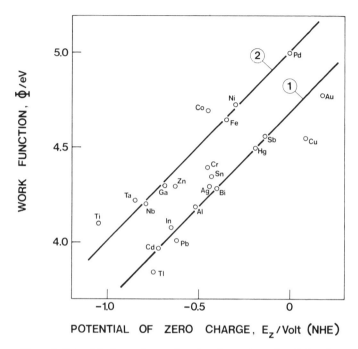

Figure 4. Potential of zero charge plotted against values of work function of metals (selected from experimental data[63]).

Deviations from the plots in Fig. 4 are then explained using equation (7). If $\delta\chi^m$ is reasonably independent of the metal, then equation (7) can be rewritten as

$$E_{pzc} = \Phi - g^s_{dipole} + K'' \tag{9}$$

The particular position of gallium has already been discussed and, in terms of these relationships, the strongly adsorbed and oriented water results in g^s_{dipole} being large in equation (9). To a first approximation, deviations from the linear plots in Fig. 4 are all attributable to the term g^s_{dipole}.

If a simple model is assumed, it is possible to derive g^s_{dipole} for metals. At sufficiently negative electrode charges, all the water molecules can be considered oriented with their positive ends toward the metal, as evidenced by the similar capacitative behavior at $q \ll 0$. In this region it is reasonable to assume that g^s_{dipole} is the same for the two metals. Hence it is possible to assume that the difference in potential at constant surface charge equals the difference in work function. From consideration of the work functions and potential shifts, we find $g^s_{\text{dipole}} = 80\,\text{mV}$ on mercury,[57] which agrees closely with the value derived from the adsorption of organic substances on mercury.[58] Water is therefore oriented with its negative end of the dipole toward the mercury, in agreement with conclusions of other observers.[59–62] Trasatti has shown that ethylene glycol, which is negligibly oriented at the interface, gives $\Delta E_z = 86\,\text{mV}$, in agreement with this derivation.[63]

A similar derivation for other metals has been presented by Trasatti,[56] for example, g^s_{dipole} for Cd is $\sim 0.23\,\text{V}$ and for Pb is $0.19\,\text{V}$. It can be seen that g^s_{dipole} increases in the sequence Hg < Pb < Cd < Ga, i.e., of increasing degree of orientation at the interface, an order which is confirmed from studies of neutral organic species, whose adsorption seems to decrease in the same sequence, and from the studies of specific adsorption of ions, which also seem to decrease according to the same sequence.[64–66] Trasatti was able to deduce the relative degrees of orientation of water on a wider range of metals, as follows:

$$\text{Au, Cu} < \text{Hg, Ag, Sb, Bi} < \text{Pb} < \text{Cd} < \text{Ga}$$

a series similar to the electronegativity scale x_m given by Pauling.[67] It also seems that if the lines of unit slope are drawn on the E_{pzc} versus Φ plots, elements with the same electronegativity seem to lie on the same line. Thus Ag, Hg, Sb, and Bi fall on the same line, $x_m = 1.9$, the same as the Pauling electronegativity of this group of elements. Equations (8a) and (8b) can now be combined in terms of the electronegativity to give

$$E_{\text{pzc}} = \Phi - 4.61 - 0.40\alpha \tag{10}$$

where $\alpha = (2.0 - x)/0.5$ and can be defined as the degree of orientation of water at the interface. α changes from zero, water not oriented, to 1.0, where water is fully oriented. Such an interrelationship is illustrated in Fig. 5. Defining the relationships in this way is equivalent to an actual contribution to the surface potential of 80 mV per 0.1 unit of electronegativity. Further discussion of the available data led Trasatti to propose a final form of $\alpha = (2.10 - x_m)/0.6$ and if the electronegativity values are now calculated from the electro-

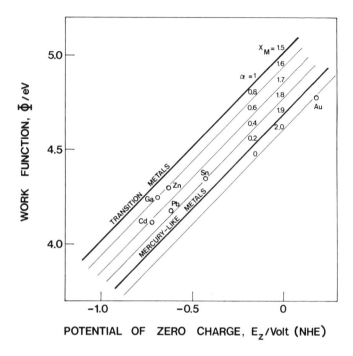

Figure 5. Plots of potential of zero charge against work function of metals; x_m is the Pauling electronegativity; α is the degree of orientation of water molecules at the interface.[63]

chemical data and Φ values, x_m seems to agree for all the metals with Pauling's values.[67] Many of the errors in earlier work seem to have resulted from inadequate consideration of the experimental data, e.g., Al must be measured with an oxide-free surface. The relationship between the intrinsic surface properties of the metal and the solution close to the surface seems well founded experimentally.

The actual orientation of water at the interface still requires further consideration.[56] The measurement of surface potentials is valuable in this context. The cell

$$Hg|N_2, H_2O|Hg$$

according to Frumkin has a potential of -0.26 V.[68] This potential may be expressed through

$$-0.26 = \chi^s + \delta\chi^m - g^s_{\text{dipole}} \qquad (11)$$

which in this case gives $\chi^s = 0.15$ V, where $\delta\chi^m = -0.33$ V. This value implies an orientation of water at the free surface with the oxygen atom pointing towards the gas phase, and agrees with the value obtained by other authors.[69,70] Further information on the relative degrees of orientation can be derived from the temperature coefficient of χ^s. This is negative. $\partial\chi^s/\partial T = -0.4$ mV °K^{-1} while at mercury $\partial\phi/\partial T = -0.57$ mV °K^{-1}.[70] In order to assess the relative magnitude of the effects at the two interfaces, $d\Phi/dT$ is required. No estimate of the relevant parameters is available for mercury. Randles and Schiffrin, on the basis of other data, suggest that the temperature coefficient of the solution in the surface region alone is -0.17 mV deg^{-1}. This lower value suggests that the water in the neighborhood of the mercury interface is about half as oriented as the water at the free solution–air interface.

According to equation (10), the difference in surface potential between Au and the transition metals due to water dipoles is 0.40 V. If this were due to a layer of completely oriented dipoles normal to the interface,

$$\chi_{\text{H}_2\text{O}} = 4\pi\mu n/\varepsilon = 0.75 \text{ V} \qquad (12)$$

where μ is the dipole moment of water, $\varepsilon \sim 7.5$, and the projected area of the water molecule is 12.5 Å^2. In order to obtain agreement with the experimental data, it is necessary to assume that the dipole is oriented at a mean angle of 32° to the surface. In terms of a model, this indicates that one hydrogen bond is still available for bulk bonding (cf. Section IV.5). This situation tends to stabilize the molecule in this orientation because the interaction of the water molecule with the bulk is also energetically favored since only one hydrogen bond is involved in reorientation phenomena.

These correlations are important since they suggest that the forces which bind the water to the metal are those concerned with the electronegativity of the metal rather than the electrostatic image forces. The monolayer adsorption on a metal is probably the simplest situation to treat theoretically, but the available treatments do not recognize the experimental facts as described here.

Macdonald and Barlow[71,72] use the model of imaging of the dipoles in the metal and include dipole–dipole lateral interactions and thermal and repulsive effects. Differences in dipole orientation are also approximated in the hexagonal lattice approximation employed. No parameter connected with the chemical bond with the surface is included. Most of the treatments of monolayer adsorption have been concerned with metal deposition in the context of thermionic emission and are not relevant in this context.[73] Since the solvent evidently interacts to different degrees with the surface, treatments based solely on dipole imaging cannot give a precise quantitative description of the system.

Since the situation discussed in this section is directly comparable with the metal–solution interface near $q = 0$, the types of interaction discussed in this section must be included in the treatments. The extent to which these interactions are involved in the models will become apparent in later sections.

III. DOUBLE-LAYER CHARACTERISTICS AT MERCURY

1. Introduction

The theoretical background to double-layer studies has been reviewed by a number of authors.[4,5,11,74,75] The most comprehensive treatment of ionic systems has been given by Parsons and Devanathan.[75] The background will be briefly outlined and some problems posed. In this section the purpose will be to describe the experimentally observable parameters at the mercury–solution interface and to indicate the relative importance of the various trends in the systems.

2. Classical Double-Layer Analysis

The thermodynamic analysis for both ionic and organic solutes is based on the Gibbs adsorption isotherm. For ionic systems this may be expressed through

$$-d\gamma = q\,dE_\pm + \Gamma_\mp\,d\mu \tag{13}$$

at constant temperature and pressure, where γ is the interfacial tension, q is the electrode charge, $E_\pm$ is the potential with respect to a hypothetical cation (E_+) or anion (E_-) reversible electrode, and $\Gamma_\mp$ is the relative surface excess of anion or cation.[4,5,76,77] Since equation (13) is a complete differential, by cross differentiation we find

$$(\partial q/\partial \mu)_{E_\pm} = (\partial \Gamma_+/\partial E_\pm)_\mu \qquad (14a)$$

$$(\partial E_\pm/\partial \mu)_q = -(\partial \Gamma_\mp/\partial q)_q \qquad (14b)$$

$$(\partial q/\partial \Gamma_\mp)_{E_\pm} = -(\partial \mu/\partial E_\pm)_{\Gamma_\mp} \qquad (14c)$$

and

$$(\partial E_\pm/\partial \Gamma_\mp)_q = (\partial \mu/\partial q)_{\Gamma_\mp} \qquad (14d)$$

Data are usually obtained in the form γ versus E, or C versus E, where C is the differential capacity defined by

$$C = (\partial q/\partial E)_\mu \qquad (15)$$

In ionic systems the specifically adsorbed charge (q^1) is defined as zF times the ionic concentration between the inner region of the diffuse layer and the metal. The charges due to anions and cations in the diffuse layer are related by diffuse layer theory (cf. Section I.2) and if one of the ions can be presumed to be absent from the inner region, the surface excess due to this ion can be equated to the diffuse layer charge of the ion and the diffuse layer theory then used to calculate the diffuse layer charge due to the second adsorbed ion. From this result the specifically adsorbed charge can be calculated,

$$q^1 = q^{D+} + q^{D-} + q \qquad (16)$$

For solutions at constant ionic strength the method of Hurwitz and Parsons can be applied[78,79]

$$(1/RT)(\partial \gamma/\partial \ln x)_{E,T,P} = -\Gamma^1 = -q^1/zF \qquad (17)$$

In this case x is the concentration of the adsorbing ion in the mixture with a salt whose ions are only present in the diffuse layer. The advantage of this method is that the diffuse layer theory is not directly involved, but there is the problem of finding a nonspecifically adsorbed salt, since recent studies have indicated that earlier results obtained using fluoride as the nonadsorbed ion are probably in error (cf. Section III.6).

The choice of electrical variable in equations (13)–(17) is not important since the thermodynamic approach does not depend on this choice. Since the diffuse layer approach involves the electrode charge q and ionic systems are better treated in terms of charge, Parsons has suggested that q should be used as the electrical variable.[80] The substitution $\xi = \gamma + qE_\pm$ is used to transform the electrical variable. From equation (14b), simple criteria for specific adsorption can be developed by comparing experimental values of $(\partial E_\pm/\partial \mu)_q$ with values of $-(\partial \Gamma_\pm/\partial q)_\mu$ derived from the diffuse layer theory.[4,6,81] Experimental values of these coefficients vary from 1.45 for I^- to 1.22 for Cl^- and 1.0 for F^- at $q \gg 0$.

The analysis of capacity–potential functions is preceded by separation of the diffuse layer contribution to give the capacity component of the inner region. This is generally analyzed using[80]

$$\phi^{m-2} = f(q, q^1) \tag{18}$$

where ϕ^{m-2} is the potential drop from the metal to OHP, to give

$$_qC^1 = (\partial \phi^{m-2}/\partial q)_{q^1} \tag{19a}$$

$$_{q^1}C^1 = (\partial \phi^{m-2}/\partial q^1)_q \tag{19b}$$

These components of capacity, C_i, $_qC^i$, $_{q^1}C^i$, are the direct result of inner layer phenomena. The data are the result of applying a model to the system and the experimental results will include any residual defects in the models assumed. The base solution capacity, frequently defined as $_qC^i$, is derived at constant q^1. As $q^1 \to 0$, the true base capacity is obtained. These values have been obtained for a number of systems. For the simplest ions, I^- to F^-, the capacities are similar but not identical (cf. Fig. 4.10, Ref. 4, p. 68). The distinctive hump is present in all cases. This suggests that the hump is the result of both ion and solvent contributions. The component $_{q^1}C^i$ is also widely reported as being independent of q. The plots from which this capacity component is derived, ϕ^{m-2} versus q^1, are seldom parallel, particularly in nonaqueous media. This may in part be due to experimental difficulties, but the systematic observation of such deviations does suggest that the simple model in which the integral capacity $_{q^1}K^i$ is used in place of the differential capacity $_{q^1}C^i$ may not be valid.

These components of inner layer capacity may be related to a simple model of the inner layer,[4]

$$_qK^i = \varepsilon_i/4\pi x_2 \tag{20a}$$

$$_{q^1}K^i = \varepsilon_i/4\pi(x_2 - x_1) \tag{20b}$$

Hence, on the basis of the experimental capacity, the inner layer distance ratio can be derived

$$_qK^i/_{q^1}K^i = (x_2 - x_1)/x_1 \tag{21}$$

By using Courtauld molecular models for molecular dimensions, the individual distances and hence ε_i, the inner layer dielectric constant, can be calculated. This is usually of the order 6–10 for most systems, a value close to the high-frequency dielectric constant of most common solvents. Further considerations of the thickness parameter are discussed in the context of specific ionic adsorption (Section III.9).

For organic systems equation (13) can be applied in the form[82]

$$-d\gamma/d\mu_A = \Gamma_A^{\text{expt}} = \Gamma_A - \Gamma_W(N_A/N_W) \tag{22}$$

In order to obtain Γ_A from Γ_A^{expt}, a model must be assumed, particularly if the coverages are high. The usual assumption is

$$\Gamma_A S_A + \Gamma_W S_W = 1 \tag{23}$$

where S_A and S_W are areas corresponding to one mole of organic substance and water in the surface layer, respectively. The analysis is then based on the Frumkin model, with the field across the inner layer being controlled by using potential as the electrical variable

$$q = q_b(1 - \theta) + q_A\theta \tag{24}$$

where the charge at $\theta = 1$ is q_A and that at $\theta = 0$ is $q = q_b$. Hence

$$C = C_b(1 - 0) + C_A\theta + (q_A - q_b)(d\theta/dE) \tag{25}$$

The assumption sometimes made, that $d\theta/dE = 0$, is only true for the simplest isotherms and does not even result from the Langmuir isotherm.[80] It is also a necessary condition that C_b and C_a are not $f(E)$ and that all the potential dependence is in θ. This is in agreement with the conclusions of Breiter and Delahay.[83]

The electrical variable is important but it is easy to show the relationship of the models to thermodynamics. Using equation (14a) for an isotherm congruent with respect to potential, we find

$$\left(\frac{\partial q}{\partial \mu}\right)_E = \left(\frac{\partial \Gamma}{\partial \ln \beta}\right)_\mu \frac{d \ln \beta}{dE} \qquad (26)$$

i.e.,

$$q - q_b = kT\Gamma_s \frac{d \ln \beta}{dE} \int_0^a \frac{\partial \theta}{\partial \ln \beta} d \ln a \qquad (27)$$

Since the isotherm is defined as being congruent with respect to one of the electrical variables, $\partial \theta / \partial (\ln \beta) = \partial \theta / \partial (\ln a)$; hence by differentiation of equation (27),

$$C = C_A - C_b = kT\Gamma_s \left[\left(\frac{\partial \ln \beta}{\partial E}\right)^2 \frac{\partial \theta}{\partial \ln \beta} + \theta \frac{\partial^2 \ln \beta}{\partial E^2} \right] \qquad (28)$$

Frumkin's relationship, equation (24), is readily derived from equation (28). For $\theta = 1$, where $C_A \equiv C_b$,

$$C_A - C_b = kT\Gamma_s \partial^2(\ln \beta)/\partial E^2 \qquad (29)$$

Hence, with equation (28),

$$C = C_A \theta + C_b(1 - \theta) + kT\Gamma_s \left(\frac{\partial \theta}{\partial \ln \beta}\right)\left(\frac{\partial \ln \beta}{\partial E}\right)^2 \qquad (30)$$

The last term in this relationship is obviously equivalent to the last term in equation (25). There is therefore no need for any model to be introduced and only the constancy of the electrical variable is involved. The general equation for the Frumkin model, equation (25), is a direct consequence of thermodynamics. In order to apply equation (25) in the usual capacitor model, either $\partial(\ln \beta)/\partial E = 0$ or $\partial \theta / \partial (\ln \beta) = 0$. The former condition exists at maximum adsorption, while the latter is applicable when $A \gg 1/\theta(1 - \theta)$ in the Frumkin isotherm,

$$Bc = [\theta/(1 - \theta) \exp(-2A\theta) \qquad (31)$$

This isotherm implies the existence of a saturation coverage with strong lateral interactions predominating. This is the principal reason why Parsons objects to the use of this model for ionic systems.[84]

The derivation of the relationships at constant charge, an approach considered unrealistic by Frumkin,[85,86] is based on equation (14b). The relationship equivalent to equation (24) is

$$E = \varepsilon_B(1 - \theta) + \varepsilon_A \theta \tag{32}$$

and the capacity is related to coverage through

$$C_A - C_B = kTT_s\left[\left(\frac{\partial \ln \beta}{\partial E}\right)^2 \frac{\partial \theta}{\partial \ln \beta} + \theta \frac{\partial^2 \ln \beta}{\partial E^2}\right] \tag{33}$$

Neither approach is at variance with thermodynamics so long as they are used in their complete forms. They do differ when applied in practice as the approximations discussed are employed and in this case the principal difference depends on the variation of the ratio ε_i/x_i, the dielectric constant to thickness ratio with coverage. For Frumkin's approach this can be expressed through

$$\varepsilon/x = K_1 + K_2\theta \tag{34}$$

while for Parsons' approach

$$x/\varepsilon = (1/K_1) - (k_2 x_A x_B/\varepsilon_A \varepsilon_B) \tag{35}$$

where the subscripts have their usual meaning and $K_1 = \varepsilon_B/x_B$ and $K_2 = (\varepsilon_A/x_A) - (\varepsilon_B/x_B)$. Comparison of these relationships with experiment seems inconclusive and recently further attempts have been made to clarify the situation by generalizing Frumkin's relationship,[87,88]

$$q = \frac{[C_b(1 - \theta)nC_A\theta]E - nC_A E_N[k(1 - \theta) + \theta]\theta}{1 + n\theta - \theta} \tag{36}$$

where E_N is the point of zero charge (pzc) where $\theta = 1$ and potentials are referred to the point of zero charge where $\theta = 0$. Both n and k are adjustable parameters dependent on the adsorption hypothesis involved. For the Frumkin case $n = k = 1$; for Parsons' case $k = 1$ and $n = C_b/C_A$; and $n = 1$ and $k = C_b/C$ for the Hansen model.[89] In terms of this approach, it is not necessary to explain the deviation of data from one approach to another since it is possible to use the adjustable parameters to indicate the degree of deviation of a specific set of data from particular models. In practice most studied systems seem to center their behavior around the Frumkin model.

It must be emphasized that this is *not* a molecular approach and does not explain why adsorption occurs or to what extent the various molecular interactions contribute in real molecular system.

3. Solvent Excesses

In this section the discussion will turn toward specific situations in which the meaning[197a] of relative surface excess becomes important. Since the experimental ionic excess is measured with respect to the solvent, it is important to recognize conditions under which the approximation of assuming it to be the real excess becomes important. This has been considered by a number of authors and evidence for the difference between the real and apparent excesses is apparent even in the classical work of Gouy.[92] The Gibbs equation (37) is applied at constant temperature and pressure. The relative excess of ionic component i, $\Gamma_{i(H_2O)}$ with respect to water, may be derived in the usual thermodynamic way using the Gibbs–Duhem equation to eliminate the chemical potentials of the reference component,[4,76,92–94]

$$-d\gamma = \sum \Gamma_i \, d\mu_i \quad (37)$$

For the relative surface excess this gives

$$\Gamma_{i(H_2O)} = \Gamma_i - (x_{salt}/x_{H_2O})\Gamma_{H_2O} \quad (38)$$

where x_z is the mole fraction of any species z. This shows that the observed surface excess is the difference between the real surface excess of component i and the excess of water present in the interphase. If m_{salt} and m_{H_2O} are the bulk concentrations of salt and water, the corresponding n's are local concentrations,

$$\Gamma_{i(H_2O)} = \int_0^\infty [n_i - (m_i/m_{H_2O})n_{H_2O}] \, dx \quad (39)$$

for x defined as normal to the interface. If the plane of closest approach is x_2, $n_1 = 0$ at $x < x_2$, then splitting the integral into parts gives

$$\Gamma_{i(H_2O)} = \int_0^{x_i} [-(m_i'/m_{H_2O})n_{H_2O}] \, dx + \int_{x_i}^\infty [n_i - (m_i/m_{H_2O})n_{H_2O}] \, dx \quad (40)$$

where the second term on the right-hand side of this equation is directly available from diffuse layer theory.

It is possible to obtain an estimate of the inner layer thickness using equations (39) and (40). If a model is assumed for the distribution of water in the inner layer, i.e., from x_w to x_2 where $x_w < x_2$ and $[H_2O] = 0$ at $x < x_w$,

$$-\int_0^{x_2} (m_i'/m_{H_2O})n_{H_2O}\, dx = -\int_{x_w}^{x_2} m_i\, dx = -m_i(x_2 - x_w) \quad (41)$$

Hence, after using the GCS theory for a 1:1 electrolyte,

$$\Gamma_{i(H_2O)} + m_i(x_2 - x_w) = 2m_i K[\exp(-e\phi_2/2kT) - 1] \quad (42)$$

where K is the reciprocal Debye length. The right-hand side of equation (42) may be derived by comparing the experimental surface excesses with those calculated from the GCS theory in the absence of specific adsorption. In dilute solutions there is no unequivocal evidence for the influence of the solvent on the measured surface excess.[95] In concentrated solutions a number of studies have shown these effects. Parsons and Zobel[96] have studied the double layer in aqueous sodium dihydrogen phosphate solutions and found evidence for an ion-free layer at $q \ll 0$. Evidence for the orientation of water with the oxygen against the mercury was also found in this study. Using electrocapillary and double-layer capacity measurements, Harrison et al.[91] determined surface excesses in concentrated solutions of LiCl, NaClO$_4$, and MgSO$_4$, while Damaskin et al.[97,98] have determined similar quantities for aqueous solutions of NaNO$_3$ and NaClO$_4$. Unambiguous values of inner layer parameters can only be estimated when specific ionic adsorption is absent. In this case n_{H_2O} in the inner layer seems to be a function of ionic strength and the nature of the cation. Interpretation of Fig. 6 in terms of a simple capacitor model would suggest a decrease in C_i of 50% when the cation changes from Mg^{2+} to Li$^+$, a conclusion not confirmed experimentally. The decrease in the thickness of the ion-free layer at high concentrations may be a result of the decrease in ion hydration due to lack of water. These results principally apply to the SO$_4^{2-}$ system. Kaganovich et al.[98] also concluded that there was an interlayer of water in NO$_3^-$ and ClO$_4^-$ solutions (cf. Ref. 107) and suggested that this was the source of various anomalies observed in solutions of these salts (cf. Section III.7). It was suggested that the capacitance maximum and anodic

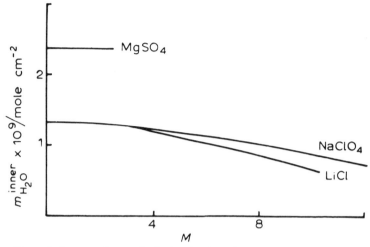

Figure 6. Amount of water in the ion-free layer for the solutions studied as a function of salt concentration (M = molality).[91]

minimum in the capacitance–potential function for these systems resulted from the formation of an interlayer of water at the electrode.

These data can be compared with the data at the air–solution interface. The ion-free layers are 1.6, 1.8, and 4.2×10^{-9} mole cm^{-2} of water for LiCl, NaCl, and MgSO$_4$ at the air–solution interface, respectively.[37] These are comparable with the data of Fig. 6 strictly only at the pzc, but at this point anion specific adsorption interferes. It is evident that for $q \ll 0$ on mercury the decreased values of thickness derive from the attraction of cations toward the electrode (cf. Section IV, Macdonald and Barlow theories).

From these studies it is evident that (a) under extreme conditions the reference species may be important; (b) cation size is less important than cation charge; (c) adsorption of oxyanions may involve specific solvent effects; and (d) comparison of behavior at the air–solution with mercury–solution interfaces is of limited value since the principal differences result from the presence of a dielectric in one case and a conducting medium in the second case.

4. Surface Excesses of Entropy and Volume

In addition to the two surfaces excesses normally encountered, namely those of charge and of ionic composition, there are two others also rigorously defined by thermodynamic means. They relate to the temperature and pressure coefficients of interfacial tension and lead respectively to the excess entropy and volume, i.e.,

$$(\partial \gamma / \partial T)_{P,\mu,E_\pm} = -\Gamma_s \tag{43}$$

and

$$(\partial \gamma / \partial P)_{T,\mu,E_\pm} = \Gamma_v \tag{44}$$

where the symbols have their usual significance, E being the electrochemical potential. If we adopt the Guggenheim approach to the interface and consider an interfacial zone σ bounded by two homogeneous phases α and β (e.g., water and mercury), then the two excesses can be further defined as

$$\Gamma_s = S^\sigma - m_{Hg}^\sigma \bar{S}_{Hg} - m_{H_2O}^\sigma \bar{S}_{H_2O} - \sum_i m_{H_2O}^\sigma (m_i^\sigma / m_{H_2O}^\alpha) \bar{S}_i \tag{45}$$

$$\Gamma_v = V^\sigma - m_{Hg}^\sigma \bar{V}_{Hg} - m_{H_2O}^\sigma \bar{V}_{H_2O} - \sum m_{H_2O}^\sigma (m_i^\sigma / m_{H_2O}^\alpha) \bar{V}_i \tag{46}$$

where the m values represent the number of moles of the component in the interphase σ or in the bulk phases α and β, and $\bar{S}$ and $\bar{V}$ are the partial molar volumes of the species concerned, assumed here to be the same in both bulk and surface phases. As is usually the case, the surface excess of ions (and other dissolved components) is expressed relative to the solvent species, but whatever convention is used to describe the interphase, it should be made clear that the surface excesses of the particular extensive quantities represent the excess values arising from the formation of the interphase from the bulk phases. In the case of the excess entropy, for example, it represents the sum of the entropies of those components in excess and the excess entropy of the other, reference components which arises because of the stresses and other forces in the interphase. It will be shown below that this latter component is of particular interest.

As observed above, the surface excesses are operationally defined by equations (43) and (44) and we should first consider how they may be determined experimentally. In principle, it is possible to measure the temperature and pressure derivatives directly, using, for example, a capillary electrometer containing a solution of fixed composition in contact with a reference electrode reversible to one of the ions. Allowance would need to be made for the concomitant variation of chemical potential and reference electrode potential with temperature or pressure. Certainly, so far as the pressure coefficient is concerned, its direct determination is experimentally difficult (and has not yet been achieved) and an alternative method which leads to almost the same range of information has been used instead. This is based on the full Gibbs equation,

$$-d\gamma = \Gamma_s dT - \Gamma_v dP + \sum_i \Gamma_i d\mu_i + q\, dE_{\pm} \qquad (47)$$

cross differentiation of which leads, in the case of the temperature coefficient, to

$$(\partial \Gamma_s/\partial q)_{\mu,P,T} = -(\partial E_{\pm}/\partial T)_{\mu,P,q} \qquad (48)$$

from which the surface excess can be evaluated by integration of the r.h.s. with respect to q.

In the case of a simple system such as

$$\mathrm{Hg}|\mathrm{NaCl\ solution}\,|\mathrm{Hg_2Cl_2}|\mathrm{Hg}$$

the relevant electrochemical potential is defined as

$$E_- = \mathscr{E}_- + [(\tfrac{1}{2}\mu_{\mathrm{Hg_2Cl_2}} - \mu_{\mathrm{Hg}})/F] \qquad (49)$$

where $\mathscr{E}_-$ is the emf of the above cell. It follows that

$$\left(\frac{\partial E_-}{\partial T}\right)_{\mu,P,q} = \left(\frac{\partial \mathscr{E}_-}{\partial T}\right)_{\mu,P,q} + \frac{S_{\mathrm{Hg}} - \tfrac{1}{2}S_{\mathrm{Hg_2Cl_2}}}{F} \qquad (50)$$

where

$$\left(\frac{\partial \mathscr{E}_-}{\partial T}\right)_{\mu,P,q} = \left(\frac{\partial \mathscr{E}_-}{\partial T}\right)_{c,P,q} + \left(\frac{\partial \mathscr{E}_-}{\partial \mu}\right)_{q,T}\left(\frac{\partial \mu}{\partial T}\right)_c \qquad (51)$$

and

$$\left(\frac{\partial \mu}{\partial T}\right)_c = \bar{S}_c \qquad (52)$$

i.e., the partial molar entropy of the salt at that concentration. Since, the capacity C is defined by $(\partial q/\partial \mathscr{E}_-)_{P,T,c}$

$$\left[\frac{\partial(1/C)}{\partial T}\right]_{c,P,q} = \frac{\partial}{\partial T}\left(\frac{\partial \mathscr{E}_-}{\partial q}\right)_{P,T,c} = \frac{\partial}{\partial q}\left(\frac{\partial \mathscr{E}_-}{\partial T}\right)_{c,P,q} \qquad (53)$$

and

$$\left(\frac{\partial \mathscr{E}_-}{\partial T}\right)_{c,P,q} = \int \left[\frac{\partial(1/C)}{\partial T}\right]_q dq \qquad (54)$$

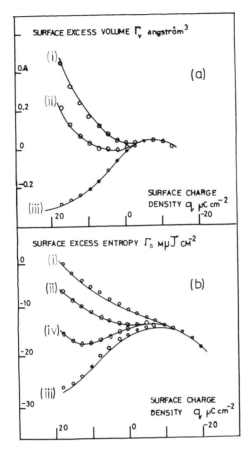

Figure 7. (a) Surface excess volume as a function of charge for 0.1 V solutions of the salts (i) $NaNO_3$, (ii) NaCl, (iii) NaF, (iv) Na_2SO_4 at 25°C and 1 atm pressure. Values are relative to the arbitrarily chosen zero at $q = 0$ for the NaF system. (b) Surface excess entropy as a function of charge for 0.1 N solutions of the salts (from Ref. 99, 100).

The second and final integration leads to Γ_s, provided that the integration constant Γ_s at $q = 0$ is known. If it is not known, then only relative values are obtainable and this is certainly the case for the excess volumes, which are derived from an entirely equivalent set of relationships.

Only two sets of measurements of Γ_s and Γ_v have been reported so far.[99-102] The results for 0.1 M aqueous solutions of NaF, NaCl, and Na_2SO_4 are shown in Fig. 7.

They are immediately interesting in that they show considerable sensitivity to the nature and degree of adsorption of the different ions present in solution. This is to be expected in principle because of the differing degrees of specific adsorption of anions, but differences are also be be observed with two different cation species neither of which is specifically absorbed. In fact, the variation at constant charge of Γ_s and Γ_v with ion type is to be expected from equations (45) and (46), which contain terms which are the products of the surface excesses and the partial ionic entropies and volumes of the ions concerned. Although absolute values of single-ion entropies and volumes are known only approximately, values relative to an arbitrary, standard ion are known exactly. If, for the

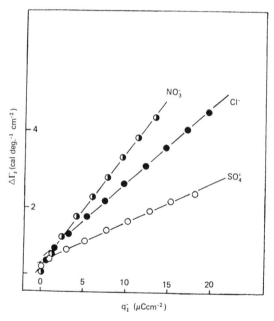

Figure 8. Surface excess entropy relative to that for 0.1 N NaF at the same value of the charge for the anion, as a function of the amount of adsorbed ion; 25°C and 1 atm.

sake of this study, we take fluoride as the reference ion, then $[(\Gamma_s)_{Cl^-} - (\Gamma_s)_{F^-}]$ should be a linear function of $\Gamma_-(q)$ with a slope equal to $(\bar{S}_{Cl^-} - \bar{S}_{F^-})$. Such a relationship is shown in Fig. 8, which bears out this simple thesis.

The question remains as to whether the ionic terms represent the major contribution to the surface excess entropies and volumes. The fact that both the air–mercury and the air–water interfaces have finite temperature coefficients of surface potential indicates that the other terms in equations (45) and (46) may be important. Indeed, once the ionic terms have been allowed for it is evident that the charge dependence of the residual excesses is a further important characteristic of the interface. The difficulty of making a proper allowance relates to the need to assume *absolute* values of the single ionic entropies and volumes. Fortunately, this is not quite the problem it once was and there are now generally agreed bases for evaluating $\bar{S}_i$ and $\bar{V}_i$, which are supported now by a number of experimental procedures.[103–104] Given such values, it is possible to calculate two other excess quantities, namely

$$S^* = \Gamma_s - \sum_i \Gamma_i \bar{S}_i \tag{55}$$

and

$$V^* = \Gamma_v - \sum_i \Gamma_i \bar{V}_i \tag{56}$$

which should reflect the nonionic contributions to the surface excesses. Values of S^* and V^* have been evaluated on the basis that $\bar{S}_{K^+} = 18.5 \text{ J mole}^{-1} \,°\text{K}^{-1}$ and $\bar{V}_{K^+} = 3.4 \text{ cm}^3 \text{ mole}^{-1}$ and that they are the same in the interface as they are in the bulk phases. They are shown in Fig. 9 as a function of q. In the case of the excess volumes they are shown relative to the value of the fluoride system at $q = 0$ and this avoids the difficulty that the absolute value of Γ_v at $q = 0$ is not known. It also allows us to neglect any other charge-independent contributions to V^*.

It is a striking consequence of this analysis that all the different S^* and V^* values fall on two common parabolas. The entropy values are the more reliable because of uncertainties in the experimental determination of $[\partial(1/C)/dP]_T$; moreover, they have been confirmed by earlier independent measurements.[105,106] The parabolic dependence is principally attributed to the variation in the contribution

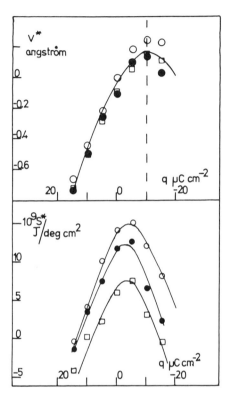

Figure 9. Reduced excess volume and temperature as a function of surface charge density.

of the solvent to the surface excess, i.e., to the difference $(S^\sigma - m^\sigma_{H_2O}\bar{S}_{H_2O})$ or $\Gamma'_{H_2O}\bar{S}_{H_2O}$, where Γ'_{H_2O} is the unknown absolute surface excess of solvent. It is clear that the orientation of the highly polar water molecules will be strongly influenced by the electric field normal to the interface and that both the entropy and the volume will decrease on either side of a point of zero preferred orientation. This point of maximum disorder evidently occurs in aqueous solution at $-6\,\mu\text{C cm}^{-2}$ and is negative of the point of zero charge because the natural preferred orientation of water on mercury (and probably on all metals) is oxygen to metal. There is independent evidence for this[61,265] and further support for the special significance of $-6\,\mu\text{C cm}^{-2}$ may be adduced from the fact that this is also near the charge value corresponding to maximal adsorption on mercury of simple organic molecules.[107,265,266] The maximal value of V^* occurs at approximately $-10\,\mu\text{C cm}^{-2}$ and although there is no reason why this should coincide with that for S^*, it is thought that further determinations of Γ_v might result in closer agreement.

Another way of describing the point of maximal S^* is to note that the Volta potential on the mercury corresponding to $-6\,\mu\text{C cm}^{-2}$ is just sufficient to overcome the natural, self-field of water on the metal surface, i.e., that the surface potential has become zero and the Volta and Galvanic potentials on the solution side of the interface are equal. It follows that between $q = 0$ and $q = -6\,\mu\text{C cm}^2$ there exists a zero of Galvanic potential difference on the solution side of the interface.

These conclusions are necessarily tentative because of the restricted range of studies that have been pursued so far. Much more data are required and they need to be supplemented by those from other methods, for example, optical techniques. Many such studies are in hand (see, for example, Refs. 108 and 109) and it is certain that they will lead to further information concerning the properties of the solvent in the interface.

5. Ionic Systems: General Characteristics

The general characteristics of the capacity–potential function were described by Grahame in his classical review.[74] On mercury an ideally polarizable range of potential of greater than 2 V is available, dependent on the anion and cation present. In aqueous media the capacity–potential function displays a characteristic minimum at potentials cathodic to the pzc, $q \sim -12\,\mu\text{C cm}^{-2}$, while at $q > 0$ the behavior is more complex. In many systems a maximum is observed when $q > 0$ and prior to the mercury dissolution reaction a further rapid rise is observed. The rate at which the final anodic rise occurs and the rational potential at which it is observed seem to depend on the stability of the mercury salt formed during the dissolution reaction for the halide series. Prior to this final increase the capacitance rise depends on the degree of specific ionic adsorption and in halide and pseudohalide solutions the series $F^- < Cl^- < Br^- < I^- \sim CNS^-$ is observed. The increase in fluoride media has been contested by Damaskin et al.[110] and discussed by Payne[111] but more recent evidence is presented in Section III.6. In other solutions the onset of dissolution and formation of a monolayer of halide salt is marked by an abrupt decrease in capacitance from values of $\sim 10^4$ μF cm^{-2}.[112] The anodic capacitance is very sensitive to temperature variation, a direct consequence of the dependence of ion adsorption energy on temperature.

The principal factor to discuss is the minimum at $q < 0$. Grahame found a striking independence of this capacitance on the cation type, although the unhydrated cation radii varied from Li$^+$ (0.07 nm) to Cs$^+$ (0.17 nm).[113,119] The temperature coefficient in this region is still significant but independent of the cation type (-0.075 μF cm^{-2} °K^{-1}).[114] This suggests that the cations must at least be hydrated and/or separated from the electrode by a layer or layers of water. There is evidence for a residual dependence on cation type[115] and this suggests that the accidental compensation of cation size by hydration is not complete.

At values of electrode charge less than the minimum the capacitance increases systematically, although the contribution of the cation to this rise has not been clearly demonstrated for ions other than Cs$^+$.[116] Gierst has derived possible surface excesses of cations assuming Li$^+$ to be nonadsorbed.[117] This is not a thermodynamic method and, since it is dependent on the capacity, neglect of the solvent and solvation contributions will invalidate these values.

The characteristics of the capacity–potential function for ionic systems in nonaqueous media have been fully described by Payne in a recent review.[115] Discussions of ionic adsorption and the hump are presented separately. For lower aliphatic alcohols the curve is rather featureless but a minimum value is attained in all cases. In methanol the temperature coefficient of the minimum capacitance is anomalous, $+0.02$ μF cm^{-2} °K^{-1}, but no rational explanation has yet been presented.

The most systematic study of the cathodic minimum was made by Payne on the formamide series of solvents. The minimum capacitance was not compared on the basis of electrode charge, so correlations with the Macdonald theory (Section IV.6) cannot be made. The minimum capacity was compared with the bulk dielectric constant of the solvents. Payne concluded that there was no direct correlation between these two quantities but concluded that the capacity does increase with increasing molecular weight of the solvent. This would indicate that the thickness of the inner layer is the pertinent variable rather than the dielectric constant of the region which was discussed in Section II. These conclusions disagree with the conclusions of Minc and Jastrzebska,[118] who suggested a functional dependence of the capacity at the minimum on bulk dielectric constant.

The data for the NH_4^+ ion are anomalous and also have to be explained.[124]

The general conclusions to be drawn from the characteristics of the minimum are twofold: (a) The capacitance minimum seems to be a characteristic of the solvent molecule, the size of the molecule being the predominant factor; and (b) the influence of the cations on the minimum is weak. Some residual influence of the difference in cation sizes is still apparent.

6. The Fluoride Ion—Is its Behavior Anomalous?

The cryptic title to this section refers to the apparently unique position of the fluoride ion as the only nonspecifically adsorbed monovalent anion. Grahame[120] found that the double-layer capacity in pure fluoride solutions could be described rather precisely by a simple capacitance model. If the diffuse layer was assumed to obey the classical GCS theory, the inner layer capacitance was found to be independent of the bulk concentration of fluoride. Since in other systems studied the inner layer capacitance was always significantly concentration dependent, Grahame suggested that in this case the fluoride ion was absent from the inner layer and was consequently defined as a nonspecifically absorbed anion. Further evidence for this behavior was inferred from the agreement of the experimental Esin and Markov coefficients with the theoretical values as derived from GCS theory. Both these methods of finding specific adsorption depend for their precision on the degree of agreement with the diffuse layer calculations. As the degree of specific adsorption tends to zero, the errors in the experimental data are magnified when GCS theory is applied and it is quite possible that there *is* specific adsorption, that is, charge in the inner layer, although it is not detectable by these techniques. The peculiar characteristics of the F^- ion were rationalized by Bockris et al.[107]

in terms of ionic solvation. The unique position of fluoride has been constantly challenged.

The capacity–potential function for $q \gg 0$ is significantly temperature dependent.[114] The capacitance–potential function shows a rise in capacitance when $q \gg 0$ similar to that found for the other alkali halides as the standard potential of the $Hg|HgF_2(Hg(OH)_2)$ couple is approached. Other evidence for the specific adsorption of of fluoride derives from the difference between the adsorption characteristics of the halides and other species when studied from solutions at constant ionic strength with fluoride and from single salt solutions.[121] Although these differences have been rationalized by various workers,[122] the arguments seem unsatisfactory.

From this discussion it is evident that the interpretation of much experimental work is relative to the "hypothetical nonadsorbed" system. Theoretical studies are also hampered by the lack of discriminating data on the fluoride ion, as Mott and Watts-Tobin[123] indicated. The anodic rise in capacitance did not agree with their dipole reorientation theory (cf. Section IV.2). The deviation was attributed in principle to three effects: (a) specific adsorption of the F^- ion, (b) specific adsorption of OH^- ions, and (c) an electrochemical reaction.

The third proposal is most unlikely since no direct current can be detected in the region of the anodic rise in capacity.[112] The degree of specific adsorption of OH^- ion was found to be very small by Payne and is probably only significant in concentrated hydroxide solutions when $q \gg 0$.[111] Of the influences proposed by Mott and Watts-Tobin,[123] only specific adsorption of F^- seems plausible.

An alternative explanation based on the adsorption of SiF_6^{2-} ions was proffered by Damaskin et al.[110] In a study that has never been repeated, they found that the anodic capacitance of "pure" fluoride media did not increase up to ~ 0.25 V versus NCE. There are other reasons to doubt the validity of this proposal,[124] as follows. (a) The results of many other workers seem to agree closely with Grahame's data;[120,124] (b) solutions which have stood in Pyrex for extended periods of time do not give significantly different capacities at anodic potentials; and (c) measurements in silicofluoride solutions do not show an increase in the capacitance but rather show a decrease as if an oxyanion had been added.[126]

Three recent studies have shown much more clearly what are the true adsorption characteristics of the fluoride ion. The work of the first two studies was concurrent and is based on the constant ionic strength technique evolved by Parsons and Hurwitz.[79,80] In this case a diffuse layer correction is avoided. The systems studied can be represented by xMKF $+ (p - x)$MKA. Verkroost chose to study the acidic fluoride system KF with additions of HF to give KHF_2. The relevant anionic species are therefore F^- and HF_2^-. In general, the observed surface excess Γ_t is given by

$$\Gamma_t = \frac{1}{RT}\left(\frac{\partial \xi}{\partial \ln x}\right)_q = \Gamma_A^i - \left(\frac{x}{p - x}\right)\Gamma_{F^-}^i \qquad (57)$$

The observed Γ_t values were anomalous and passed through zero as $q \gg 0$. The precise charge depends on the concentration ratio $[F^-]/[HF_2^-]$. If fluoride is nonadsorbed, then thermodynamically Γ_A^i must be equal to Γ_t and in this case be negative. This is most unlikely on an electrode at $q \gg 0$ and one is forced to the conclusion that the second term in equation (57) must be significant, ie., $|x/(p - x)\Gamma_{F^-}| > \Gamma_A$. A similar approach was adopted by Hills and Reeves.[121] In this case mixtures of F^- and PF_6^- were employed and, as shown in Fig. 10, the observed specifically adsorbed charge ($= zF\Gamma_t$) passes through a maximum and then decreases to negative values. This phenomenon is the direct result of fluoride adsorption and is seen to increase with increasing fluoride ion concentration. These authors attempted to estimate the degree of specific adsorption of F^- by assuming a limiting slope for PF_6^- adsorption and then applying equation (57) to the system. The limiting slope chosen was not correct, as a recent study of PF_6^- from KPF_6 solutions has shown.[128] In spite of the approximations, the range of adsorption is clear and the values are of the correct order of magnitude.

This is confirmed in the detailed study made by Schiffrin[126] of the adsorption of F^- from KF solutions at 15 and 0°C. Using the procedures developed by Parsons (cf. Section III.2), a careful analysis showed that the F^- ion is specifically adsorbed when $q \gg 0$, the degree of adsorption being strongly dependent on temperature. The data at 15°C are shown in Fig. 11. At both temperatures, for $q \sim 0$ there is no specific adsorption of F^- and this is in agreement

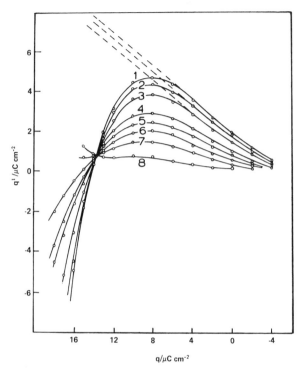

Figure 10. (———) Charge due to specifically adsorbed hexafluorophosphate ions (q^1) as a function of charge on the metal (q) and bulk concentration of KNF_6; (— —) estimated "true" adsorbed charge due hexafluorophosphate ions. x = (1) 0.3; (2) 0.2; (3) 0.1; (4) 0.075; (5) 0.050; (6) 0.035; (7) 0.020; (8) 0.010.[121]

with earlier observations. The adsorption does not become significant until $q \sim 6 \,\mu C \, cm^{-2}$, which agrees with the observations of Hills and Reeves.[121]

From this evidence it must be concluded that there is no monovalent anion for which specific adsorption is unambiguously absent over the entire range of polarization. As a consequence, much of the data for mixed electrolyte systems must be reanalyzed taking into account this fact, a task which is already being undertaken.[121,127]

7. Capacitances over the Entire Concentration Range

There are several systems which are susceptible to study over the entire range of concentration from infinite dilution to molten salt.[129-130] These, as well as displaying behavior typical of both dilute and concentrated solutions, can give information as to which parts of the capacitance–potential function are most sensitive to solvent effects. As shown in Fig. 12, the following trends are important.

(a) The hump in the capacity–potential function when $q \sim 0$ at first increases and subsequently decreases with increasing salt concentration.

(b) The hump moves to more cathodic potentials with increasing salt concentration.[131]

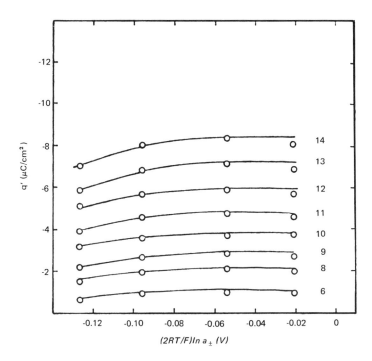

Figure 11. Adsorption isotherms for fluoride ions adsorbed from aqueous KF at 15°C.[126]

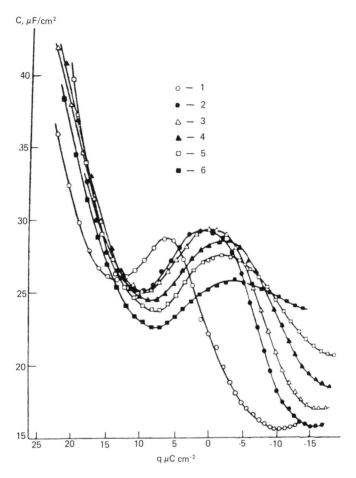

Figure 12. Capacitance of the double layer for $LiNO_3$ in aqueous media at 60°C: (1) 0.1; (2) 1.0; (3) 2.0; (4) 4.0; (5) $LiNO_3 \cdot 6H_2O$ (~7 M); (6) $LiNO_3 \cdot 3H_2O$ (~11 M).[131]

(c) The anodic minimum gradually disappears.

(d) The hump increases in width with increasing salt concentration.

(e) The anodic capacitance for $q > 15\ \mu C\ cm^{-2}$ is no longer a function of salt concentration at high concentrations.

(f) The capacity for $q < -10\,\mu C\,cm^{-2}$ increases with increasing concentration.

(g) The cation size seems important in determining the hump characteristics at higher concentrations.

(h) An apparently linear relationship exists between the charge at the hump and the concentration.[137]

For most of these characteristics no explanation has been offered and, although it is possible to discern trends due to solvation effects, ion size, etc., the only valid conclusion that can be drawn from these data is that the solvent must be involved to some extent in the formation of the capacitance hump in aqueous media.

8. Maxima in the Capacitance–Potential Function

The capacitance–potential function displays many features, among which the incidence of maxima is probably one of the most discussed. The hump is found for aqueous solutions of most salts and is dependent on the anion present. The fluoride system is one of the simplest superficially since the inner layer capacity, with which this characteristic of the system is associated, is independent of anion concentration in the region of the hump.[120] It is clear from the previous discussion that fluoride is specifically adsorbed as $q \gg 0$. The capacitance at the hump, which increases as the temperature is lowered, is evidently the result of fluoride adsorption.[126] No specific adsorption of fluoride is detectable at the hump and this explains its concentration independence. If the capacitative analysis is carried out and values of $_qC^i$ are derived, these values are significantly concentration dependent when $q > q_{hump}$. This capacitance decreases as the fluoride concentration is increased and the formation of the hump is therefore the result of fluoride adsorption. It is not due to an inflection in the q^1 versus q relationship, as Bockris et al. has proposed.[107] If $_qC^i$ is extrapolated to $q^1 = 0$, the so-called solution capacity is obtained and this is shown in Fig. 13. There remains a small hump which is significantly temperature dependent and this may be due to either the water reorientation phenomena or to changes in inner layer thickness dependent on F^- adsorption. This type of analysis has been applied to many aqueous and nonaqueous systems. In aqueous media two types of humps are generally observed, those in the halide systems and those in the oxyanion systems.

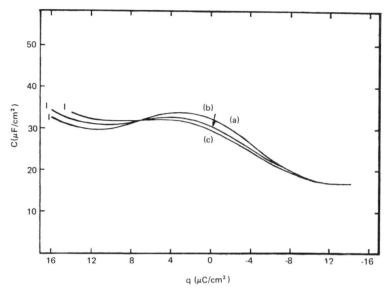

Figure 13. Apparent inner layer capacitance for the mercury/aqueous KF interface calculated assuming no adsorption at any potential. The degree of scatter at different concentrations is indicated. (a) 0, (b) 15, and (c) 25°C.[126]

In halide and pseudo-halide systems the hump decreases in size as the degree of specific adsorption increases and hence for the simple halides the hump decreases and occurs at higher capacity values according to the series Cl < Br < I. In the latter system no hump can be observed; only a slight inflection in the steeply rising adsorption capacitance[133] can be observed. As shown for the fluoride system (Fig. 13) and for dilute aqueous solutions of simple salts, the hump invariably occurs as $q > 0$. For the halide ions and some of the oxyanions there is a direct relationship between the inflection in the q^1 versus q relationship and q at the hump in the capacity–potential relationship. This is further discussed in Section V.

In aqueous oxyanion systems the hump is more pronounced and occurs nearer $q = 0$ but still at $q > 0$.[122,134] There is apparently a direct relationship between the hump and the properties of anions of the structure-breaking type, e.g., BF_4^-, PF_6^-, ClO_3^-, ClO_4^-, NO_3^-, CF_3COO^-, etc., the humps in all these cases having similar characteristics although the molecules have widely differing shapes,

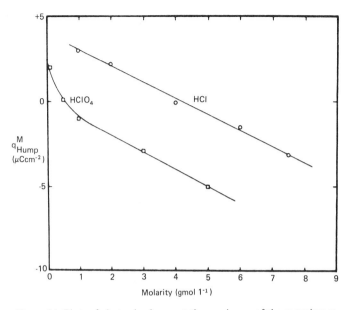

Figure 14. Plots of electrode charge at the maximum of the capacitance hump, q_{hump}, versus acid concentration.[132]

sizes, and adsorbabilities. For the systems where analysis has been possible, the hump is apparent in the $_qC^i$ plots.[122,126] In acidic media, when $[H^+] > 1$ M plots of q_{hump} versus $[H^+]$ are evidently linear and the slope seems rather independent of the anion (Fig. 14).[132] This behavior is not associated with the Faradaic process but may be associated with the presence of a partial ion-free layer at the interface. This effect is more fully discussed in Section III.3.

The capacity hump has also been observed in a number of nonaqueous systems. No hump is found in ionic solutions of the lower aliphatic alcohols, even at low temperatures, where the hump is likely to be amplified.[114,134] Humps in nonaqueous solutions were first identified in the capacity curves for CsCl and $SrCl_2$ in formamide.[135] Studies of the adsorption of diethylether in formamide indicate that the hump occurs at the maximum of adsorption of the diethylether and this indicates that the solvent itself must be involved in the formation of the hump in this case.[136] The hump occurs at $q \sim -8 \,\mu C\, cm^{-2}$, but comparative studies of water–

formamide mixtures with the pure solvent at the air–solution and mercury–solution interfaces suggest that only small orientation effects are involved in this solvent.[137] The hump is distinctly temperature dependent, $-0.04\ \mu\text{F cm}^{-2} \text{deg}^{-1}$ (formamide) and $-0.10\ \mu\text{F cm}^{-2} \text{deg}^{-1}$ (water) for solutions of 0.1 M $NaNO_3$ in these two solvents.

A similar symmetric hump is observed in N-methylformamide but the hump is better defined in this system. The maximum in organic adsorption also occurs at the hump but the temperature coefficient of the capacity at the hump is larger, $-0.06\ \mu\text{F cm}^{-2} \text{deg}^{-1}$.[138]

No hump can be detected in dimethylformamide. The interphase in this system seems deficient in ions and as a consequence the surface tension of the interphase increases with increasing salt concentration.[139,140]

In N-methylacetamide two humps are found, one at the same electrode charge as the single hump found in formamide and N-methylformamide, and a second, anodic to the pzc, as is usually found in aqueous systems. A similar situation appears to exist in N-methylproprionamide. In both solvents each of the humps has a negative temperature coefficient.[115] From this comparison of the series of amide solvents it is apparent that the hump only occurs in solvents which possess "unsaturated" hydrogens on the nitrogen atom. These particular solvents seem to be associated in the bulk phase through the "unsaturated" hydrogens.

Humps are also found for the two related solvents dimethyl sulfoxide (DMSO) and sulfolane. In both cases the hump is well formed when $q > 0$ in the presence of PF_6^- or ClO_4^- ions, for DMSO at $q \sim 8\ \mu\text{C cm}^{-2}$ and for sulfolane at $q \sim 12\ \mu\text{C cm}^{-2}$.[95,141,142] The humps with the cyclic carbonates, ethylene and propylene carbonate, are of an entirely different character. In both cases they occur when $q > 0$ but the hump is now more of a flat maximum in PF_6^- solutions, occurring at $q \sim 10\text{–}12\ \mu\text{C cm}^{-2}$. The decrease in capacity anodic to the hump is of the order of $3\ \mu\text{F cm}^{-2}$. There is also evidence for a well-defined hump with 4-butyrolactone and 4-valerolactone.[143] The humps in these latter solvents are not well understood since little evidence for ionic adsorption or molecular orientation has been found. Further studies are necessary to clarify the contribution of these factors in these solvents.

9. Anion Adsorption

A further problem in ionic systems which has frequently been considered and about which there is no real consensus of opinion is the problem of why specific adsorption occurs at all at the mercury–solution interface. Before trying to build a model for the system it is useful to be able to understand the diverse contributions to the total free energy of adsorption of the ion. In this context correlations of the type presented in Sections II.5, III.5, III.7, and III.8 are useful. For the ionic systems an extra set of correlations must be involved since the possibility of some degree of chemical bonding cannot be excluded.

Attempts to correlate the degree of ionic adsorption on mercury with free energies of formation of the ionic mercury salts are not successful.[107] A similar correlation with the covalent bond energies on the basis of the reaction

$$M(g) + \tfrac{1}{2}X_2(g) = MX(g) \tag{58}$$

is unsatisfactory.[74,144] It was therefore considered that solvation phenomena must play a large part in determining the free energy of the adsorption process.[107,144] Two approaches have been suggested by Barclay.[145,146] There seems to be a good correlation between the degree of specific adsorption and the degree of chemical softness (tendency toward covalency) and hardness (tendency toward ionicity). Chemically "soft" ions (I^-, S^{2-}, CN^-, CNS^-) are strongly adsorbed, intermediate cases (Br^-, N_3^-, NO_2^-) are less strongly adsorbed, while "hard" ions (Cl^-, ClO_4^-, NO_3^-, F^-) display weaker specific adsorption. This latter category is the most difficult to justify within this correlation. If solvation energy is included, however, and the scale of *solution* softness developed by Klopman[147] is introduced, a description using orbital energy and solvation can be semiquantified and good agreement with experiment obtained for the halide series. The alternative correlation is with the strength of the coordinate bond, which Barclay argues is a better description of a bond at mercury. In the first instance this can be made by comparison with the energetics of the reaction

$$M^+(g) + X^-(g) = MX(g)$$

These correlations also seem satisfactory, as do similar correlations with work functions for different metals, and indicate that some degree of orbital overlap may be possible in the adsorbed state, although the pure solvent layer shows a similar dependence on the nature of the metal. A simple separation of these contributions is not possible.

In view of the involvement of the ion solvation energy in the adsorption process and the availability of precise data in non-aqueous media, consideration of these data in this context is essential. Grahame found that the adsorption characteristics of ions were different in methanol.[148] This behavior was later confirmed by Garnish and Parsons.[149] A number of these systems were recently reviewed by Payne.[115] Damaskin has attempted to quantify the difference between the adsorption characteristics in dimethyl formamide (DMF) and water. The principal difference between the two solvents, in addition to the different degree of specific adsorption of ions in DMF, is the position of SCN^-, which in aqueous solution is adsorbed to about the same extent as iodide, while in DMF it is adsorbed less than chloride.[150-152] The total free energy of adsorption of the ion may be expressed as a sum of free energies of interaction,

$$\Delta G_{ads}^X = \Delta G_{A^-,Hg} - \Delta G_{A^-,X} - \Delta G_{X,Hg} + \Delta G_{X,X} \quad (59)$$

where X is the solvent. By analyzing the contribution from the terms for the two solvents, Damaskin was able to show that the different solvation of the ion by the two solvents was decisive in determining the difference in adsorption characteristics. This involved models and approximations that are open to criticism, but this type of explanation can also be applied to other systems, e.g., DMSO, where the anion solvation is weaker than in water.[153] The free energy difference for the adsorption of ions from two solvents is given by

$$\Delta(\Delta G_{ads}^{X-H_2O})_{Hg} \approx (\Delta G_{A^-,H_2O} - \Delta G_{A^-,X}) + [(\sigma_X - \sigma_{H_2O})/\Gamma_\infty] \quad (59a)$$

which involves the difference in interfacial tension between the solvents and the difference in anion solvation energy.

A further factor which might have been included in these calculations is the degree of accommodation of the ion in the solution, the structure-making or -breaking effect of the ion in Gurney's notation[154] treated by Conway and Gordon.[264] It was pointed out by Parsons that the hump seemed to correlate with this property of the ions and since the relationship of the hump to specific adsorption is reasonably well founded in aqueous systems, it is reasonable to introduce a similar factor here, especially since it is not possible to relate the ionic activites in the two solvent systems.[155]

A number of semiquantitative and quantitative attempts have been made to describe these systems. There is one clear group of workers who attribute the free energy of adsorption to the interaction of the ion with its electrostatic images in the metal and OHP (this matter is discussed further in Section IV.2) and, in general, neglect the essentially molecular nature of the entire inner region. Andersen and Bockris[156] considered the adsorption of ions from solution semiquantitatively. Initially, they reject the possibility of covalent, coordinate, ionic, or imaging phenomena as the principal source of the adsorption energy. Image forces are included but the contribution to the total energy is considered to be small.[107] Metal-ion dispersion forces and repulsion forces (cf. Section IV.6) are included. The dependence of the degree of specific adsorption on the primary hydration number of the ion is crucial to the model. Consideration of the adsorbed water monolayer is included, using the reorientation model (Section IV.5). On the basis of lateral repulsion calculations, θ_w was shown to be ~ 0.9, which agrees with data derived from the adsorption of water vapor from the gas phase.[157] The calculations of ion adsorption involve the number of water molecules replaced on the electrode by the adsorbing ion and the number of water molecules on the electrode accepted into the primary solvation sheath of the ion. This is equivalent to desolvating the side of the ion adjacent to the electrode and calculating the interaction of the primary hydration sheath in the inner layer with the inner layer field. The latter contribution is set to zero by Bockris. Using this model, the entropy, free energy, and heats of adsorption were estimated. Some of the specific calculations used in estimating the components of the entropy, for example, must be questioned. Is it reasonable to invoke vibrational states for the adsorbed system as if it were a gas phase only making allowance for the lack of freedom in one dimension? As with most models, it would seem that too few data are available to test the ideas. The only precise measurements of ΔH are for I^- and here the agreement is not good, -11.5 kcal mole^{-1} as compared to the experimental value of -16.4 kcal mole^{-1}.[158] There is a difference in the standard states here. For the halide series the calculated ΔH values are $F^-(+14.1)$, Cl^- (-9.6), Br^- (-12.9), and I^- (-16.4). In this case fluoride could not be specifically adsorbed, in contrast to the conclusions of Section III.6. The differences between the free energies of adsorption of I^- and Br^- and of Br^- and F^- are experimentally 1.7 and 2.4 and theoretically 1.4 and 0.6 kcal mole^{-1}, respectively. Similar calculations were made according to this model, with minor modifications, for water and ions by Bodé.[159,160]

An alternative approach was adopted by Lorentz and his co-workers which suggested partial covalent bonding of ions.[165,166] This involved determining the double-layer capacity as $\omega \to \infty$ and $\omega \to 0$. They derived a parameter which was considered to represent the partial covalent character of the bond at the electrode, but both Parsons and Damaskin have clearly shown the equivalence of this parameter to the distance ratio $(x_2 - x_1)/x_2$, a function which can be readily obtained from alternative analysis (Section III.2).[155,161] Vetter has approached the same parameter from a slightly different standpoint.[162–164] by defining an electrosorption valency v where

$$vF = (\partial \mu_s/\partial \phi^{m-2})_\Gamma = -(\partial q/\partial \Gamma)_{\phi^{m-2}} \quad (60)$$

The dependence of the adsorption of a substance in the inner layer is treated as a function of ϕ^{m-2} and depends on v.

The reaction for electrosorption is written as

$$A_{aq}^z + vM\text{-}OH_2 \rightleftarrows M_v\text{-}A^{z+\lambda} + vH_2O + \lambda e^- \quad (61)$$

Thermodynamic measurements only give the total excess of A^z and $A^{z+\lambda}$ in the metal-to-bulk-solution region. In the diffuse layer A^z is therefore eliminated using diffuse layer theory (cf. Section III.2).

The interpretation of v, which can be shown to be equivalent to $-(x_2 - x_1)/x_2$, is based on the separation

$$v = zg - \lambda(1-g) - vH_w + H_A - \frac{1}{F}\int \left(\frac{\partial C_D}{\partial \Gamma}\right)_{\phi^{m-2}} d\phi^{m-2} \quad (62)$$

where g is a geometric factor, H_A and H_w are dipole factors for ion (or organic species) and water, and the final term is capacitative. By estimating the contributions to v, Vetter was able to interpret the degree of charge transfer at metal solution interfaces for a number of ionic and organic solutes. In particular, $\lambda_{Cl^-} < \lambda_{Br^-} < \lambda_{I^-}$, which agrees with the order of specific adsorption of these ions from aqueous media.

A number of interesting conclusions can be drawn from this study. Since $dv/d\phi^{m-2}$ is negative for halides, $(\partial C_D/\partial \Gamma)_{\phi^{m-2}} > 0$, the capacity increases with increasing adsorption. The most important conclusion is, however, that the calculated specifically adsorbed charge is not equal to the charge at present calculated by diffuse layer separation (Section III.2). The metal charge q cannot be used in electrostatic calculations since

$$q = q_e + \lambda F \Gamma_i \tag{63}$$

where q_e is the real electron excess of the electrode. For iodide $\lambda F \Gamma_i > 0$ hence $q_e < 0$. This should be compared with equation (17). It means that the calculations of image forces, isotherms based on q^1 fits, and all the associated parameters are possibly suspect, particularly in the case of I^-, where $\lambda \sim 0.2$. For chloride the interactions are probably purely electrostatic within experimental error.

10. Organic Systems

These systems have been fully described in two recent reviews by Frumkin and Damaskin[165] and Damaskin et al.[166] The systematic study of the influence of organic compounds on electrocapillary curves derives from the work of Gouy, who found that the electrocapillary curve was depressed in the region of the pzc.[167] The curves frequently coincide with the base solution capacity at the anodic and cathodic extremes of polarization, indicating desorption of the organic species. This discussion will be restricted to a few important trends that are important for the discussion of models.

The shift in the pzc on adsorption of organic molecules on the mercury surface was associated by both Frumkin and Williams[168] and Butler[169] with their dipole moment. When the dipoles are adsorbed at the pzc they are oriented and their normal dipole components set up an adsorption potential corresponding to the shift in pzc. This is seen to be confirmed by Gouy's work.[170] Comparison of the adsorption of various aliphatic oxygen compounds at the air–solution and the mercury–solution interfaces at the pzc showed that the adsorption potentials at the two interfaces were almost identical.[171] It was suggested that this was due to the oxygen in both cases being oriented toward the air or mercury.[171] Comparison of these adsorption potentials for aromatic compounds gives considerable discrepancies in both magnitude and sign. The tendency of the benzene ring to lie flat on the mercury surface with the consequently enhanced π-bonding interaction with a positively charged interface can explain these differences. The effect is so strong that in the case of $C_6H_5NH_3^+$ ions electrode ion repulsions are insufficient to compensate for π-bonding interactions.[172] At both the air–solution and mercury–solution interfaces a large contribution to the free energy of adsorption results from the squeezing out effect, from the bulk solution to the interface, of the organic molecule.[173–175]

The characteristic dependence of surface coverage on the electrical variable for simple aliphatic compounds displays a peak which is approximately independent of electrode charge as the bulk concentration of the organic species increases.[107] The symmetry of the function varies from symmetric to very asymmetric.[107]

The orientation of the adsorbed species has received much attention. Apart from the "flat" orientation of aromatic compounds,[265] a similar position seems to be assumed by saturated aliphatic compounds with two or more functional groups.[166,171,176] The presence of high dipole moments does not seem to be important since amino acids lie flat on mercury.[177] The presence of substituents can modify the adsorption behavior, e.g., benzene-sulfonic acid, for which two position adsorption can be observed.[178]

The influence of metal is also important. Significant data could, until recently, only be obtained with liquid electrodes, but with recent improvements in technique data for solid electrodes are becoming more reliable. On liquid gallium the degree of adsorption seems lower, probably due to the specific interactions of water with the metal.[179] The general characteristics are similar to those for mercury with most organic compounds.[180]

IV. MODELS OF THE DOUBLE LAYER

1. Introduction

As indicated in Section I.1, the description of model-type approaches requires some background of earlier theories on which many of the more recent advances are based. It is not possible to describe fully these in this article and only an outline is given. The recent advances are clearly divisable into a number of separate approaches. Much attention has been given to the role of the solvent in the interface and since any treatment of ionic and organic systems should specifically involve the solvent model, this is presented first. Most ionic models do not include this contribution and this is reflected in the treatments of ionic adsorption. Any realistic model of this system must include both these contributions, and the extent to which this has been achieved is described. Few truly molecular theories of organic adsorption have been described and the attempts that have been made are described.

The influence of the metallic phase is usually neglected on a molecular scale in these theories and a description of the gallium–solution interface which displays specific solution–metal interactions is included.

It is not possible to give full details of all the theories but this treatment points out the deficiencies and advantages of the particular approaches and attempts to assess the relative importance of each theory. This discussion is extended in the final section.

2. Earlier Theories of Adsorption

Having identified the principal effects which require explanation by any theory of the interfacial region, the remainder of this discussion will be concerned with the models which have been proposed to describe these features and their applicability. The fundamentals of the Frumkin theory have already been described and only the later developments of this theory will be considered. The remainder of this section will be concerned with early treatments of ionic systems in the absence and presence of specific adsorption.

Gouy was the first to suggest that the finite size of ions would result in the diffuse layer not extending all the way up to the electrode surface.[181] The developments of the model on which the ensuing theoretical considerations are based have been reviewed by Grahame[74] and Parsons[11] and since they are of an essentially macroscopic nature, will not be further discussed here. The earliest attempt to give a quantitative description of the double layer is due to Esin and Shikov,[184] who were attempting to explain the anomalous shift in the point of zero charge with salt concentration observed by Esin and Markov[182] ($>kT/e$). They attribute to Frumkin the suggestion that this anomaly arises from the discreteness of charge of the specifically adsorbed ions.[183] The usual representation of q^1 as a smeared charge distribution at the inner Helmholtz plane was replaced by an array of discrete point charges, the true potential at the site of the adsorbed ion being the micropotential ψ_a. Consideration of the discreteness alone led to an overcorrection for the Esin and Markov effect. Ershler suggested that this theory does not include imaging of the ion in the double layer and the mercury.[185] Calculations including this effect were later repeated by Grahame.[134]

The ions, due to mutual repulsion and in the absence of thermal agitation, might be considered as being arranged in an hexagonal array. The production of a hole in the array, assuming the array does not rearrange itself, affords a calculation of ψ_a. The work done to produce these holes in two arrays in parallel of opposite charge sign is negligible. The work done on the system to bring ions into the plane from infinity is

$$ze(\psi - \psi_a) - z^2 e^2/\varepsilon x \tag{64}$$

where the second term represents the work of mutual attraction of the ions. In this expression ψ is the smeared macropotential. ψ_a can be derived for two such hexagonal arrays by considering finite arrays and summing concentric circles of charge. ψ_a is the potential difference between the two charge sheets, a value which is readily obtained from this summation as

$$\psi_a = 0.805 \psi^*_{\text{cont}} \tag{65}$$

where ψ^*_{cont} is calculated by assuming a continuous smeared charge density equal to that of the discrete layer. A circular hole is removed from these planes equal in area to that of one discrete charge, the concentric summations involving areas whose charge density equals the area of the discrete charge sheets. For an infinite array

$$\psi^*_{\text{cont}} = (4\pi q/\varepsilon)[(r_1^2 + \delta^2)^{1/2} - r_1] \tag{66}$$

which leads to, after substitution for $\psi = 4\pi q/\varepsilon$,

$$\psi_a/\psi = 0.766(\delta/r) - 0.697(\delta/r)^3 \tag{67}$$

Since δ/r, the ratio of the distance between the planes to the hole

Models of the Double Layer

radius, is seldom greater than 0.2, equation (67) reduces to the result obtained by Esin and Shikov, within the errors due to differences in computational procedures. These considerations may now be transposed to a model system. Equation (67) can be immediately applied to the two discrete layers of charge to obtain the average potential gradient between the two layers,

$$\frac{d\psi}{dx} = 0.766 \frac{\delta}{r} \frac{4\pi q^2}{\varepsilon\varepsilon_0} \tag{68}$$

where ε is the dielectric constant of the region and ε_0 is the permittivity of free space. In this case q^1 is the ionic array or adsorbed charge density. From geometric relationships for the hexagonal array, $r = 4.3 \times 10^{-10}(q^1)^{1/2}$; hence

$$d\psi/dx = -(2.24 \times 10^{-10} \delta/\varepsilon\varepsilon_0 \chi q^1)^{3/2} \tag{69}$$

The potential drop across the inner layer is $(d\psi/dx)x_2$ if an approximately linear relationship of ψ and x is assumed. In order for this to be true, $\varepsilon \neq f(q^1)$ and $\delta \neq f(q^1)$. Irrespective of any functional interdependence, $\partial \psi^{m-2}/dq^1 \to 0$ as q^1 tends to zero. This behavior is not found in either aqueous iodide systems, other aqueous systems (cf. Ref. 90), or nonaqueous systems (cf. Ref. 186). The direct plot of equation (69) is also evidently nonlinear (cf. Fig. 16), but the nonlinearity of the plot is not conclusive because of the other functional dependencies.

Ershler noted that at reasonable electrolyte concentrations the outer Helmholtz plane (OHP), in the presence of specific anion adsorption, will be populated by cations which will form a thermally smeared charge region and hence this may be considered as an ideal conductor.[185] If a reasonable dielectric change is postulated for the OHP, imaging at the OHP and electrode is possible. Imaging of the charge array will give a further contribution to the micropotential. Since the diffuse layer is entirely described by the Poisson–Boltzmann distribution, imaging will only contribute to the potential drop across the inner layer. The imaging situation in the real system is as shown in Fig. 15, where $\beta = x_1$ and $\delta = x_2 - x_1$. The multiple reflections are summed using the inverse square law and the resulting contribution to the field is half the potential difference

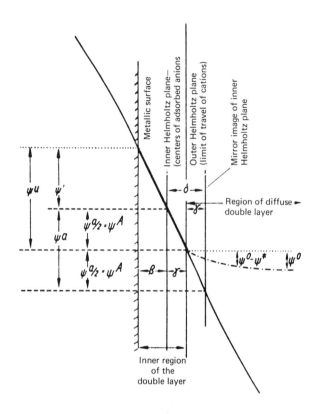
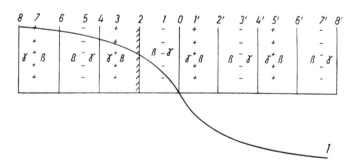

Figure 15. (a) Schematic of potential distribution within the double layer, (b) imaging diagram for ions (1) in the OHP (0) and metal (2) assuming perfect imaging in the OHP and metal.[134]

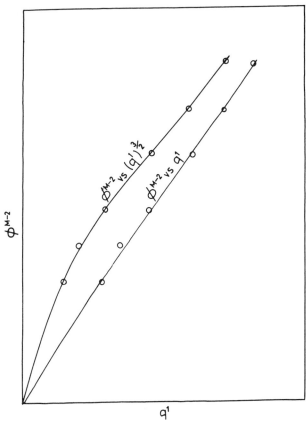

Figure 16. Potential generated by specifically adsorbed iodide ions on mercury, plotted as a function of surface charge concentration q^1.[134]

generated by two layers of charge, one placed at the real and one at the image position in the solution separated by a uniform dielectric.

The potential gradient in the region between the metal and OHP is nearly constant since it is the sum of a large number of concave and convex potential segments from the imaging system. This potential is given by $-2\pi q^1 \delta/\varepsilon\varepsilon_0$, which after substitution gives a linear dependence of potential drop across the inner layer on q^1. Such a relation is found in Fig. 16. This suggests that if the approximations regarding ε and q^1 are reasonable at $q = 0$, a multiple imaging model seems a better description of the system than the single imaging system. The solvent is not considered explicitly and this type of treatment can only help in the understanding of systems with specific adsorption.

The consequences of the assumption of a linear isotherm in terms of the distance ratio for the inner region are described elsewhere.

The simplified approach of Ershler, Grahame, and later Levich et al.[187] has been criticized by Macdonald and Barlow.[188] The infinite imaging theory certainly seems to give reasonable agreement with experiment, but the model on which it is based is still weak. Both imaging phases are considered to have an infinite dielectric constant and the dielectric constant of the inner layer is not considered to be dependent on q^1. The dielectric constants of the individual parts of the inner layer are also considered to be equal, $\varepsilon = \varepsilon_\beta = \varepsilon_\gamma$, where ε is undefined. Under the usual conditions of study of ionic adsorption, the actual surface coverage with ions is usually less than 0.3.* A principal contribution to the observed phenomena under these conditions might be expected to derive from the solvent, which is not considered. Thermal equilibrium of the array configuration is assumed and since the micropotential is derived as a departure from the smeared approximation, only low values of coverage can be used. The interactions between ions and solvent are also neglected in all the treatments at the molecular level.

A second quantitative attempt to describe the electrical double layer in electrolytes was formulated by Macdonald.[189] This was particularly directed at the nonadsorbed electrolyte systems, e.g., fluoride in water. The diffuse layer is assumed to be described entirely by the GCS theory, although dielectric saturation in the inner part of the diffuse layer was considered using Grahame's empirical formula[190]

$$\varepsilon_1 = \varepsilon_\infty + \frac{\varepsilon_0 - \varepsilon_\infty}{[1 + (b/m)\varepsilon^2]^m} \tag{70}$$

where ε_0 is the dielectric constant at zero field ($\varepsilon = 0$), ε_∞ is the

---

*This would correspond in three-dimensional solutions to high-concentration conditions where ε is much less than the normal static value.

limiting high field value of dielectric constant, and m, on the basis of Malsh's experiments,[191] is 0.5. This equation, as Macdonald noted, is based on measurements which did not really approach the limiting high-field value. The constant b, assigned 0.5 by Grahame, was recalculated by Macdonald and Barlow. The influence of this correction to the total diffuse layer capacity is likely to be small, but the influence of this effect on the imaging situation at the OHP may be more important.

The concepts behind this theory are basically to express the characteristic properties of the inner layer in terms of "bulk" parameters, averaged macroscopic properties such as dielectric constant and compressibility.

Assuming a charge-free inner layer, the field strength $\mathscr{E}^i$ for $x_m \to x_2$, metal to OHP, is given by $V^i/x_m - x_2$, where V^i is the inner layer potential drop, assumed linear across the inner layer. A second effect that must also be considered is the problem of compression of the inner layer by the field. The heuristic formula

$$x_2/x_2^0 = (1 + \beta p)^{-1} \tag{71}$$

is assumed, which gives an initial linear compressibility and tends to a limiting value at high pressures p. The inner layer thickness at $p = 0$ is x_2^0 and β is an isothermal modulus of linear elasticity. The electrostatic pressure across the inner layer is given for a field $\mathscr{E}$ by classical electrostatic theory as[192]

$$p = \varepsilon \mathscr{E}^2/8\pi \tag{72}$$

whence

$$x_2/x_2^0 = (1 + \alpha \varepsilon \mathscr{E}^2)^{-1} \tag{73}$$

and

$$\alpha = \beta/8\pi \tag{74}$$

Equations (71)–(74) are difficult to apply since in general ε is a function of $\mathscr{E}$ and p.

These three considerations can now be combined, diffuse layer saturation, dielectric saturation in the inner layer, and compression effects, to give an expression for the total capacity. The total mean potential difference across the double layer is

$$V_0 = 4\pi q x/\varepsilon_1 + (2kT/e)\sinh^{-1}(q/2A) \qquad (75)$$

The second term is the normal diffuse layer term, and $x = x_2$ in the first term. Fitting these data to Grahame's data for NaF at one solute concentration gives the undetermined parameters of the system, which should be independent of the bulk electrolyte concentration.[120] In the absence of field this gives $\varepsilon_1^0/x_0^2 = 3.3$ Å^{-1}. For a water monolayer $x_2 = 4.6$ Å, which gives $\varepsilon_1^0 = 15$. If ε_∞ is taken as five, from independent high-frequency measurements, the total capacitance versus potential function can be fitted quite accurately over a wide range of electrode charges for $q \leq 0$. It would, of course, be expected that the fits would be good since C versus electrode potential is concentration independent for NaF and so long as the basic fit to the experimental function is good, all subsequent fits will be satisfactory.

It is necessary to introduce a value of β into equation (74) to fit the experimental data, but the value of β required is of the order of five times the bulk value of β. This is reasonable, since the ions are not smeared over the outer Helmholtz plane but are discrete. The compression is therefore a local phenomenon dependent on the presence of the ion at the OHP. The ion will therefore compress a small number of water dipoles whose compressibility will bear little resemblance to the bulk value of β, which will depend extensively on the bulk structure of water. A more refined treatment will be presented in a later section.

The solvent itself has been considered essentially as a continuum in these treatments. Mackor, however, suggested that the molecular condenser, at cathodic potentials, is made up of solvent molecules.[193] The independence of capacitance at the cathodic minimum was suggested as being due to the presence of at least a monolayer of solvent molecules between the electrode and the ions. At high negative and anodic potentials it was suggested further that the orientation of the surface layer is complete. No mathematical model involving these effects was, however, proposed by Mackor. These ideas on orientation were not directly involved in the early Macdonald and Grahame models. Bockris and co-workers introduced a quantitative treatment of water dipoles to explain the anomalous Tafel slope for hydrogen evolution at nickel electrodes as the pH

of the solution changes. The field was considered to orient the water dipoles, thus increasing the activity of water in the inner layer reaction zone. The introduction of such a contribution gave quantitative agreement with the data as a function of pH.[194]

Grahame considered the role of the solvent in a different light.[195] The hump in the capacity potential function was not featured in early theories, since its significance and widespread occurrence was not at that time recognized. Grahame studied systematically the mercury–aqueous solution interface as a function of temperature and concentration. In order to explain the significantly temperature- and potential dependent capacity of the inner layer, he suggested that an "icelike" layer which was capable of "melting" as the temperature increased was involved. As a consequence, the hump decreased in size. The water molecules were also considered to be oriented with their oxygen atoms, i.e., their negative end, toward the mercury when $q \gg 0$. No quantitative calculations on the basis of this idea were made, but the concept which embodies the idea of interactions between water molecules, is valuable. Such interactions have been inferred from experimental data. If one considers an isotherm of the Flory–Huggins type[197a]

$$\beta a = [\theta/p(1 - \theta)^p] \exp A\theta \tag{76}$$

where p is the number of solvent molecules replaced by the adsorbing species. It is found experimentally* that p is seldom greater than unity. If a system can be approximated by this relationship with such a value of p and this has been found for a range of molecules of widely differing sizes, it is possible to interpret the data in terms of the replacement of groups of solvent molecules by the adsorbing species. In Damaskin's studies[197b] associated groups of solvent molecules up to 40 $Å^2$ in size were suggested. However, it is difficult to see how the water structure adjusts itself to provide replaceable groups equal in size to that of various organic molecules.[197a]

The first attempt to treat the general case of a system with ionic adsorption over the entire range of polarizability was put forward by Devanathan.[198] The treatment is based on a simple electrostatic representation of the Stern model. Continuous charge distributions are assumed at the IHP and OHP and the medium

---
*This requires evaluation of the second derivative of the isotherm and is thus very demanding of the accuracy of the experimental adsorption results.

between the OHP and metal is considered to be continuous. The potential drop across the entire double layer is expressed by

$$\phi^{m-s} = \frac{4\pi q x_2}{\varepsilon_1} + \frac{4\pi q^1 (x_2 - x_1)}{\varepsilon_1} + \phi^2 \quad (77)$$

Assuming that ε_1, x_1, and x_2 are constant, the capacity is given by

$$\frac{1}{C} = \frac{d\phi^{m-s}}{dq} = \frac{1}{K_{m-1}} + \frac{1}{K_{1-2}} + \frac{1}{K_{1-2}} \frac{dq^1}{dq} + \frac{d\phi^2}{dq} \quad (78)$$

The final term in this expression is derived from diffuse layer calculations, i.e.,

$$d\phi^2/dq = (1/C_d)[1 + (dq^1/dq)] \quad (79)$$

Values of q^1 were calculated from equation (78), employing the cathodic minimum capacity as the criterion for zero specific adsorption. At the minimum $dq^1/dq = 0$ and the inner layer capacity becomes $K_{m-1}K_{m-2}/(K_{m-1} + K_{m-2})$. Values of x_1 and x_2 were taken from the radii of ions and the solvent. Equation (78) was solved using an iterative technique for dq^1/dq. Equation (77) is correct but equation (78) assumes a constancy of K_{m-1} and K_{1-2} and consequently requires that x_2/ε_1 be charge independent. In the absence of specific adsorption $dq^1/dq = 0$; hence the first two capacity components on the rhs of equation (78) must represent the "solvent" capacity. Since Devanathan does not use the diffuse layer criterion for specific adsorption of ions, data interpreted on this model must disagree with other data. This model has also been strongly criticized by Payne,[199] Macdonald and Barlow,[188] and Frumkin et al.[200] Devanathan assumed that $\varepsilon_1 = 7.2$ and that the inner zone was homogeneous. None of the consequences of the inner layer field acting on solvent dipoles was considered.

In spite of the criticisms, the model has been widely used. It must be noted that this is not a suitable theory on which to base a molecular model. It does raise an interesting question specifically as to the cathodic branch of the capacity–potential functions. Is the cathodic capacity minimum a result of zero specific adsorption? One of the best criteria for specific adsorption is the temperature dependence of capacitance, which is small in the region of the minimum. Cathodic of the minimum, the capacity remains rather independent of temperature for the lower alkali metals and consequently both this model and Gierst's seem suspect.[117,201] In terms of "macroscopic" models, Macdonald's model seems more appropriate in this region.

3. Organic Systems: Classical Treatments

The basis of the Frumkin theory has already been discussed. In terms of the above discussions, this theory is essentially a macroscopic model. At the microscopic level, the Butler theory must be considered.[169] Although it does not give quantitative agreement with experimental data, it does contain the elements of present-day dipole-based theories. The basic problem Butler identified was the question of the form of the potential-dependent function in the expression for the lowering of surface tension,

$$\Delta \gamma = \gamma_0 f(V) \tag{80}$$

The electrical work done in moving a molecule B of known polarization from region I in which the field strength is zero to a region II in which the field strength is $\mathscr{E}$ and displacing a molecule A occupying the same volume δV but of different polarization per unit volume in the opposite direction is given by

$$w = \tfrac{1}{2}(\alpha_A - \alpha_B)\mathscr{E}^2 \, \delta V \tag{81}$$

where $P = \alpha \mathscr{E}$ for moderate field strengths.

According to the Boltzmann distribution equation, if $n_B{}^0$ is the number of molecules of B in region I, n_B in region II is

$$n_B = n_B{}^0 \exp[-\tfrac{1}{2}(\alpha_A - \alpha_B)\mathscr{E}\,\delta V/kT] \tag{82}$$

where $\mathscr{E} = KV$. It was found that phenol adsorption on mercury closely followed this relationship if $V = 0$ at the maximum of the surface tension versus potential curve.

The maximum in the adsorption–potential function frequently occurs when $q \neq 0$ of the base solution. This was accounted for by allowing the adsorbed molecules to have a polarization in the absence of the electric field arising from dipole orientation at the interface. The work of adsorption would then be

$$w = \tfrac{1}{2}[(\alpha_A - \alpha_B)\mathscr{E}^2 + (p_A - p_B)\mathscr{E}]\,\delta V \tag{83}$$

hence
$$\Delta\gamma = \Delta\gamma_0 \exp[-(aV^2 + bV)] \qquad (84)$$
where a is the constant part of the exponential in equation (9). The maximum now occurs at $V_m = -b/2a$, where $\Delta\gamma_m = \Delta\gamma_0 \exp(b^2/4a)$. The capacity at the interface was assumed to be 18 μF cm^{-2} and the polarizability was derived using

$$\alpha = (\varepsilon - 1)/4\pi \qquad (85)$$

Butler used the *bulk* dielectric constant value for ε. If the correct value is employed, the fits to the data are not so satisfactory.[107] Butler did recognize this deficiency in the theory but did not offer a quantitative explanation. Bockris has also pointed out that if the correct values of the polarizabilities are used in this system of equations, the polarizability term is insufficient to produce the characteristic parabolic dependence of the adsorption on electrode charge until $q \sim 20$ μC cm^{-2}.[107] The asymmetry in these curves is also not explained by this theory.

4. Summary of Basis for Recent Developments

1. The basic Gouy–Chapman–Stern theory for the distribution of ions across the double layer is generally accepted and a linear separation of inner and diffuse layers is generally a first step in model-type interpretations, although, in a definitative theory, their mutual interactions must be involved.

2. The ions must be considered as discrete charges of finite size.

3. Some degree of imaging of the ions in the metal (and the solution) can be included in the models to at least partially explain the Esin and Markov effect.

4. In most treatments the solvent is treated as a continuum and molecular treatments neglect many factors. Little quantitative progress has been made in this area.

5. Models for organic systems display a similar lack of dependence on molecular parameters. No specific solvent effects are considered in Frumkin's model, but they are implicit in Butler's theories, although they were not correctly treated at that time.

6. Of the features of the capacity versus potential function, the cathodic minimum can be explained on the basis of macroscopic parameters.

7. The quasi-thermodynamic approach of Devanathan, while correct in principle, neglects the variation of the constants with potential. This approach does not immediately help in the formulation of a molecular model.

5. Recent Developments

The recent developments in the capacitance model result not so much from advances in measuring techniques as from the recognition of the wide range of factors that must be included in a realistic double-layer model. The concepts advanced by Macdonald, Watts-Tobin, and Mott are the starting point from which many of the recent advances stem. Watts-Tobin[202] was the first to analyze quantitatively the role of water in the double-layer model. The average field necessary to orient a water molecule of dipole moment μ is kT/μ, which at 300°K gives $\mathscr{E}$ as 6.6×10^6 V cm^{-1}. This is compatible with double-layer fields. The energy of the hydrogen bond in bulk water is 4.5 kcal mole^{-1}. This is much greater than kT, so in the bulk the hydrogen bonding will be virtually complete at any instant in time. The strength of the electrostatic field necessary to break such hydrogen bonds is given by $N\varepsilon\mu \sim 4.5$ kcal, which gives $\mathscr{E}$ a value of 5×10^7 V cm^{-1}. This calculation assumes the value for μ in vacuum. The water molecules in contact with the metal cannot be hydrogen bonded in the same way as in the bulk system. The high heat of adsorption of water vapor on mercury measured by Kemball suggests that the surface molecular configuration is very different from the bulk structure and it has been suggested that hexagonal packing prevails independent of the bulk structure.[204] The data on adsorption of monolayers of water on mercury have recently been reviewed by Smith,[205] who has shown the wide divergence between the different sets of data available. Most of the theoretical work has been based on Kemball's data but, as is shown in Fig. 17, these data are not in agreement with the more recent results of Law.[206] The interfacial tension of mercury in contact with a monolayer of water is very similar to the interfacial tension of mercury in contact with pure bulk water, which adds support to the separation of interfacial water from the bulk phase as being a reasonable first approximation. Such a qualitative discussion does not help the model. More quantitative correlations are necessary. The entropy of adsorption suggests that the water

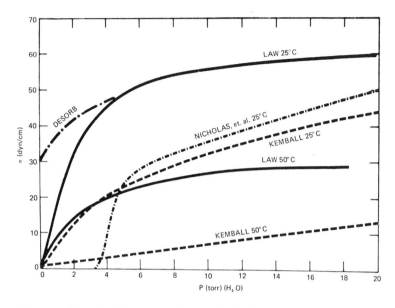

Figure 17. Plot of surface pressure due to adsorbed water vapor on mercury as a function of the partial pressure of water vapor.[205]

molecules are adsorbed in pairs.[204] A possible model might be that there are three bond-forming directions toward the mercury with one free for bond pair formation. Such a natural orientational situation would, of course, be true only in the natural or field-free environment. In the field of the double layer, it was suggested that the molecules could have one of two orientations, with the dipole either toward or away from the metal. The relative energies of the two positions can be estimated at thermal equilibrium and hence the relative populations of the two states can be estimated. The water dipole is in contact with an imaging medium and Watts-Tobin considers that the forces between the dipole and interface are entirely electrostatic. The interaction energies W between two such systems can be described in terms of the dipole μ and quadrupole moments θ_i, the distance between the systems R, and the unit vectors between the origins of the two systems λ_i. If ε_i is the integral dielectric constant of the intervening media, the interaction energy of the two systems is

$$\tfrac{1}{2}W = -\mu^2(1 + \cos^2\theta)/2\varepsilon_i R^3$$
$$- 3\mu(\cos\theta)[(\theta_3 - \theta_1)(1 - \lambda^2) + (\theta_3 - \theta_2)(1 - \lambda_2^2)]/2\varepsilon_i R^4$$
$$+ O(1/R^5) \qquad (86)$$

This system is essentially the image situation at the mercury–solution interface. The treatment requires that R be greater than the spread of charges around the origin of the two systems. If accurate values of three quadrupole tensors were available, the orientation could be calculated, but the two orientations discussed above must be assumed. For these orientations $\lambda(\pm\sqrt{\tfrac{2}{3}}, 0, -\sqrt{\tfrac{1}{3}})$ with the proton pointing away from the metal and $\lambda(0, \pm\sqrt{\tfrac{2}{3}}, \sqrt{\tfrac{1}{3}})$ when the lone-pair orbital points away from the metal should be substituted in equation (86). The components of the dipole moment normal to the metal in the two orientations are respectively $\mu/\sqrt{3}$ and $-\mu/\sqrt{3}$. Substituting into equation (86) gives the difference between the two interaction energies in the two positions,

$$G = 2\mu(2\theta_3 - \theta_2 - \theta_1)/\sqrt{3}\varepsilon_i R^4 \qquad (87)$$

The first term in equation (86) is the dipole–dipole energy $-2\mu/3\varepsilon_i R^2$, in both positions. If $R \sim 3$ Å and Buckingham's estimate of $2\theta_3 - \theta_1 - \theta_2 = 3.9 \times 10^{-26}$ esu is assumed, $\varepsilon_i G \sim 1.5$ kcal mole^{-1}. In order to have a value of the heat of adsorption of water that agrees with Kemball's value, we must have $\varepsilon_i = 1$ for this interaction. Such a calculation relies heavily on Buckingham's value,[207] which is an order of magnitude larger than Duncan and Pople's value.[208] The model is not strong since it neglects entirely the lateral interactions in the layer and chemical interactions with the metal. Dispersion interactions and dipole–dipole interactions must make substantial contributions to the total energy of the layer. Since the dipole–dipole interactions are considered small by Watts-Tobin, a nonnormal orientation of the dipole to the metal is not unreasonable (cf. Section II.5). The principal theoretical objection to this approach is the use of the quadrupole moment. Not only, as pointed out above, does it rely heavily on one theoretical value, but in order to use this value, the molecular system must be defined in the same coordinate system as the theoretician assumed in the model from which it was calculated. This can be illustrated if the center of mass of the system, instead of the electrical center as Buckingham used, is employed as the coordinate center of the system.[209] In this case a quadrupole moment of -0.13×10^{-26} esu was calculated. Such a value would not help the Watts-Tobin theory! The quadrupole moment is also vectorial and the pertinent component must be that normal to the interface for interaction with the double-layer field. A better theoretical description is necessary before the quadrupole moment can be involved realistically in theories of this region.

Since hydrogen bonding with water is favorable, the oriented dipole will interact with the next layer of water molecules and so on until the chain either terminates at the metal or an ion. If the chain interacts with the electric field between metal and bulk, i.e., ϕ^{m-s}/nd where n is the number of molecules and d is the molecule length, the dipole moment of the chain, assuming the nonnormal orientation, will have a component $\mu^* = \pm n\mu/\sqrt{3}$ normal to these planes and the interaction of the chain with this field is $-\mathscr{E}\mu^* = \mp\mu\phi^{m-s}/\sqrt{3}d$.

If there are N free molecules per unit area of the interface, L with their dipole moment one way and M with their dipole moment the other way, the energies of the chains associated with these molecules are

$$A = C' - \mu\phi^{m-s}/\sqrt{3}d \qquad (88a)$$

$$B = D' + \mu\phi^{m-s}/\sqrt{3}d \qquad (88b)$$

where C' and D' are equal to $W/2$ in equation (87). Since $N = L + M$, the relative numbers of molecules in the two states are

$$L = N \exp(-A/kT) \qquad (89a)$$

$$M = N \exp(-B/kT) \qquad (89b)$$

There is a surface charge q on the mercury which will contribute $4\pi qd/\varepsilon_0$ to the potential drop. The total potential drop is therefore

$$\phi^{m-s} = (4\pi qd/\varepsilon_0) - (4\pi N\mu/\sqrt{3}\varepsilon_0) \tanh[\mu(\phi - V)/\sqrt{3}\,dkT] \qquad (90a)$$

In this presentation the contribution of adsorbed ions has been omitted. The differential capacity of the layer is $\partial q/\partial \phi^{m-s}$, i.e.,

$$C = (\varepsilon_0/4\pi d) + (N\mu^2/3d^2kT) \operatorname{sech}^2[\mu(\phi - V)/\sqrt{3}\,dkT] \qquad (90b)$$

A second exponential term was included to express the adsorption of complexes, i.e., specific adsorption. The dielectric constant was taken as two but the possible error in this value was recognized. The value of G of 1.5 kcal mole^{-1} results in a hump 0.4 V anodic to the experimentally observed value. If Duncan and Pople's value of the quadrupole moment were accepted,[208] the hump would occur in the region of the experimental hump, namely 0.66 V (cf. 0.5 V observed experimentally). On the cathodic branch Watts-Tobin accepts the compression model for the increasing capacity as proposed by Macdonald.

This theory was later extended to treat other features of the capacitance–potential function.[211] The interaction of the first layer of molecules with the second layer will give at least partial saturation of the second layer. It might suggest that some degree of long-range structure is not unreasonable, the degree of structuring by the nth layer being about half the $(n-1)$th layer. The dielectric constant rises rapidly in the region of the nonadsorbed cations.

From this model a number of pointers can be drawn.

1. The tendency of the water molecules to be oriented in the absence of external field is weak, but significant (cf. the dipole p.d. at the pzc).

2. The hump is due to a high effective dielectric constant when the field in the inner layer is not strong enough to give saturation.[210] In terms of equation (90b), this is when the final term gives a small contribution to the total capacitance.

3. The dependence of the capacitance hump on temperature is roughly reproduced in water for the cathodic branch. The hump significantly increases with decreasing temperature (cf. Fig. 18). The agreement is good for the pure solvent hump in N methylformamide for $q \ll 0$[212] (cf. Sections III.8 and V).

An approach similar to that proposed by Watts-Tobin was suggested by Damaskin et al.[197,213] The parallel plate capacitor model [equation (24)] is assumed. Between the plates of one capacitor the water has one orientation and between the plates of the second it has the reverse orientation. The work to reorient the water dipoles between the plates of the capacitor per mole of adsorbed solvent (Γ_s) is

$$\Delta E = K_0 \psi_A \Delta \psi_N / \Gamma_s \qquad (91)$$

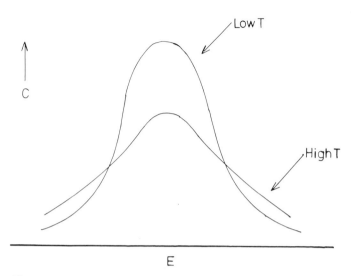

Figure 18. Capacitance of a water monolayer as a function of temperature according to Watts-Tobin.[210]

where $\Delta\psi_N$ is the difference between the potential generated by the dipoles in the two orientations, ψ_a is the potential with dipoles all in one orientation, and K_0 is the real integral capacitance of the double layer.

If an adsorption isotherm of the form

$$\theta/(1 - \theta) = \beta \exp(-\Delta E/kT) \quad (92)$$

is employed, where ΔE is given by equation (91), the capacity due to water is given by

$$C_{H_2O} = K_0 + K_0 \Delta\psi_N \, d\theta/d\varphi \quad (93)$$

If $K_0 = \varepsilon/4\pi x_2$, $\Delta\psi_N = [8\pi\mu(\cos\theta)/\varepsilon]N$ and since $\theta = 1/\sqrt{3}$, the final equation of the Watts–Tobin theory can be obtained [equation (90)],

$$C_{H_2O} = (N\mu^2/3\,d^3kT)\,\mathrm{sech}^2[\mu(\psi_A - V)/\sqrt{3}\,dkT] \quad (94)$$

Instead of calculating C_{H_2O} on the basis of assumed parameters, Damaskin attempts the reverse procedure. In particular, N can be estimated and this seems to be $\sim 40\,\text{Å}^2$, far larger than the geometric area of the water molecule. This suggests, in agreement with with Macdonald and Barlow and the results of fits to the Flory–Huggins isotherm, that water may be aggregated in neutral groups on the surface.

The temperature dependence of the hump agrees of course with that of Watts–Tobin. The asymmetry is then introduced by defining the two integral capacity components as K_0^+ and K_0^-.[214] This means that ΔE in equation (92) is given by $(K_0^- \phi'_N - K_0^+ \phi''_N)$, where different potentials are generated in the two orientations in the limit of complete orientation of the layers. This does not really help the model since, although it gives a symmetry to the C versus q plot from $q = 0$, it still suffers from the weakness of the earlier macroscopically based theories. K must be dependent on the inner layer field and since no saturation function has been introduced, the cathodic limits of the C versus E function will not be reproduced. In general, the same comments apply to this theory as apply to the Watts-Tobin theory and the types of theories proposed to calculate the interaction terms for the Frumkin isotherm (Section IV.8).

The first description of the inner layer by Macdonald and Barlow (Section IV.2) was basically heuristic in approach and the subsequent description is of the extensions that were proposed to avoid, as far as possible, the arbitrary nature of some of the earlier assumptions. The principal advance of these extensions was a description of the system in terms of molecular parameters.[215] The dielectric constant cannot be used since the medium is not isotropic (Gauss). The approach must therefore be through the displacement $D = \varepsilon \mathscr{E}$,

$$D \equiv 4\pi q = \mathscr{E}_1 + 4\pi P \qquad (95)$$

where P is the polarization. This can be expressed as the sum of induced dipole and orientational polarization contributions, $P = P_1 + P_2$, where

$$P_2 = \mathscr{E}_{\text{eff}} \alpha_1 \qquad (96)$$

and

$$P_2 = N_v \langle \mu \rangle \qquad (97)$$

In these expressions $\mathscr{E}_{\text{eff}}$ is the effective field at the dipole site and α_1 is the volume polarizability; $\langle \mu \rangle$ is the average value of the permanent dipole moment in the direction of the field $\mathscr{E}$ and N_v is the density of water molecules in the compact layer. Of these parameters N_v is readily calculated for a hexagonal lattice with spacing r_1. From geometric calculations $r_1^{-3} = (\frac{3}{4})^{3/4} N^{3/2}$. The value of $\mathscr{E}_{eff}$ at the dipole site is also easily calculated for a hexagonal array,

$$\mathscr{E}'_{eff} = 4\pi q - (\tfrac{3}{4})^{3/4} \sigma N_0^{3/2} [\langle \mu \rangle + (P_1/N_v)] \tag{98}$$

where σ is taken from Topping's work, 11.034; and $N_0 = N_v d$, where d is the layer thickness.[216] Combining equations (96) and (98) gives

$$P_1 = [4\pi q \lambda \alpha_1 - s\lambda N^{3/2} \langle \mu \rangle]/\varepsilon_{eff} \tag{99}$$

Substituting for P_1 leads to

$$\mathscr{E}_{eff} = [4\pi q - (\tfrac{3}{4})^{3/4} \sigma N^{3/2} \langle \mu \rangle]/\varepsilon_{eff} \tag{100}$$

where the effective dielectric constant, $\varepsilon_{eff} = 1 + sd_0 N^{1/2} \simeq 1\text{-}3.$[188] It is still necessary to derive an expression for $\langle \mu \rangle$ as a function of $\mathscr{E}_1$, the average field acting to orient the dipoles.

In order to accomplish this, the overall effective field tending to orient the dipoles must be considered. The charge-independent part of the field is included by introducing an effective natural field $\mathscr{E}_n$. This component is considered to be temperature independent and the additivity of the extra field when $q \neq 0$ is assumed. Such a contribution can be derived in the following way. A dipole is first considered in a field $\mathscr{E}$. It has an energy of $-\mathscr{E}\mu \cos \theta$ for the dipole oriented at an angle θ to the field. To this must be added the anisotropy energy W_a, which can be expressed by

$$W_a = W_0[\sin^2(\tfrac{1}{2}\theta) - 1] = -\tfrac{1}{2}W_0(1 + \cos \theta) \tag{101}$$

where the anisotropy energy depends on θ. The interaction energy of the dipole with the field is given by Boltzmann statistics as

$$\langle \mu(\mathscr{E}) \rangle = \mu \int_0^\pi z \cos \theta \sin \theta \, d\theta / \int_0^\pi z \sin \theta \, d\theta \tag{102}$$

where

$$z = \exp[-(1/kT)(-\mu \mathscr{E} \cos \theta + \tfrac{1}{2}\omega_0 \cos \theta)]$$

The natural field is then given by $W_0/2\mu$. The required dependence of $\langle\mu(\mathscr{E})\rangle$ is then given by

$$\langle\mu(\mathscr{E})\rangle = a(\mathscr{E} + \mathscr{E}_n)h(\mathscr{E} + \mathscr{E}_n) \quad (103)$$

where $h(\mathscr{E}) \equiv (3/x)L(x) \equiv (3/x)(\coth x - x^{-1})$ (Langevin function) and $x = \mu\mathscr{E}/kT$, $a \equiv 4\pi N_v \mu^2/3kT$. If infinite imaging of the dipoles were included in the calculation the fields would be increased, resulting in an earlier onset of dielectric saturation.

Removing a single dipolar molecule from the hexagonal sheet gives a field at the vacancy $\mathscr{E} = \mathscr{E}_{\text{eff}}$. If the dipole is now returned to its position in the array having left all the other charges and distributions in the system unaltered, the field is now altered by the polarization of the reentrant molecule. This contribution depends on the specific orientation of the reentrant molecule in some uncertain manner, and could lead to further asymmetry in the saturation function. Macdonald considers that shielding of this dipole by the electronic polarization depends on the square of the optical refractive index (n) through

$$\mathscr{E} = \mathscr{E}_n + \{\mathscr{E}_{\text{eff}}/[1 + \lambda(n^2 - 1)]\} \quad (104)$$

In addition to the anisotropy energy W_a, the permanent dipole-image attraction, $W_i = -\mu^2(1+\cos^2\theta)/2S^3$, where S is the dipole-image distance, must be included. The relative magnitudes of the two contributions is easily obtained at $q = 0$. The image energy is ~ -0.6 kcal mole^{-1} for single imaging and ~ -1.4 kcal mole^{-1} for infinite imaging, while the anisotropy energy may be about -0.3 kcal mole^{-1} at absolute zero. The image energy at $q = 0$ will result in some dielectric saturation occurring even at $q = 0$, in the absence of the externally impressed field. This leads to a value of the dielectric constant ~ 15 at the pzc. Depending on how the anisotropy energy is defined in terms of θ, the degree of asymmetry in the total function can be varied. The electrostriction effects a capacity increase cathodic to the minimum in the capacity–potential function. Since equation (104) involves the dielectric constant, it must be replaced if the treatment is to be consistent. Thus the electrostrictive pressure p can be redefined through

$$p = 2\pi q[q - P + \lambda(\partial P/\partial \lambda)_q] \quad (105)$$

The inclusion of λ is to correct for changes in the dipole moment

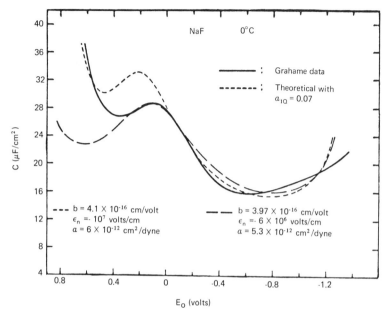

Figure 19. Comparison of the microscopic theory and experiment for NaF–water at 0°C.

after compression. Since the polarization can be fully described, it is possible to describe the total field across the double layer and hence the capacity. The fitting technique involves iteration to find values of m [equation (70)] and the other variables. This has been clearly described by Macdonald and Barlow.

The earlier theory gave a symmetric capacity versus potential function. This is now clearly asymmetric (Fig. 19), the degree of asymmetry being sensitive to the values of the parameters chosen. The fit has been made in practice in terms of the original Booth–Grahame saturation formula, which involves b.[196] In addition to the successful treatment of the general experimental relationship, there is also the temperature dependence of the capacity function, which is clearly defined experimentally. Both this theory and the earlier theory of Watts-Tobin give realistic temperature coefficients of capacity (cf. Fig. 18 in the region of the capacity hump).

The possible ramifications and extensions of this type of approach have been outlined by Macdonald and Barlow, but since

Models of the Double Layer

they have not yet been implemented they will not be treated at this juncture.

An alternative, strictly molecular approach has been devised by Levine *et al.*[217] It still involves a simplified model and only the first layer of water molecules are considered. These molecules are regarded as a plane monolayer of electrostatically interacting polarizable species in vacuum between two plane interfaces beyond which the media are continuous (mercury and aqueous phases). The water molecules are considered to be in a two-dimensional hexagonal lattice and the dipoles of the molecules are taken as point charges situated at the lattice sites. If the number of sites is N per unit area and d is the distance between nearest neighbors, $d = (2/\sqrt{3}N)^{1/2}$. However, Levine also considers d to be the thickness of the inner region, although in practice both N and d may be expected to vary with the electrode charge q. In addition, the inner layer thickness variation with q need not be directly related to the changes in N and d with q. These interrelationships are neglected in this treatment. A further limitation of the basic model is the use of a continuum model for the aqueous phase and the neglect of specific interactions between the monolayer and adjacent phases.

Only the normal components of polarization are involved, $p^\downarrow$ and $p^\uparrow$, in the assumed two-state model. These components are not necessarily equal, depending on the orientation of the dipole to the interface in the two positions. $p^\downarrow$ would correspond to the hydrogens of water toward the mercury and $p^\uparrow$ vice versa. It is useful to relate these components through

$$p = \tfrac{1}{2}(p^\uparrow + p^\downarrow) \quad \text{and} \quad \delta = (p^\uparrow - p^\downarrow)/2p \quad (106)$$

i.e.,

$$p^\downarrow = p(1 - \delta) \quad \text{and} \quad p^\uparrow = p(1 + \delta) \quad (107)$$

If the permanent dipole moment of water is p_w and the model in Fig. 20(b) is assumed, we have

$$p^\uparrow = p_w \quad \text{and} \quad p^\downarrow = p_w \cos\theta \quad (108)$$

and hence

$$p = p_w \cos^2 \tfrac{1}{2}\theta \quad \text{and} \quad \delta = \tan^2 \tfrac{1}{2}\theta \quad (109)$$

If all the sites are occupied

$$N = N^\uparrow + N^\downarrow \quad (110)$$

and the mean dipole moment per site can be derived

$$m_p = (p^\downarrow N^\downarrow - p^\uparrow N^\uparrow)/N = p(\lambda - \delta) \tag{111}$$

where

$$\lambda = (N^\uparrow - N^\downarrow)/N \tag{112}$$

As can be seen from equation (112), λ measures the degree of preferential orientation of the molecule and hence in equation (111) the term $-p\delta$ represents the contribution of permanent polarization while $p\lambda$ is due to preferential orientation. In addition to m_p there will be an induced dipole moment which can be defined as m_i per molecule.

The ensuing treatment is based on the approximation of random distribution of the dipoles independent of their orientation over the surface sites.[218] The consequences of this assumption are crucial to the validity of the treatment. Any dipole in a particular orientation would be expected to influence the orientation of dipoles on adjacent sites both by electrostatic and by higher-order interactions. Such local order phenomena will depend on the solvent system and will be relatively more important in highly structured solvents such as water or the formamides than in the less structured aliphatic alcohols. In the worst case long-range order within the monolayer is conceivable. The model employed here neglects *all* such correlations and the mean field at each site is taken to be the same for all sites and is calculated by assuming the same polarization for each monolayer molecule. Nonnormal dipoles will give horizontal components of dipole moment, but the random approximation averages all such contributions to zero in an isolated system.

In the normal direction, there will be in general a mean external field due to the charge on solution and mercury equal to $4\pi q$; to this must be added the residual field due to the two orientational states expressed in the form of residual energies $E_r^\downarrow$ and $E_r^\uparrow$. These cannot be equal since they include specific nonelectrostatic interaction energies, dipole and quadrupole image energies, and dipole–dipole interactions with the second layer of water molecules (hydrogen bonding). These terms depend on the orientation of the dipole and can be considered as the interactive part of the energy controlling dipole orientation.

Models of the Double Layer

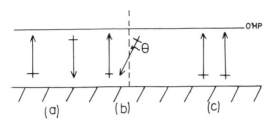

Figure 20. Possible dipole orientations in the inner layer.[217]

The total dipole component normal to the surface at each site is $m_i + m_p$ and hence the field acting at any site to its neighbors is

$$\mathscr{E} = 4\pi q + (C_e/d^3)(m_p + m_i) \tag{113}$$

where C_e is an effective coordination number. If the hexagonal lattice were situated in vacuum, in the absence of the interface, the field at the site due to nearest-neighbor interactions is $(m_p + m_i)/d^3$ and $C_e = 6$. Topping[216] has shown that the model in Fig. 20(c) with identical dipoles on all sites is given by putting $C_e = 11.0$. In the present situation the water dipole is adjacent to a metallic conductor and on the other side is a medium whose conductance can vary over wide limits depending on the salt concentration. The dipoles are therefore imaged to some extent in these phases, these images in planes parallel to the original dipole diminishing in magnitude by multiples of $f = (\varepsilon - 1)/(\varepsilon + 1)$, where ε is the dielectric constant of bulk water. The field at the dipole site is increased effectively making C_e larger than Topping's value. Macdonald and Barlow[219] have estimated these contributions to the site potential and if only the first three planes of images are retained in each phase, an arbitrary but reasonable approximation for C_e is

$$C_e = 11.0[1 + 0.1612(1 + f) + 0.0522f + 0.0067(1 + f)] \tag{114}$$

from which $C_e = 15.2$ for water at 25°C.

The induced polarization at each site is given by

$$m_i = -\alpha\mathscr{E} \tag{115}$$

Hence

$$\mathscr{E} = [4\pi_q + (C_e m_p/d^3)]/Z \tag{116}$$

where $Z = (1 + \alpha C_e/d^3)$ and represents the extent to which the polarizing factor is reduced by the polarization effect at $q = 0$. It is easy to obtain a relationship between the relative populations of the two orientational states, from the familiar Boltzmann law. The energy in the down state is $p^\downarrow \mathscr{E} + E_r^\downarrow$ and in the up state is $-p^\uparrow \mathscr{E} + E_r^\uparrow$ and hence

$$\frac{N^\downarrow}{N^\uparrow} = \frac{\exp[-(p(1-\delta)\mathscr{E} + E_r^\downarrow)/kT]}{\exp[(p(1-\delta)\mathscr{E} - E_r^\uparrow)/kT)]} \tag{117}$$

Both m_i and the van der Waals energies are assumed to be the same in the two states. The same relationship can be derived from the equilibrium conditions for the system, minimizing the free energy of the system, $F = E - kT \ln g$, where g is the number of configurations of the dipoles on the lattice sites per unit area. It is convenient to define the quantities

$$\Delta e_r = (E_r^\downarrow - E_r^\uparrow)/kT \quad \text{and} \quad e_w = C_e p^2/d^3 kTZ \tag{118}$$

which are respectively the orientational anisotropy in the residual energy and the intensity of dipole–dipole interactions reduced by the depolarizing factor Z. In addition v can be defined,

$$v = 4\pi p/Z \tag{119}$$

In terms of these parameters, equation (117) leads to

$$\ln[(1 + \lambda)/(1 - \lambda)] = -\Delta e_r - 2vq + 2e_w(\delta - \lambda) \tag{120}$$

This equation relates q to λ. These relations can readily be examined at the ecm, using equation (120). From the lhs of equation (120), expanded as a power series, $\lambda = \lambda_0$ at $q = 0$ is given by $-\Delta e_r/34.2$ if the dipole moments in the two states are assumed equal. This means that, from the equation

$$m_p + m_i = [p(\lambda - \delta) - 4\pi q\alpha]/Z \tag{121}$$

The mean dipole moment is $\lambda_0 p_w/1.42 \ll p_w$. The large value of C_e resulting from the long-range dipole–dipole interactions leads to a small mean dipole moment at the lattice site. This result is obtained by using only the first term of the series expansion. If, on the other hand, a dipole orientation of 52.5° is assumed, $\delta - \lambda_0 = 0.022$. This small value means that the field at a site due to the permanent polarization ($-p\delta$) causes a preferential orientation which gives a polarization term which largely compensates for the permanent polarization (a feedback factor resulting from the strong dipole–dipole interactions).

These relationships can now be employed to derive values of the potential drop χ across the water monolayer at the ecm and the inner layer capacity C_i. For $q = 0$

$$\chi = 4\pi N(m_p + m_i) \tag{122}$$

Using equations (121) and (122), χ is given to a first approximation by

$$\chi = \frac{2\pi Np(1 - \delta^2)\{\Delta e_r + \ln[(1 + \delta)/(1 - \delta)]\}}{Z[1 + e_w(1 - \delta^2)]} \tag{123}$$

With identical dipole contributions ($\delta = 0$) and $\theta = 0$ and using the same constants as were used in interpreting equation (120), i.e., $N^{-1} = 12.5$ Å^2, $d = 3.80$ Å, $p_w = 1.84$ D, $\alpha = 1.5$ Å^3, $C_e = 15.2$, and $T = 25°$, gives $\chi = -115\Delta e_r$ mV. If $|\Delta e_r| \approx 1$, this is of the correct order of magnitude. However, if at $q = 0$, $E_r^\downarrow < E_r^\uparrow$, then $\Delta e_r < 0$ and $\chi > 0$. This is clearly incorrect but if $\delta \neq 0$, then using $\theta = 52.5°$ or $\delta = 0.243$,

$$\chi = -182(\Delta e_r + 0.496) \quad \text{mV} \tag{124a}$$

and

$$\lambda_0 = \delta - 0.0436(\Delta e_r + (0.496) \tag{124b}$$

If $-0.496 < \Delta e_r < 0$, $\chi < 0$, and $0.243 > \lambda_0 > 0.221$, the dipoles are still preferentially in the "down" state but although this preferential orientation makes a positive contribution to χ, this is more than counteracted by the negative contribution due to the permanent upward polarization $p\delta$ per molecule, due to the fact that $p^\uparrow > p^\downarrow$, i.e., $p^\uparrow N^\uparrow > p^\downarrow N^\downarrow$ in spite of $N^\uparrow < N^\downarrow$.

For $q \neq 0$

$$V = 4\pi qd + 4\pi N\frac{p(\lambda - \delta) - 4\pi\alpha q}{1 + \alpha C_e/d^3} \tag{125}$$

which, after differentiation, yields

$$\frac{1}{C_i} = 4\pi d\left[1 - \frac{4\pi N\alpha}{d(1 + \alpha C_e/d^3)}\right] + \frac{4\pi Np}{1 + \alpha C_e/d^3}\frac{d\lambda}{dq} \tag{126}$$

where

$$\frac{d\lambda}{dq} = -\frac{v(1 - \lambda^2)}{1 + (1 - \lambda^2)e_w} \tag{127}$$

Two anomalies can be explained by this theory: the hump in the capacity–potential relationship occurring at $q > 0$ and the temperature coefficient of the capacity. The former can be derived

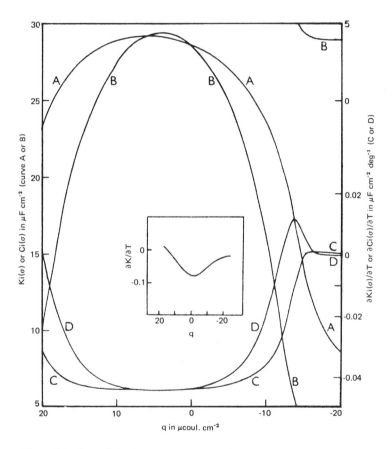

Figure 21. Plots of the integral and differential capacities of the inner layer capacity due to solvent dipoles and the temperature coefficient of these capacities according to the theory of Levine et al. Calculations used $p^\uparrow = p_w, p^\downarrow = 0.804 p_w$, $\Delta e_r = -0.4$, and $N^{-1} = 10 \text{ Å}^2$.[217]

by finding the condition for the occurrence of a stationary point in the capacity–charge relationship. By choosing a value of $\delta \neq 0$ and a small negative Δe_r, a maximum in K_i is possible at $q > 0$. With the parameters assumed to describe the potential due to the water dipole monolayer, Levine et al.[217] were able to show that the hump would occur in water at $\sim 3.1 \,\mu C \, cm^{-2}$. The second dependence may also be derived theoretically by differentiating equation (126) with respect to temperature. The calculated temperature coefficient of capacity is not identical with the experimental value, as can be seen from Fig. 21, but the theory predicts that not only the maximum temperature coefficient occurs at the maximum in the capacity–potential function, but that at the extremes of charge the dependence decreases. The example presented assumes $p^\uparrow \neq p^\downarrow$. The result should really only be compared with experiment for the charge range $8 > q > -10 \,\mu C \, cm^{-2}$ since outside this range specific cation or anion effects may be expected even in fluoride media at 25°C.

The capacity–charge behavior also does not seem to be in agreement with experimental data, but over the charge range discussed above the basic shape of the relationship is correct and the magnitude of the capacity is approximately correct. The curves can have variable degrees of symmetry by choosing $E_r^\uparrow \neq E_r^\downarrow$ or $p^\uparrow \neq p^\downarrow$. The theory makes a number of important approximations. It assumes that N is constant. If N is allowed to increase with increasing q, an increase in the capacity results, e.g., at $q = 20 \,\mu C \, cm^{-2}$, N^{-1} decreasing from 12.5 to 8.33 $Å^2$ causes C_i to increase by 5.5 $\mu F \, cm^{-2}$. The capacity is now known to contain a contribution from fluoride ion adsorption at $q \gtrsim 8 \mu C \, cm^{-2}$ and consequently it is difficult to estimate the validity of such calculations. In addition, for this model it is assumed that the inner region thickness could, for simplicity, be equated to the nearest-neighbor lateral separation of the dipoles. Changing N, which the substitution of more reasonable values would imply, would also change $E_r^\uparrow$ and $E_r^\downarrow$. The lack of consideration of lateral interactions, which is the principal feature of the Bockris approach, will tend to lead to large deviations from the proposed relations at large cathodic and anodic potentials, where the orientation of the dipoles in a specific direction may be considered to be complete. The inclusion of this term would give a contribution to the capacity which would tend to counteract the low values of capacity predicted by this theory at the extremes of polarization. Further examination of the applicability of this model will be presented in the final discussion section.

Bockris et al.[107] in their theory of ionic adsorption, include consideration of the influence of solvent orientation on the properties of the double layer. The two-position dipole model, "up" and "down," is again used for the water orientations. The equilibrium between water in the bulk and water adsorbed on the electrode in a particular orientation is described by

$$k_w c_w (1 - \theta^\uparrow - \theta^\downarrow) = k^\uparrow \theta^\uparrow \tag{128}$$

$$k_w c_w (1 - \theta^\uparrow - \theta^\downarrow) = k^\downarrow \theta^\downarrow \tag{129}$$

where the k values are rate constants. If n is the coordination number of the dipole and E is the lateral interaction energy between a pair of adjacent dipoles, the energy of the "up" dipoles is expressed by Bockris as

$$\mu \mathscr{E} + (EnN^\downarrow/N) - (EnN^\uparrow/N) \tag{130}$$

and for the "down" dipoles equation (130) with the sign reversed. Since $\theta^\uparrow/\theta^\downarrow = e^{2x}$, where kTx is the sum of the energies arising from field and mutual interactions of the dipoles, if a simple Langevin model is assumed,

$$R = \frac{\theta^\uparrow - \theta^\downarrow}{\theta^\uparrow + \theta^\downarrow} = \frac{N^\uparrow - N^\downarrow}{N^\uparrow + N^\downarrow} = \tanh x \tag{131}$$

the relationship between the energies of the dipoles and the orientation ratio R is thus derived as

$$R = \tanh[(\mu \mathscr{E}/kT) - (REn/kT)] \tag{132}$$

and since the potential across the dipole layer is given by

$$\chi = 4\pi\mu(N^\uparrow - N^\downarrow)/\varepsilon \tag{133a}$$

hence

$$\chi = 4\pi\mu NR/\varepsilon \tag{133b}$$

Bockris uses a factor 2π is equation (133a). The capacity due to dipole reorientation is then simply given by differentiating equation (133b) with respect to q and substituting for the field, $\mathscr{E} = 4\pi q/\varepsilon$. The data shown in Fig. 22 have been calculated for the Bockris

Models of the Double Layer

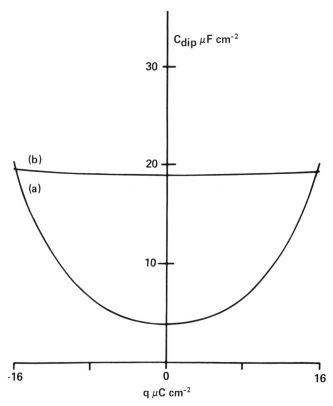

Figure 22. Comparison of capacities due to dipoles in the inner layer; calculated according to the theory of Bockris et al[107a] (a) assuming no lateral interactions (b) assuming lateral interactions ($E/kT = 0.5$).

theory. A plot of the Watts-Tobin theory with no lateral interactions was included in the original incorrect plot of Fig. 22. The curves from the two theories are rather similar in shape although the actual magnitude of the capacity differs at the minimum. The similarity between these curves is not helpful since there are some essential differences in the models involved in the Watts-Tobin and Bockris treatments. According to equation (90b), the dipolar capacitance contribution in the Watts-Tobin model is added to an essentially potential-independent contribution according to a *parallel* capacitor configuration. The hump in this model is therefore the result of adding a *positive* dipole capacitance contribution to the second capacitative component. The BDM model, on the other hand, indicates that the dipole capacity contribution passes through a minimum near $q = 0$. In order for this contribution to give a hump in the total capacitance–potential function, the dipole capacitance contribution must be added in *series* to any other capacity components and must be a *negative* capacitance contribution. The data as given in Fig. 22 must be considered in the light of these assumptions related to the models.

The validity of these calculations can be questioned since a number of factors which may be important seem to have been ignored or treated in an oversimplified manner. Of particular importance are the approximations involved in the derivation of the lateral interaction energies. The water dipoles are assumed to have only two orientations which are identical in moment except for sign. A simplified, nonassociative formalism for the dipole interactions is assumed and this may have a strong influence on the lateral interaction energy. This would effectively change the coordination number of the solvent dipole, a quantity of prime importance in these calculations. This is expecially so in the case where $N^\uparrow \simeq N^\downarrow$. The dipoles in one orientation will be effectively shielded by the surrounding dipoles and although the field with which the dipole interacts is formally given by $4\pi q/\varepsilon$, the microfield at the dipole site may well be considerably different. Although equilibrium is assumed between the bulk and surface layer, the treatment is still essentially one of a vacuum monolayer and the possibility of longer range interactions into the bulk of the solution is neglected (as they are in all other theories). The role of the metal is considered using equations (138)–(141) in terms of the different distance of the dipole from the metal in the two orientations of the dipole. The difference in the image contributions calculated for a difference in distances of 0.05 Å gives an energy of 0.9 kcal mole^{-1} for the difference between the two orientations of the dipoles. There is no consideration of any other interaction between the solvent molecule and the metal. The inclusion of a nonelectrostatic term or nonnormal dipole orientation in the two directions would give a very different lateral interaction energy. In addition, the possibility that the total dipole population of the inner layer may change (due to electrostriction) with electrode charge is ignored, a factor pointed out by Conway and MacKinnon.[267]

Models of the Double Layer

It is possible to estimate the surface potential of a dipole monolayer from equation (133b). Although the equation indicates that for $q = 0$ we have $\chi = 0$, the inclusion of a term asymmetric in dipole orientation, e.g., the different image contributions in the two orientations, or some nonelectrostatic term, results in $\chi \neq 0$ at $q = 0$. However, if a completely oriented monolayer model is assumed for the water layer, equation 133b yields a value of 1.2V for the surface potential.[107b] Conway, MacKinnon and Tilak have calculated this value on the basis of a 3 Å lamina of normal water but if a depolarization factor were introduced, this anomalously large value would be reduced.

In order to treat organic adsorption, Bockris assumes a more sophisticated model of the solvent which attempts a calculation of specific intermolecular forces.[239] The model still makes certain assumptions. Basic among these is the choice of the electrical variable. Bockris argues that charge should be chosen because in the calculation of field strength, the pertinent calculation involves only one unknown, i.e.,

$$\mathscr{E} = 4\pi q/\varepsilon \qquad (134)$$

the dielectric constant ε. Although this seems superficially correct, the inclusion of a dielectric constant for what is essentially a non-isotropic medium over which, according to Macdonald and Barlow, $\varepsilon = f(\mathscr{E})$, can lead to complex theoretical problems. In addition it is assumed that lateral interactions of the solvent depend only on the relative numbers of molecules in the two orientations.

The energies of the molecules in the two positions are now expressed by the more complex relationships

$$\langle \Delta G_s^\uparrow \rangle = \frac{N_s^\uparrow}{N_s}(nE^{\uparrow\uparrow}) + \frac{N_s^\downarrow}{N_s}(nE^{\uparrow\downarrow}) + \mu \cos \alpha^\uparrow \mathscr{E} + \Delta G_c^\uparrow \qquad (135)$$

and

$$\langle \Delta G_s^\downarrow \rangle = \frac{N_s^\downarrow}{N_s}(nE^{\downarrow\downarrow}) + \frac{N_s^\uparrow}{N_s}(nE^{\downarrow\uparrow}) + \mu \cos \alpha^\downarrow \mathscr{E} + \Delta G_c^\downarrow \qquad (136)$$

where the average over all the solvent molecules is

$$\langle G_s \rangle = (N_s^\uparrow/N_s)\langle \Delta G_s^\uparrow \rangle + (N_s^\downarrow/N_s)\langle \Delta G_s^\downarrow \rangle \qquad (137)$$

In this system of equations, as in equation (130), n is the effective coordination number of the dipoles which includes both nearest neighbors and the effect of neighbors further away and the E values are particular nearest-neighbor interaction energies. The field with which the dipole interacts is derived from equation (134) and the normal gas dipole moment is assumed* for μ. The final terms $\Delta G_c^{\uparrow}$, $\Delta G_c^{\downarrow}$ are the equivalents of Macdonald and Barlow's anisotropy energy and include all other "chemical" contributions to the dipole energy. These terms do not imply chemical bonding with the electrode but probably contain a large dispersion contribution. This, it is suggested by Bockris, arises from the interaction of the lone pair of electrons on the oxygen or from the OH bond. The dispersion interaction between the adsorbed molecule and the metal can be calculated from[220]

$$U_{\text{disp}} = \pi NC/6r^3 \qquad (138)$$

where N is the number of atoms per cubic centimeter of adsorbent and C is a constant for the system,[221]

$$C = 6mc^2 \frac{\alpha_1 \alpha_2}{(\alpha_1/\chi_1) + (\alpha_2/\chi_2)} \qquad (139)$$

where α_1 and α_2 are the polarizabilities of water and the metal and χ_1 and χ_2 are the diamagnetic susceptibilities of water and metal. Values of χ_i are not precisely known ($\sim \pm 10\%$ errors are common). Bockris was able to show that if the molecules were considered to be in two orientations and the molecule's center was at a different distance from the electrode in the two positions, the differences in ΔU_{disp} for different metals were of the correct order of magnitude for different metals but did not differ sufficiently from metal to metal.[222] In addition, image interaction should be included in this term. This could be expressed by

$$U_{\text{image}} = -\mu^2/8r^3 \qquad (140)$$

The difference between the energies of imaging in the two positions can again be estimated if the distance between the dipole and surface differs in the two positions, namely

$$\Delta U_{\text{image}} \approx 3\mu^2 d/4v^4 \qquad (141)$$

where r in equations (138) and (141) is the distance between the center of the dipole and the conductor and d is the asymmetry of the dipole. The sum of these contributions for mercury is of the order

---

*Actually the value in the liquid for H_2O is probably 2.4 D rather than 1.87 D as in the gas phase.

of 2.3 kcal mole^{-1} for an energy contribution which may be considered equivalent to $\Delta G_c^{\uparrow} - \Delta G_c^{\downarrow}$ in equations (135) and (136). The lateral energies of interaction of the dipoles are estimated using the classical relationship

$$E_{ij} = -(\mu_s^2/\gamma^3)[2\cos\alpha_i\cos\alpha_j - \sin\alpha_i\sin\alpha_j\cos(\beta_i - \beta_j)] \quad (142)$$

where the special orientation of the dipoles is characterized by α and β, r is the distance between the centers of the dipoles, and μ_s is the total dipole moment of solvent molecules. Since $N_s = N_s^{\uparrow} + N_s^{\downarrow}$ and R is defined by equation (131), from the Boltzmann distribution

$$R = \frac{\exp[-(\Delta G_s^{\uparrow})/RT] - \exp[-(\Delta G_s^{\downarrow})/RT]}{\exp[-(\Delta G_s^{\uparrow})/RT] + \exp[-(\Delta G_s^{\downarrow})/RT]} \quad (143)$$

From these equations the procedure followed in the initial treatment might be adopted to derive the capacity–potential relationship and the surface potential due to dipoles. This potential is given by equation (133b) with equation (143) substituted for R. This latter equation is no longer a symmetric function of q and it will have a finite value at $q = 0$. The residual "chemical" contribution can now result in the maximum in the entropy of the inner layer being at $q > 0$, depending on the largest energy contribution to ΔG^c. Since the dispersion energy contribution seems largest when the oxygen is nearest to the electrode, the surface potential at $q = 0$ will be negative, in agreement with experimental observation. The magnitude of χ and the capacity have not yet been calculated by this model. The results will depend on the detailed nature of the models used to estimate the various energy contributions. A number of useful observations may be made. In addition to the problem of the dielectric constant that should be involved in the dipole–dipole interaction calculations, the field with which the dipole interacts must be defined precisely. The mean field is not correct since it neglects the depolarizing effect of the dipole at the site and the potential generated by the rest of the dipoles in the immediate environment. This requires consideration of the local order. If the dipole interacts with the surface, is it still correct to use a "vacuum" value for the dipole moment? To these and the other simplifications and problems raised by the solvent aspects of this theory, no immediate solution can be foreseen. The model has given impetus to the search for an adequate description of the behavior of the inner layer water and hopefully other workers will be able to continue from the basis of these ideas.

6. Recent Ionic Models

The principal contributions to the field were considered by Macdonald and Barlow[223] and Levine et al.[224] in recent extensive reviews. The theoretical work discussed in these reviews is based on the viewpoint that the principal part of the free energy of adsorption of an ionic species can be attributed to the image of the discrete ion charge at the electrode. Bockris et al.[107] reject this approach on the basis of a simple calculation of the probable magnitude of the imaging energy of the ion which is inconsistent with the experimental free energy of adsorption of ions. Ion size, hydration, and adsorbability are suggested to be the principal causes of ions being specifically adsorbed.[107] There seems, as in the organic and solvent descriptions, to be a polarization of views over the extent to which these different parameters contribute to the total energy of the system. In addition, all the treatments either neglect or underestimate the limiting cases of their models. At low ionic concentrations the model must tend to the solvent model, which is usually excluded.

While they are reasonably successful in explaining the Esin and Markov effect, the discreteness-of-charge theories have until recently been very difficult to apply in practice since the results were not presented in a form suitable for comparison with experimental data. The number of parameters which have to be determined independently is larger than the experimental data will allow to be evaluated. It is frequently impossible to fit adsorption data to better than a one- or two-parameter isotherm and theoretical studies are necessarily hampered by the macroscopic experimental approach. Little attempt has been made to transform experimental or theoretical data into forms that are directly comparable. This discussion will outline the two possible approaches. The discreteness-of-charge approach will only be treated in terms of developments later than Refs. 223 and 224.

A number of different methods of calculating the discreteness-of-charge effect have been proposed. The earliest work of the Soviet authors has already been discussed. This approach used the two-dimensional hexagonal lattice approximation for the ionic distribution in the inner layer. Although a useful technique, this approach does require consideration of the stability of the array with respect to ionic concentration and temperature. For both metallic and dielectric nonaqueous phases the rigid lattice cannot be maintained when the effective ionic hard-sphere co-area exceeds 70 Å^2.[225,226] (If this were to correspond to the effective co-area derived from the virial fit to the experimental data, the behavior of few ions could be described by this model. Such a comparison is not reasonable.)

An alternative approach is via the "cut-off disk" model, in which the area around a given ion is considered to be free from other ions on the IHP.[134] The remaining charges are then considered smeared over the IHP. Levine et al.[227] were able to show that the effective area Grahame chose, equal to the mean area per adsorbed ion, was too large.

Buff and Stillinger[228] and Hurwitz[229] attempted to treat the system rigorously using cluster theory to approach the problem of ion distributions at the interface. At the IHP the ion is considered to be in a region of interaction of radius equal to the distance of nearest approach of the ions (i.e., a "cutoff disk" of this radius). Image forces are shown to weaken the lateral electrostatic interactions between the adsorbed ions so much that only short-range forces are effectively operative. This represents the opposite extreme to the hexagonal lattice model, in which electrostatic repulsive forces predominate. The real situation obviously lies in the middle, with both short- and long-range forces adequately treated in conjunction with the solvent dipole array problem.

There is a tendency to consider the inner layer in isolation, but the theoretical work on discreteness-of-charge effects has clearly shown the importance of a correct treatment of the diffuse layer. Some of the theoretical problems in treating this region have already been presented (Section I.3).

The presence of ions in the diffuse layer will modify the self-atmosphere potential of the adsorbed ion, and to quantify this influence requires a model for the dielectric constant as a function of distance from the metal. Earlier models tended to assume a dielectric discontinuity at the OHP and either perfect or some form of truncated imaging at this and the solution–metal discontinuity. This theory may be appropriate at high diffuse layer charge densities, but since these are not frequently encountered, it cannot be generally valid. Recently Buff and Goel were able to show that the introduction of a variable dielectric constant in the neighborhood of the OHP would influence the calculation of the self-atmosphere potential.[7,8] In particular they showed that with decreasing curvature of the dielectric constant-distance function (degree of discontinuity at the OHP) the self-atmosphere potential decreased. Neglect of this contribution leads to an overestimate of the self-atmosphere potential of the ion and the potential distribution at the ion site.

An anisotropic model of the dielectric constant of the inner layer proposed by Macdonald and Barlow[223] has been criticized by Levine et al.[230] This model involves a dielectric constant ε_L ($= 1$) parallel with the interface and the usual constant ε ($= 6$–15) normal to the interface. These values were introduced to obtain agreement of the proposed theory with experiment. Levine suggests that the ε_L should exceed ε because the average field in the lateral plane must be zero. Isotropic dielectric constants are assumed by Levine et al.[230]

Two zones divided by the IHP are identified. They have dielectric constants ε_1 and ε_2, respectively.[230] If the fraction $(\varepsilon_2 - \varepsilon_1)/(\varepsilon_1 + \varepsilon_2)$ is small and the OHP is approximated by a perfectly conducting plane, at either high OHP potentials or high electrolyte concentrations the discrete charge calculations are simplified. In this case the dielectric is assumed to be discontinuous at the OHP. The discreteness of charge potential for the ion adsorbed on the IHP is the sum of three terms; (1) the self-atmosphere of the adsorbed ion obtained by assuming that no change has taken place in the mean charge density on the IHP; (2) the disk potential due to the removal of mean adsorbed charge on the IHP from the exclusion area of the adsorbed ions; and (3) the contribution from the perturbation of the mean adsorbed charge from outside the exclusion disk when the ion is adsorbed at the disk center on the IHP.

This can be expressed as

$$\phi_1 = e_i \Delta\phi_0 + \phi_a + \phi_d \qquad (144)$$

where e_i is the adsorbed ion charge. The first term in equation (144) is considered by Levine to be small enough to be ignored and the theoretical treatment involves finding the form of the terms ϕ_a and ϕ_d which are dependent on position in the double layer. For the simple model outlined above ϕ_1 at the adsorbed ion site is given by

$$\phi_1 = -8\pi\beta\gamma q^1 g/(\varepsilon_1\gamma + \varepsilon_2\beta) \qquad (145)$$

with the factor $g = 1$. If the infinite disk model is retained, the next stage is to replace the perfectly conducting OHP by a diffuse layer. With the infinite disk model the diffuse layer behaves as if it were an uncharged medium of thickness $1/\kappa_e$ at whose outer boundary is a perfectly conducting wall, where κ_e is the local Debye–Hückel parameter defined by

$$\kappa_e = \kappa \cosh(e\psi_d/2kT) \qquad (146)$$

where κ is the bulk Debye–Hückel parameter and the dielectric constant is considered to be that of the bulk system, ε. The factor g is now not unity,

$$g = \frac{1 + (\varepsilon_2/\varepsilon\kappa_e\gamma)}{1 + (\varepsilon_2/\varepsilon\kappa_e\gamma)[\gamma\varepsilon_1/(\gamma\varepsilon_1 + \beta\varepsilon_2)]} > 1 \qquad (147)$$

Fitting these data as described in Section V shows that the model assumed for equation (147) may be incorrect since g is too large. This seems to be a consequence of the disk size assumed and the model for the diffuse layer. The ion areas on the IHP are measured by the distance of closest approach of the ions, a. The principal object of Levine's recent studies has been to investigate the approximation for the diffuse layer, which is one of Bockris' principal objections to this type of approach, i.e., to what extent can the OHP be considered a conducting plane.[231]

Returning to equation (144), the cutoff disk potential at a distance ρ from the ion on the IHP is

$$\phi_a(0, \rho) = -2\pi q^1 \int_0^a \rho' M(\rho, \rho') d\rho' \qquad (148)$$

where M is the "Green's function" which expresses the potential at any point on the IHP due to a ring of unit change of radius ρ' on the IHP. The third contribution, the perturbation potential, is also required at the ion site ($\phi_d(0, 0)$). The generalized potential $\phi_d(0, \rho)$ is given by the solution of an integral equation which in its linearized form is

$$\phi_d(0, \rho) = \Phi(0, \rho) - (2\pi q^1 e/kT) \int_a^\infty \rho' M(\rho, \rho') \phi_d(0, \rho') \, d\rho' \quad (149)$$

$$\Phi(0, \rho) = -(2\pi e q^1/kT) \int_a^\infty \rho' M(\rho, \rho')[e\phi_0(0, \rho') + \phi_a(0, \rho')] dp' \quad (150)$$

The principal problem is to obtain solutions to equations (148)–(150) in order to obtain g in equation (145). These methods are rather complex and will not be discussed further. For the model of a perfectly conducting OHP the potential contributions ϕ_0 and ϕ_a are expressed as slowly convergent series which describe the multiple electrostatic imaging in the mercury wall and the OHP. These series may be converted into rapidly convergent series by considering Fourier transforms and applying the Poisson summation formula. The third term requires solution of equations (149) and (150).

The upper and lower bounds to equation (148) are respectively obtained by replacing the diffuse layer by pure water and a perfectly conducting OHP. This can be visualized in the following manner. In the absence of salt in the aqueous phase a point charge q^1 on the IHP "sees" an image charge fq^1, where $f = (\varepsilon_2 - \varepsilon)/(\varepsilon_2 + \varepsilon)$ at a distance 2γ on the opposite side of the OHP. Since ε_2 is in the range 10–20 and $\varepsilon \sim 80$, f in this case is ~ -0.5. If the metal wall is ignored, the potential on the IHP is the sum of potentials due to charges q and fq. Since $\varepsilon > \varepsilon_2$, q is screened by its image charge, the magnitude of the screening increasing as the value of $\varepsilon \to \infty$, $f = -1$. The true situation lies between these limits and for the contribution ϕ_a, obtained by integrating equation (148) for $\rho = 0$, Levine derives

$$\phi_a(0, 0) = -\frac{4\pi a q^1}{\varepsilon_1 + \varepsilon^2} \int_0^\infty \frac{1}{k} J_1(ka) H(k) dk. \quad (151)$$

which in the limit of the perfectly conducting OHP may be approximated by the sum of $\phi_a{}^0(0, 0) + f_{21}\phi_a{}^1(0, 0)$. The first term has already been introduced in a less general form, namely equation (145) with $g = 1$,

$$\phi_a{}^0(0, 0)$$

$$= -\frac{8(\beta + \gamma)q^1}{\pi(\varepsilon_1 + \varepsilon_2)} \left[\frac{\pi^2 \beta \gamma}{(\beta + \gamma)^2} - \frac{\pi a}{\beta + \gamma} \sum_{n=1}^\infty \frac{a_n{}^0}{n} K_1 \left\{ \frac{n\pi a}{\beta + \gamma} \right\} \right] \quad (152)$$

Models of the Double Layer

As $a \to \infty$ equation (145) is obtained with $g = 1$. f_{21} is the dielectric ratio term $[= (\varepsilon_2 - \varepsilon_1)/(\varepsilon_2 + \varepsilon_1)]$ and is nonzero. $\phi_a{}^1(0, 0)$ must be derived from the relationship for the potential at a radial distance from the charge. Levine formulates this as a correction to the exclusion disk potential with perfectly conducting planes bounding the inner region $\phi_a{}^0(0, 0)$ and this is a complex expression from which the relative magnitude of the contribution of the term

$$f_{12}\phi_a{}^1(0, 0)/\phi_a{}^0(0, 0) \sim 0.1$$

should be noted.
Calculations have been made on the basis of the full expressions for these contributions and a number of trends are apparent. For a fixed inner layer thickness $d(=\beta - \gamma)$, as $\beta \to d$ the screening increases and $\phi_a(0, 0)$ decreases. For pure water the differences in screening efficiency of the metal and solution result in local screening minima. In the perfect conducting case, minimum screening occurs at $\beta = \gamma = \frac{1}{2}d$, where the first term in equation (152) for $\phi_a(0, 0)$ is maximal. If g is defined through equation (145), its value usually increases as ε_2 increases. In the presence of a diffuse layer an intermediate model applies and the minimum screening might be expected in the case where $\beta \simeq \gamma$.

The most important result of these calculations is that the contributions of terms ϕ_0 and ϕ_d in equation (14.4) are small and that the $\phi_a{}^0(0, 0)$ is of the order of $10\,kT$, varying between the extremes of a perfectly conducting system and pure solvent system by $\sim kT$.

The objections of Bockris to the imaging theory are dependent to some extent on the relative inefficiency of the OHP as a conducting plane, and this assumption of a perfectly conducting OHP does lead to overestimation of the screening effect of the diffuse layer. The correction for this is simply obtained by modifying the imaging function $H(k)$ in equation (151) by adding the local Debye–Hückel length onto the distance between the ion and the effective OHP imaging plane (now at $\gamma + 1/\kappa_e$). A complete treatment of the diffuse layer has been suggested but it would seem that approximation of the diffuse layer by a dielectric slab and a conducting plane into the bulk of solution only introduces about a 1% error in the calculations. The relative efficiency of the OHP as a conducting plane now depends on the value of κ_e, but the magnitude of the discreteness-of-charge potential was based on a calculation which included the influence of the diffuse layer through $H(k)$, and consequently these objections to this approach may in part be answered. The influence of ionic concentration on the contribution of the diffuse layer is now clear through the concentration dependence of κ_e. Smith and Levine have recently extended these calculations.[230b] In this revised approach they use a continuous dielectric constant-distance function. The results indicate good agreement with the previous discontinuous model for the discreteness of charge potential under similar double layer conditions. This at least partially removes the objections of Bockris and his coworkers to the introduction of discontinuous imaging planes in the earlier models. Levine has also refuted many of the objections raised by Wroblowa and Muller which are described in Section V. As little or no comparison with experiment has yet been described for these recent advances, objective judgement of the practical success of these theories is not yet possible.

An alternative approach to the problem of ionic adsorption was suggested by Bockris et al.[107] This approach evolved from Devanathan's model of the double layer (Section IV.2). The principal element of this theory is the introduction into a theoretical formulation of the earlier suggestion of Parsons that the hump in the capacity–potential curve arises primarily from the maximum in the value of dq^1/dq.[232] Such a result was first experimentally observed in the case of the adsorption of sulfonate ions at the mercury–water interface. The peak in this system occurs at the half-saturation value of the adsorbed charge.

In order to describe the interface, Bockris identifies six important trends in systems involving ionic adsorption. Other aspects, such as the cause of specific adsorption, are discussed elsewhere. This theory sets out specifically to describe (a) the cathodic minimum in the capacity–potential function, (b) the specific adsorption of ions, (e) the inflection in the q^1 versus q relationship, which, it must be noted, is not universally observed, (d) the maximum in the capacity–potential functions, (e) the maximum in the adsorption of simple aliphatic molecules which occurs at $q \sim -2\,\mu\text{C}\,\text{cm}^{-2}$ for some systems (see Section IV.7), and (f) the contribution of water dipoles.

In earlier theories the potential across the inner layer was taken as the electrical variable, but Parsons[232] has consistently advocated the use of charge as the electrical variable in ionic systems. On the other hand, potential has been advocated by Frumkin and Damaskin[165] and seems more satisfactory for organic systems. Bockris et al.[107] maintain that if one considers a dipolar layer at an electrode, the field cannot be meaningfully defined by $\Delta V/\delta$, where δ is the double-layer thickness and $\Delta V = V_\theta - V_{\text{pzc}}$, where V_θ is the potential measured at a dipole coverage θ.

Since V_{pzc} is itself dependent on the number of dipoles adsorbed, which is a function of the electrode charge, ΔV is not clearly defined. These interrelationships are complex. Electrode charge, on the other hand, is directly determinable and removes, in principle, the necessity for the types of complexities introduced by the constant-potential approach. In ionic systems the use of charge seems well founded, but for organic systems the θ versus E relationship frequently seems more symmetric.

The dielectric constant of the inner layer, which is required for calculation of the field on the constant-charge basis, is frequently assumed to be in the range 4–6. In fact, even for nonaqueous solutions the inner layer dielectric constant still seems to be in the same range irrespective of the dipole size, molecule size, or observed capacitance value. This corresponds to the high-frequency dielectric constant for the solvent measured at $\omega \sim 10^{11}$ Hz, arising from polarization of electronic shells and nuclei. This is in contrast to Macdonald's treatment, which involves the optical value of ε corresponding to electron polarization only. Both these components should really be considered in making up the distortion dielectric constant of water free from contributions due to the permanent dipole moment. For this layer, Bockris suggests that the dielectric constant is six (ε_L). The second layer of water molecules is obviously linked to the first but probably has a higher dielectric constant ε_H. This model of the inner layer suggested by Bockris differs significantly from that of Stern and Grahame in that the first layer of water molecules is not considered at any electrode charge to be part of the solvation shell of cations.[211] This is an essential part of both Macdonald's and Grahame's treatments. This change was introduced because the independence of the capacitance at the cathodic minimum is not a result of the Grahame–Stern theory since the changes of radii of solvated cations should lead to a cation-dependent capacitance.

The potential drop is given by[107]

$$-\Delta\phi = 4\pi q \sum (x_i/\varepsilon_i) \tag{153}$$

where x_i is the distance from the electrode and ϕ is the potential; hence

$$\frac{1}{K} = 4\pi\left[\left(\frac{1}{\varepsilon_L} - \frac{1}{\varepsilon_H}\right) + \left[\frac{t}{2}(2 + \cos 60) + r_i\right]\bigg/\varepsilon_H\right] \tag{154}$$

thus if $\varepsilon_H/\varepsilon_L$ is sufficiently large, $K \neq f(x_i)$ and hence K does not depend on the ion radius r_i. Using this simple model with $\varepsilon_L = 6$ gives $\varepsilon_H = 30$. In equation (154) t is the water molecule diameter. These values of ε_L and ε_H can be compared with the value chosen by Watts-Tobin,[202] which, if substituted in these equations, would result in negative capacity values.

The reasons why ions are specifically adsorbed is still not clear, as was shown in Section III.9 Bockris bases his theoretical interpretations on three factors: (1) For halide ions the degree of adsorption seems to correlate with the cube of ionic radius; (2) cations are adsorbed less than anions; for ions of the same radius cations are adsorbed to $\frac{1}{4}$–$\frac{1}{3}$ the extent of anions[107]; and (3) the influence of solvation of the ions which must contribute to the differences between adsorption of ions from different solvents is widely recognized. In aqueous media, where solvation phenomena seem better understood, ions with high primary hydration numbers, Li^+, Na^+, and F^-, are significantly less adsorbed.

The possibility of covalent bond formation has been considered by various authors, but Bockris suggests that the degree and type of ionic hydration is the principal factor determining the degree of specific adsorption.[156] This excludes chemical interaction energy from the model.

Image energy is also considered, but Bockris regards this energy contribution as small compared with the hydration effect. The complete model must obviously include all interactions and further discussion of the approximations involved in the models is given in Sections IV.5 and V.

The apparent dependence of adsorption on the ion hydration number suggests that the ions with weak hydration can be stabilized by contact with the metal, gaining energy by partial contact with the infinite dielectric constant of the metal. Thus for half the ion

$$\Delta F^\circ = -e^2/4r_i\varepsilon \tag{155}$$

which gives a value of the free energy of adsorption of 11 kcal mole^{-1} for typical values substituted in equation (155). There is one glaring inconsistency in equation (155), the inverse dependence of the free energy of ionic adsorption on ionic radius. The inverse of this relation would be more reasonable. Bockris justifies this by suggesting that ions of larger radii, which are the only ones specifically adsorbed, are situated partially in regions of higher dielectric constant. This seems consistent with the observation that specific adsorption of ions is only detectable if the ionic radius is greater than 1.4 Å. For ions of smaller radius the anionic hydration energy falls below the cationic hydration energy.

The basic adsorption isotherm can be written in the usual form,

$$n_a = 2r_i n_i \exp(-\Delta F^\circ/RT) \tag{156}$$

where three contributions to the free energy of adsorption are identified, the free energy change for the adsorption of an ion at $q = 0$, $\theta = 0$, ΔF°, the Coulombic energy of the dipole formed by the adsorbed ion and its image, based on the Langmuir model[234]—this gives attractive and repulsive contributions.

The order of imaging in this final contribution is considered to be single by Bockris, although it is evident from earlier discussions that the correct order must be between single and infinite. According to Levine et al.,[224] for single imaging, ε must be the same from the electrode outward otherwise, multiple imaging is inevitable.

If R is the distance between the centers of the adsorbed ion and the image of the adjacent ion and $2r$ is the distance between the center of the ion and its own image, the total electrostatic energy of the system is classically

$$U_{\text{rep}} = (e^2/\varepsilon r) - (e^2/\varepsilon R) \tag{157}$$

where $R^2 = r^2 + (2r_i)^2$ and $\pi n_a(\tfrac{1}{2}r)^2 = 1$. To this must be added the influence of the field acting across the OHP to the IHP on the free energy of adsorption of the ion. The potential drop from solution to the OHP is apparently neglected. Substituting for these contributions to ΔF° in equation (157) gives the complete BDM isotherm,[250]

$$\ln[\theta/(1-\theta)] = \text{const} + \ln a_i + Aq - B\theta^{3/2} + D\theta^{5/2} \quad (158)$$

where

$$\text{const} = \ln(2r_i F/10^3 |q_{\text{sat}}|) - \Delta G_c^\circ/RT,$$

$$A = 4\pi e(x_2 - x_1)/\varepsilon_H kT$$

$$B = \frac{\pi^{7/2} e^2 r_i^2}{4\varepsilon kT}\left(\frac{q_{\text{sat}} N_a}{F}\right)^{3/2} = \frac{\theta^{7/2} e^2 r_i^2 n_T^{3/2}}{4\varepsilon kT}$$

$$D = \pi^3 r_i^2 n_T B/20$$

In these equations n_T is the total number of sites at the IHP, q_{sat} is the saturation value of the charge density at the IHP, and a_i is the activity "of the contact adsorbing species." B includes a two-dimensional analog of the Madelung constant. If the ratio of areas occupied by the ion on Bockris' theory to the number of water molecules replaced on ion adsorption on the surface is p, it is clear that $p = 1$ in this theory.

If the ions do not interact by a type of electron sharing and if the crystal radii can be assumed for the adsorbed ions at the interface, p should be a function of ion size, since it is a simple physical parameter. The ionic radii increase in the order Cl^- (1.8 Å) $< Br^-$ (1.94 Å) $< I^-$ (2.15 Å) and the effective area occupied as r_i^2. However, as Levine et al. note,[230] in the lateral plane of the electrode one should use the partially hydrated ion size, in which case the ionic radii will tend to be equalized. The value of p should still be greater than unity. Levine suggested recently on theoretical grounds that $p \geq 2$.

The consequences of the isotherm, equation (158), can be further analyzed. If the derivative $d\theta/dq$ is obtained, then the equation for inflection points is given by[250]

$$(1 - 2\theta)(1 - \tfrac{3}{4}B\theta^{3/2}) - \tfrac{3}{4}B\theta^{7/2} + (15/4)D\theta^{5/2}(1-\theta)^2 = 0 \quad (159)$$

If $D = \pi^2/20$, $B = \alpha/2$, $\alpha = 0.982$, and $n_T = 1/\pi r_i^2$, the equation for inflection points is satisfied for values of θ related by

$$1 - 2\theta = \tfrac{3}{4}B\{[\theta^{3/2}(1-\theta)^2(1-2.45\theta)]\} = \tfrac{3}{4}Bf(\theta) \quad (160)$$

A minimum value for B can be estimated if it assumed that $\varepsilon r_i < 5 \times 10^{-7}$ cm. This gives $B > 27.6$. If values of θ are now substituted in equation (160), the results indicate that there are three roots, three inflection points in the θ versus $f(q)$ curve. The values of the roots depend on the value of B, but will lie within the ranges

$$\theta_1 < 0.2, \quad 0.3 < \theta_2 < 0.41, \quad \theta_3 > 0.67$$

If B is large (55), $\theta_1 \sim 0.1$, $\theta_2 > 0.38$, and $\theta_3 \sim 0.86$. The sign of $f'(\theta)$ for the three roots gives the maxima and minima in the capacitance–potential function. The first and third roots correspond to maxima and the second to a minimum in the C—$f(q)$ curve. There are a number of other interesting conclusions that can be drawn from this formulation of the ionic isotherm. For the three roots to occur, $B > 24.84$, and since B is inversely proportional to temperature, the hump disappears as the temperature increases. In general terms the maximum and minimum features of the capacity curve are explicable in the presence of adsorbed ions. If the limiting cases of the isotherm are considered, there can be no hump in the absence of specific adsorption. The third hump has not been observed but the hump does depend on the anion concentration and on anion size in the experimentally predicted manner. Other comparisons with experiment and alternative theories are presented in the next section and in Section V.

7. Intermediate Models

The justification for this section will be apparent after the preceding sections, which have all polarized attention on one aspect of the problem of describing systems including or not including ionic adsorption. At the outset it will be recognized that any purely theoretical approach will be immensely complex in view of the number of interactions that must be considered. No quantitative approach has yet been made to this problem and the models to be described are generally qualitative and at best semiquantitative. The ions alone (Section IV.6) and the solvent alone (Section IV.5) are partially understood, but the sort of complexity actually involved was recognized by Frumkin et al.[235] In studies of the temperature dependence of capacity in a wide range of salts they showed that the peak in the integral capacitance curves can only be explained by assuming orientation of the water in the inner layer. They suggested, in addition, that if the surface coverage with anions were not too small, the ions would sufficiently polarize adjacent water molecules and effectively lower the contribution of the water dipoles to the observed capacity. For this to occur, the dipole must be effectively parallel to the interface.

This type of approach has been developed by Payne,[122] Devananathan and Tilak[236] and Hills.[237] In the case of strongly adsorbing ions the hump is probably the result of quasichemical interactions which will override the electrostatic interactions of inner layer water. In the case of weaker specific adsorption as the electrode charge increases ($q > 0$), the field will increase across the inner layer, orienting the water molecules, and the number of specifically adsorbed ions will also increase. If the ions are considered hydrated on the IHP, this increased adsorption means that a larger fraction of the water molecules is associated with the adsorbed ions. This ion-associated water is essentially disoriented with respect to the normalizing influence of the field. Since disorientation of water leads, in general, to an increased capacitance, the rising capacitance at $q > 0$ can be explained. The capacitance cannot increase indefinitely, because as the electrode charge increases the inner layer field also increases and competes with the local field to orient the water molecules normal to the electrode. This automatically increases the order in the layer and consequently the capacity will tend to decrease.[238] This field is, however, being interpreted on the basis of a model not involving adsorption, and although the electrode charge increase, the total charge on the solution side ($= -q$) includes a large contribution from the specifically adsorbed charge. In the presence of weak adsorption, q_{diffuse} due to cations is small and it is trivial to show that the total diffuse layer charge is small. Since it is the *diffuse* layer charge which gives rise to the inner layer field and this builds up much more slowly on the anodic branch than on the cathodic branch, where specific adsorption is generally less, this model does not seem particularly satisfactory. A second effect might also be inferred, namely the decreasing rate of anion adsorption due to mutual repulsion of the ions at higher coverages. This effect is not particularly apparent in the published plots of q^1 versus q. The workers who advocate this theory invoke the usual Grahame–Stern model and propose that these latter influences result in the observed decrease in capacity which results in the capacity hump. If an alternative model of the inner layer were proposed—the inclusion of a second layer of water molecules between the weakly adsorbed ion and the electrode—then this model might be more reasonable.

Payne[122] has attempted to incorporate the influence of the replaced solvent dipoles on the adsorption of an ionic species on the capacitance analysis of the inner layer. If Grahame's model of the inner layer is adopted and a linear potential drop is assumed,[133]

$$_{q^1}\phi^{m-2} = -4\pi q^1(x_2 - x_1)/\varepsilon_i \qquad (161)$$

If the solvent dipoles are replaced by a monovalent ion and $q^1 = zn_i e$, then

$$_{q^1}\phi^{m-2} = -4\pi n_i[e(x_2 - x_1) - n_w\bar{\mu}]/\varepsilon_i \qquad (162)$$

where n_w is the number of solvent dipoles and $\bar{\mu}$ is the mean dipole moment of dipoles replaced by ions. In this case the capacity ratio is given by

$$_q K^i/_{q^1} K^i = [(x_2 - x_1)/x_2] - (ln_w/x_2) \qquad (163)$$

where l is the mean dipole length in the direction of the field. It is well known that there are a number of simple aqueous systems for which $(\partial \phi^{m-2}/\partial q^1)_q \sim$ const. This means that equation (161) would seem to apply. However, if the system is described in practice by equation (162), this means that $1/\varepsilon_i[e(x_2 - x_1) - n_w\bar{\mu}] =$ const. It seems likely that $n_w\bar{\mu}$ should increase as the electrode charge increases when dipole orientation is complete, so that $x_2 - x_1$ must also increase. This suggests that the number of water molecules in the region $x_2 - x_1$ increases, and is compatible with the ideas proposed as a result of the pressure and temperature dependence studies in Section III.4. Since the fluoride study[126] seems to give only a small hump, it is possible that $n_w\bar{\mu}$ is small and to explain this, nonnormal dipole orientation or reduced dipole moment due to chemical interaction with the electrode would have to be assumed. An alternative interpretation of the inner layer thickness has been presented in Section III.9.

The approach described above claims to explain a number of effects.

1. The stronger the ions are hydrated, the weaker is the degree of specific adsorption. The partially dehydrated ion, when adsorbed, occupies a surface area in agreement with that found in Damaskin's interpretation of the Watts-Tobin theory (Section IV.5).

2. The stronger the ion hydration, the less pronounced is the hump. The freedom of rotation of the water molecules is reduced near these ions to a greater extent than it is near more weakly adsorbed ions.

3. The maximum in the adsorption of simple aliphatic compounds attains a maximum when $q < 0$ in the region of ion desorption.

4. The adsorption of weakly adsorbed ions should give a hump when $q > 0$ if the water molecules have, as is generally agreed, a negative residual orientation at $q = 0$. At higher concentrations a hump occurs at negative values of q due to the increased probability of adequate ion adsorption required to produce hump effects.

8. Organic Systems

The basic theoretical descriptions of the adsorption of organic species at the mercury–solution interface are also polarized into two types. The Frumkin school[82] considers that the elementary step in the adsorption of the organic species from solution is the same as localized adsorption of a gas on the surface of a solid. Interaction forces between the adsorbed solute particles are considered (the constant A in the Frumkin isotherm), as is the displacement of more than one molecule of the solvent on adsorption of the solute (n). This model differs from that of Bockris et al.[239] in that it considers an alternative set of interactions, namely that solute–solute and electrode–solute interactions predominate. The assumptions Bockris makes are presented in conjunction with this theory.[107] A priori, no choice can be made between the validity of the models since they will be applicable only to systems in which the assumptions implicit in the model are valid. If these assumptions can be matched by the system under study, the theory can be tested and will be as adequate as the models used in calculating the various interactions included in the formalism adopted. Obviously a generalized model will include all the possible interactions within the system, but such a description will in general not be helpful since the large number of parameters which must be determined, and of which the majority are not amenable to independent determination, results in fits to the experimental data which are at best fortuitous. The parameters will, in general, only be valid as "the most likely value" and the acceptable range of these values will give little information regarding defects in the model.

In terms of such a description, both the Bockris[107] and Frumkin approaches[82] must clearly be inadequate. These models do give a realistic starting point from which further, more refined models can be evolved.

A theoretical description of the Frumkin model[82,165] has been attempted by Kir'yanov et al.[241-243] The model proposed is based on the general propositions of statistical thermodynamics, so that at equilibrium the partial free energies of the adsorbed and solution phases are equal. The principal point is to interpret the shift in point of zero charge (φ_N) on adsorption of the organic species, which they consider is not explained on the Bockris theory,[107] although in the later theory they are adequately accounted for since the organic dipole is allowed to reside in the inner layer.[239] The Frumkin school interprets this shift as being due to the replacement of water dipoles oriented with their negative ends toward the mercury by dipoles of the organic substance. The potential shift in the case of aliphatic oxygen compounds is due to the orientation of the "CO" bond which gives a positive displacement of up to ~ 0.5 V.

Damaskin et al.[82] suggest that the evidence for the inclusion of the polar group within the compact part of the double layer derives from (1) the lowering of interfacial tension being proportional to the length of the hydrocarbon chain for an homologous series, and (2) the fact that Traube's rule is obeyed, which, according to Langmuir, means that for fatty acids, amines, and alcohols the hydrocarbon chains lie flat on the surface. This means that the polar groups of the adsorbed molecules must lie within the compact layer, the dipole forming some fixed angle with the interface.[240] This angle will be almost independent of coverage.

Interaction forces between adsorbed molecules must depend on the solvent properties in the compact layer. At low coverages Kir'yanov suggests that the contribution of the solvent can be accounted for with sufficient accuracy using a dielectric continuum approximation. A molecular model should attempt a proper description of this region, possibly along the lines of Levine's model (Section IV.5). In the continuum model the dielectric permittivity of the compact region must be taken as a function of the degree of coverage. The assumption that the organic molecule will have a very different permittivity in the inner region does not seem justified in practice, as many studies of nonaqueous solvents have indicated. The dielectric constant of the region is seldom found outside the range $4 < \varepsilon < 6$. Studies by MacDonald and Barlow suggest that a lateral dielectric constant of unity might be more appropriate in calculating the contribution of the lateral interactions (cf. Section IV.6).

The first step is to derive an expression for the energy of the discrete dipole layer and the treatment adopted closely follows that adopted for gallium (Section IV.9). Only dipole–dipole interactions are considered at not too high degrees of coverage and it is assumed that all dipoles have the same orientation ($\mu_a > 0$) when the dipoles have their negative ends toward the metal. If the same nomenclature is assumed as was used for gallium, i.e., $x_2 = \beta + \gamma$, with the dipole centers located at β, the free energy contribution to the system due to dipole–dipole interactions can be expressed through

$$-G_s^{\text{dip}} = S \int_0^\phi q \, d\phi + \chi(\mu_a^2 N/\varepsilon_i) - f(\mu_a, \varepsilon, x_2, N) \quad (164)$$

where ϕ is the potential drop at the interface, and $v = Sx_2$ is the volume occupied by N dipoles in a dipole monolayer. N is the total number of dipoles and μ_a is the normal component of the dipole moment of the adsorbed molecule. The first two terms are self-explanatory, while the final term requires a detailed model. This can be derived from the hexagonal lattice model for higher coverages or the smeared-out disk model at lower coverages. The partial free energy of the system is determined by differentiating equation (164), $\zeta_s^{\text{dip}} = (\partial G_s^{\text{dip}}/\partial N)_{\psi,S}$. A large number of partial derivatives can result from this differentiation, particularly from the last term. The equations are given in the original paper.[241–243]

The organic molecule can in general replace more than one solvent molecule and consequently the entropy factors can be calculated. This can be derived using statistical volume fraction theory (Flory and Huggins),[244,245] which gives for the surface

Models of the Double Layer

phase partial free energy ξ_s'

$$\xi_s' = \xi^0 + \xi_s^{dip} + kT \ln[vv_\infty^{n-1}/(v_\alpha - n_v)^n] \tag{165}$$

where $v_\infty = N_\infty/S$ is the number of accessible sites per unit area and ξ^0 is a constant.

Van der Waals forces of interaction between adsorbed solvent molecules are included but the empirical approach is adopted. In an expansion in powers of density of sites occupied, the lowest quadratic term is retained, $-\bar{a}v^2$, where $\bar{a}$ is a positive constant characterizing the van der Waals dispersion forces. The differential of this term must be added to equation (165) to give the total partial free energy of the surface phase. The chemical potential of the organic species in the bulk phase is in general given by (at equilibrium $\xi_s = \xi_v$)

$$\xi_v = \xi_v^0 + kT \ln a \tag{166}$$

Using equation (165) and the discussion of van der Waals forces and assuming unity activity coefficients, the isotherm takes the form

$$B_0 c = \frac{\theta}{n(1-\theta)^n} \exp\left(-2\bar{a}\theta + \frac{\xi_s^{dip}}{kT}\right) \tag{167a}$$

where

$$B_0 = \exp[(\xi_v^0 - \xi_s^0)/kT] \tag{167b}$$

The other terms have their usual significance.

In order to solve the problem of the complex ξ_s^{dip} contribution and also $\bar{a}$ and B_0, Kir'yanov et al. assume the Frumkin model, approximating these contributions.[242,243] They retain the discreteness calculations for $\xi(\theta)$ in equation (168) using the hexagonal lattice model since the contribution needs to be precise at high coverages, while at low coverages the difference between models does not give appreciable errors. The isotherm then takes the form

$$B(\phi)c = [\theta/(1-\theta)] \exp[-2\bar{a}\theta - \eta(\theta) + \xi(\theta)] \tag{168}$$

where

$$B(\phi) = B_0 \exp\left\{-\left[\int_0^\phi q_0 \, dp + C'\phi(\lambda\phi_N - \phi/2)\right]/A\right\} \tag{169}$$

Having formulated the interactions systematically, the introduction of the model in the final calculations seems disappointing, but since in the initial hypotheses only two types of interactions are considered and a continuum model is assumed for the solvent, such an approach is rational.

The Frumkin model was employed as a first approximation to derive equation (168) from equation (167). As Kir'yanov et al. note, the Frumkin model gives a constant x_2. In order to avoid this approximation, the Frumkin–Hansen model was assumed with the approximation, as before, that μ is constant. The exponential in equation (168) now becomes $[-2a\theta - 2\alpha\phi\theta - \eta(\theta) + \xi(\theta)]$. $\xi(\theta)$ is derived from the hexagonal lattice model for the adsorbed dipoles and $\alpha = C'\phi_N(\lambda - 1)/A, A = kTv_\infty$.

If the isotherm is fitted to data which fits the Frumkin–Hansen model, values of B_0 can be derived and the parameters have the same value as the practical isotherm except for the adsorption equilibrium constant B_0. Its value calculated from the procedure indicated here is larger than the experimental value, due to the contribution of image forces and dipole–dipole interactions at larger degrees of coverage.

The system chosen for fitting in the case of the adsorption of surface-active organic species which increase the double-layer capacity was thiourea/sodium fluoride.[246] This can be treated in a way analogous to the treatment of the gallium interface (Section IV.9). The lattice function $f(\mu_a, \varepsilon, \delta, \theta)$ is derived from the usual hexagonal lattice or smeared charge models. It is assumed that to a first approximation, $\varepsilon \neq f(\theta)$, $\varepsilon \neq f(q, \phi)$, and $x_2 \neq f(\theta)$. The classical formula for q is employed,

$$q = (\varepsilon\phi/4\pi x_2) + (\mu_a v_\infty \theta/\delta) \tag{170}$$

The capacity is then derived by differentiating with respect to potential,

$$C \approx \frac{dq}{d\phi} = \frac{\varepsilon}{4\pi x_2} + \frac{\mu_a^2 v_\infty}{kTx_2^2} \frac{\theta(1-\theta)}{1 - 2a\theta(1-\theta) + [\theta(1-\theta)/kT]\partial^2 F/\partial\theta^2} \tag{171}$$

The second term is due to the adsorption of dipoles. The term $\partial^2 F/\partial\theta^2$ is the electrostatic repulsion term for dipoles. In order to derive this simplified version, the Frumkin model has been applied and the method again reduces to a calculation of the dipole–dipole contribution to the interactive term in the isotherm. Since the adsorbed layer is considered to be dilute, even at the maximum adsorption observed, the van der Waals term a is set to zero. Using $\varepsilon = 8$ and $x_2 = 4.4$ Å, a good fit to the experimental curves is obtained and since the coverage is not high, the type of model assumed for the adsorbed layer is not important. The problem with this type of approach is that is uses a predetermined formulation for the interactions and then attempts to force the theory into this framework. A truly molecular theory should give the desired θ versus q or C versus q relationship from molecular calculations alone and not require the introduction of a macroscopic function, however well justified experimentally, to give the basic form of the calculated functions.

Conway et al.[268] have shown that neglect or incorrect evaluation of the relative size of adsorbate (organic) and solvent in the adsorption isotherm can be equivalent to introduction of a substantial interaction effect in an isotherm of the type of equation (167a). When the relative size factor is neglected, further sophistication in the treatment of interaction and orientation effects is meaningless,[268] at least with regard to comparison between theoretical predictions and experimental results.

The Bockris, Gileadi, and Müller (BGM) theory of organic adsorption[239] was proposed at the same time as the BDM theory of ionic adsorption.[107] The theory was developed to explain the adsorption of simple aliphatic compounds at the mercury–solution interface. The approximations incorporated in the solvent model described in Section IV.5 are retained. In addition, in order to describe the mixed solvent–solute inner layer system, the following approximations are also made.

1. The Butler model,[169] which explains the quadratic dependence of the free energy of adsorption on the electrical variable through the interaction of the differences in polarizabilities $\Delta\alpha$ and dipole moments $\Delta\mu$ with the field, $\Delta G_{ads} = \frac{1}{2}\Delta\alpha\xi^2 + \Delta\mu\xi$, is rejected by Bockris because $\frac{1}{2}\Delta\alpha\xi^2 \ll \Delta\mu$ at the field strengths usually employed. Only the $\Delta\mu$ term is considered in the BGM theory.

2. Lateral interactions of the solvent molecules are independent of coverage by the organic adsorbate and only depend on the relative numbers of solvent molecules in the "up" and "down" positions (cf. Section IV.5 and the random approximation introduced by Levine et al.[217])

3. The dependence of coverage on the bulk solute concentration is given by a modified Langmuir relationship in which allowance is made for the areas occupied by solute and solvent molecules.[197a]

4. Lateral interactions between the adsorbed solute molecules are neglected since their polar groups are either far apart or interact in a part of the double layer where ε is high. This model therefore considers that in general for simple aliphatic molecules the hydrocarbon chain alone is located in the inner layer. The model does not exclude the organic dipole from the inner layer and it will become clear that the presence of the dipole in the inner layer is essential to explain the adsorption behavior of, e.g., butanol. The objections of Damaskin et al.[82] do not seem to be valid since the large shift in the pzc on adsorption of the organic species is readily explicable.[269]

If the adsorption of the organic solute (A) is measured at equilibrium, then

$$A_{sol} + n(H_2O)_{ads} \rightleftharpoons A_{ads} + n(H_2O)_{sol} \quad (172)$$

and $f(\theta)$ can be defined by

$$f(\theta) = x_A K_A / x_S^n K_S^n \quad (173)$$

where S refers to the solvent. The equilibrium constants for adsorption of solute and solvent are related to the corresponding free energies of adsorption through

$$K_S = \exp(-\langle \Delta G_S \rangle / RT) \quad (174b)$$

$$K_S = \exp(-\langle \Delta G_S \rangle RT) \quad (174b)$$

$\langle \Delta G_S \rangle$ has already been derived for the solvent using the two-position adsorption model (Section IV.5). The term $\langle \Delta G_A \rangle$ has yet to be defined in terms of the model outlined above. Since lateral (dipole–dipole) interactions have been ignored, only the interaction of the dipole with the field and the "chemical" contribution to the adsorption energies are considered. These are given for the two possible dipole orientations as

$$\langle \Delta G_A^\uparrow \rangle = \mu_A (\cos \alpha_A^\uparrow) \mathscr{E}_A^\uparrow + \Delta G_N^{c\uparrow} \quad (175a)$$

$$\langle \Delta G_A^\downarrow \rangle = \mu_A (\cos \alpha_A^\downarrow) \mathscr{E}_A^\downarrow + \Delta G_A^{c\downarrow} \quad (175b)$$

$\Delta G_A^{c\uparrow} \neq \Delta G_A^{c\downarrow}$ since $\mathscr{E}_A^\uparrow \neq \mathscr{E}_A^\downarrow$, the polar part of the molecule being in different fields in the double layer in the two orientational states.

Models of the Double Layer   345

If the free energy is averaged over all solute molecules,
$$\langle \Delta G_A \rangle = (N_0^\uparrow/N)\langle \Delta G_A^\uparrow \rangle + (N_0^\downarrow/N)\langle \Delta G_A^\downarrow \rangle \quad (176)$$
where $N = N_0^\uparrow + N_0^\downarrow$.
From equations (173)–(176) a relationship between the number of molecules in the "up" and "down" states can be derived. This relationship is equivalent to equation (137) for the solvent system. Substituting for equations (175a) and (175b) in equation (173) gives the complete adsorption isotherm. It is still necessary to derive explicit forms for the interactions involved in the isotherm. The treatment of the adsorption parameters for the solvent has already been discussed in Section IV.5. The description of the polarity of the dipole adopted agrees with normal chemical convention; the arrowhead indicates the negative end of the dipole. Bockris derived explicit forms for the adsorption of phenol and n-butanol from water and methanol with HCl as electrolyte.

The total isotherm contains the term $(\Delta G_A^{c\uparrow} - \Delta G_A^{c\downarrow})$ for both solvent and solute ($\equiv \Delta\Delta G_{\text{solv}}^c$ and $\Delta\Delta G_A^c$). In addition to these, there are six other identifiable contributions to the energy of the layer. The experimental systems are then fitted by summing the specific sets of interactions for various models of the inner layer. The simplest case that can be identified is when the solvent contribution to the energy is symmetric with respect to both its lateral and field interactions and the solute field interactions are zero [$Z_A^{\uparrow \text{o} \downarrow} = \mu_A^{\uparrow \text{or} \downarrow}(\cos \alpha_A^{\uparrow \text{or} \downarrow}) \varepsilon_A^{\uparrow \text{or} \downarrow}$ and similarly $Z_S^{\uparrow \text{or} \downarrow}$ for the solvent]. If a value for $\Delta\Delta G_{\text{solv}}^c$ is assumed, the θ versus q relationship is symmetric about and electrode charge given by the sign of $\Delta\Delta G_{\text{solv}}^c$ principally. The behavior of butanol adsorbed from aqueous solution seems to correspond to this model. The general shape of the maximum seems to agree, while at charges remote from the maximum in adsorption the deviations between experiment and theory are large.

The contribution of the solvent lateral interactions in the absence of solute interaction can be complex, depending, as it does, on the specific model assumed for the behavior of solvent molecules in contact with the metal. A choice of a suitable model must not only be consistent with providing the best fit to the experimental data, but also with all other evidence, e.g., surface potential of a monolayer of the solvent. Bockris assumes, in general, that $Z_A^\uparrow = 0$. Considering $Z_A^\downarrow$ to be a finite and a symmetric solvent contribution, in this case the organic dipole interacts in the down position with the inner layer field. In the phenol system the molecule can lie flat on the electrode $(Z_A^\downarrow)$ and in Bockris' view, with the dipole in the reverse orientation $(Z_A^\uparrow)$ the polar group does not interact with the inner layer field, although if the planar orientation, with π-bond interaction, is still retained, the dipole should still interact with the field. It is also suggested that $\Delta\Delta G_A^c \sim \Delta G_A^{c\uparrow}$, which is not reasonable if the two flat orientations were assumed, since the contribution $\Delta G_A^{c\downarrow}$ is intuitively nonzero. The adsorption of phenol from aqueous solutions is suggested as a system corresponding to this model.

By changing the solvent to methanol, it is possible to study a system in which solvation contributions are very different, and to explain the adsorption of phenol from methanol, a solvent model which gave the desired characteristics was chosen and is essentially the same model as that for the comparable aqueous system, the different characteristics of the θ versus q resulting from the different model assumed for the solvent. Again similar comments must apply.

For the butanol–methanol system the model including solvent interactions was again chosen but a contribution from $\Delta G_A^{c\downarrow}$ was included. Of all the fits to the experimental data in the original papers, this seems best, but the coverage by the organic solute is low and rather characterless and consequently the fit is not sensitive or discriminating.

From these observations a number of general conclusions can be drawn. If only dipole–dipole and dipole–field interactions of the solvent are considered and $\Delta\Delta G_A^c \sim \Delta G_A^{c\uparrow}$, then, as discussed by Conway and Dhar,[265] the width of the θ versus q curve depends on n [equation (172)], the dipole–field and lateral interactions of the solvent. The width increases with repulsive interactions $E^{\uparrow\uparrow}$ and $E^{\downarrow\downarrow}$ and attractive $E^{\uparrow\downarrow}$ but decreases if either $E^{\uparrow\uparrow}$ and $E^{\downarrow\downarrow}$ are attractive. The maximum in the θ versus q relationship is related,[265] on the charge scale, to the stable orientation of the solvent molecule, which is determined by both chemical interaction and dipole–dipole (repulsive) and dipole–field interactions. The asymmetry in the theoretical curves is weak and at charges remote from the adsorption maximum the experimental and theoretical curves differ. This is partially due to the presence of halide ions in the inner layer which are not considered in the basic theory. For the asymmetry to be due to the solvent alone requires interaction energies larger than are reasonable experimentally and in order to explain the observed experimental asymmetry, Bockris suggests that the solvent asymmetry must be reinforced by the solute asymmetry.

When the solute molecules are involved in attractive interactions with the field, the adsorption maximum shifts toward this energetically favorable region. This may significantly alter the symmetry of the function. In the worst cases, continuously increasing adsorption can be observed with increasing polarization in this direction. Bockris suggests that studies in different solvents are valuable in this case since they indicate that solvent is responsible for the asymmetry in the adsorption curves. This is only observed in the presence of suitable solutes and consequently field interaction of the solute may be involved. Why this should be different in the two solvents is not clear unless the solute assumes different positions in the inner layer in the two cases.

This theory has been subjected to criticism by a number of workers, in particular at a thermodynamic level, but it is clear that this approach does not attempt to use thermodynamic formulations and must be considered as an attempt to describe the adsorption phenomena in "absolute" terms. As with the other models discussed, based at a similar level, the theory is only as good as the fit and this is in terms of θ and is not very sensitive in general. It also depends on a large number of calculations of molecular interactions which have not been theoretically well characterized.

9. The Gallium–Solution Interface

The experimental characteristics of this electrode–solution interface were briefly described in Section II. The principal difference between this and the mercury–solution interface is the strong chemisorption of the water molecule at gallium. This is indicated not only by the difference in ion adsorption characteristics, but by the large free energy of hydration of the gallium interface (18–19 μJ cm^{-2}) as compared to the mercury interface (12.5 μJ cm^{-2}). This difference must be accounted for in terms of a model which must include not only the characteristic "electrostatic" dipole–dipole, dipole–ion, dipole–electrode, ion–ion, and their associated image contributions, but should also treat the chemical interaction of the solvent dipole with the metal and then consider the change in the characteristic properties of the dipole this entails.

A theoretical description of this system was suggested by Kir'yanov et al.[50] The problem was set up along the lines suggested earlier by Levich et al.[187] This model assumed that the inner layer can be considered in isolation, the diffuse layer contribution is negligible, and only one layer of dipoles is considered on the electrode. The centers of water dipoles are located at a distance $(\beta + x_d)$ from the metal ($x_d < 0$) and the ions are considered to be centered on a plane at a distance β from the electrode surface. The total inner layer thickness is $(\beta + \gamma)$. This may be equated with x_2 in the usual theory. The potential distribution in the inner region can be simply described by the volume densities of free ion charges ρ_{free} and bound dipole charges ρ_{bound} through Poisson's equation,

$$\nabla \Phi^{(1)}(\mathbf{r}) = -4\pi(\rho_{\text{free}} + \rho_{\text{bound}}) \quad (177)$$

If the inner layer is considered to contain *point* charges, then

$$\nabla \Phi^{(1)}(\mathbf{r}) = -4\pi e_a \delta(x - \beta) \sum_{m,n} \delta(y - y_{m,n}) \delta(z - z_{m,n}) \quad (178)$$

where e_a is the charge of the specifically adsorbed ion, x is the distance from the electrode in a normal direction, the subscripts m, n refer to the ion positions, and δ is the Dirac delta function. The dipole moment considered is the projection of the normal component onto the plane $(\beta + x_d)$. The boundary conditions of equation (178) can be easily derived. When these are substituted into equation (178), the equation can then be solved for $\Phi^1(\mathbf{r})$. The detailed variation of potential and field in the inner layer at any position $\mathbf{r} = (x, y, z)$ can be calculated.

The solution involves three terms, one from the average potential drop due to charge on the electrode, one due to e_a, and one due to μ, the water dipole. The second and third terms are summations over a zeroth-order Bessel function.

For the average values of the potential in the inner layer the solution takes the form

$$\psi = \psi_q + \psi_{q^1} + \psi_\mu \quad (179)$$

where ψ_q represents the normal electrostatic contribution ($= 4\pi q x_2$), ψ_{q^1} is the component due to the charge on the specifically adsorbed ions ($=4\pi q^1 \gamma$), and ψ_μ is due to the dipoles of the solvent ($=4\pi\mu n$). Similarly, the micropotential ψ_i at the site of the adsorbed ion can

Models of the Double Layer

be expressed through

$$\psi_i = -R\psi + \psi_a + \psi_{dip} + \psi_{rep} \qquad (180)$$

where R is the distance of approach of the ions measured from the OHP relative to the inner layer thickness and the first term represents the mean inner region potential at the ion site. ψ_{dip} is the dipole component contribution to the micropotential, ψ_{rep} is a quantity independent of the degree of filling and is the charge-imaging contribution and ψ_a is the ionic component of the micropotential calculated using Levich's method,[187] the limitations of which have been described by both Macdonald[223] and Levine.[224] This method only treats the dipoles in the immediate neighborhood of the ion and the remaining charges and dipoles are considered smeared out. For the micropotential the components are as follows:

$$\psi_a = -\frac{8ze}{\gamma_a} \sum_{k=1}^{\infty} \left(\frac{-1}{k}\right)^{k+1} \sin\left(\frac{\pi k \beta}{x_2}\right) \sin\left(\frac{\pi k \gamma}{x_2}\right) K_1\left(\frac{\pi k \gamma_a}{x_2}\right) \qquad (181)$$

$$\psi_{dip} = \frac{8\mu}{n_2 r_{dip}} \sum_{k=1}^{\infty} (-1)^{k+1} \sin\left(\frac{\pi k \beta}{x_2}\right) \cos\left(\frac{\pi k (\gamma - x_d)}{x_2}\right) K_1\left(\frac{\pi k r_{dip}}{x_2}\right) \qquad (182)$$

$$\psi_{rep} = \frac{2ez}{x_2}\left[\ln 2 + \left(\frac{\gamma - \beta}{2x_2}\right)^2 \sum_{k=1}^{\infty} (2k-1)^{-1}\left(k - \frac{\beta}{x_2}\right)^{-1} \right.$$

$$\left. \times \left(k - \frac{\gamma}{d}\right)^{-1}\right] \qquad (183)$$

All the terms in equations (179) and (180) can be derived on the basis of suitable model-type calculations and Kir'yanov et al.[50] claim that the resulting system of equations takes into account the discrete nature of the adsorbed ions and the dipolar layer of adsorbed solvent molecules. There is no attempt to describe the interaction of the water molecules with the interface and consequently any numerical values of, for example, water orientational angle or dipole moment must be speculative. Such values cannot be derived from the types of measurements so far completed at gallium. The hexagonal lattice model which Levich's theory implies for the calculation of the ionic component of the micropotential ψ_a must be questioned in this case. The ion in this system is not necessarily in direct contact with the surface and as a consequence, the potential well in which the ion will reside will be shallower than in the comparable mercury case. The validity of this calculation should be reexamined in this system, with particular reference to the thermal stability of the array. From this treatment it is possible in principle to derive (1) the ionic adsorption isotherm, (2) the capacity–potential function, and (3) the surface potential generated by the surface-dipole layer.

The isotherm can be generated from the relation for the micropotential with the Boltzmann relation,

$$q^1 = k_0 a_\pm \exp[(-zF/RT)\psi_i] \qquad (184)$$

where ψ_i can be obtained through equation (180). The term ψ_{rep} does not depend on the degree of filling or coverage of the surface, so that equation (184) can be written as

$$\theta = Ka_\pm \exp\left\{-\frac{zF}{RT}\left(\frac{\gamma}{x_2}\psi_{pzc} + \psi_a + \psi_{dip}\right)\right\} \qquad (185)$$

where K contains all coverage-independent terms. If the approximate model based on simple site filling is introduced, i.e.,

$$-\psi_{pzc} = 4\pi\chi\gamma\theta + b(1 - \theta) \qquad (186)$$

then from equations (185) and (186) the "Esin and Markov" coefficients may be calculated after substitution for the unknowns from the earlier relationships.

The capacity can be expressed through

$$C_\psi = C_0 + \frac{\gamma}{x_2}\frac{\partial q^1}{\partial \psi} + 4\pi C_\psi{}^0\frac{\partial(\mu n)}{\partial \psi} \qquad (187)$$

where $C_0 = 1/4\pi x_2$. This is a simplified relationship which neglects terms in $d(\text{distance})/d\psi$, i.e., electrostrictive terms which automatically appear when $dq/d\psi$ is derived from the preceding relationships. This method is considered superior to the treatment of the mercury interface both of Watts-Tobin and of Bockris in that in this case the dipole potential is properly included.

If $b = b_0$ [equation (186)] and $b_1 = 4\pi\chi\gamma$, where χ is the potential due to the dipole monolayer, equation (186) reduces to

$$-\psi_{pzc} = b_1\theta + b_0(1 - \theta) \qquad (188)$$

Hence b_0 characterizes the position of the pzc in the absence of specific adsorption, $b_0 = (-\psi_{pzc})_{\theta=0}$. The gallium electrode differs from the mercury electrode in that $b_0(Ga) \gg b_0(Hg)$. It can be shown experimentally that the difference between these values is ~ 0.50 V. As was shown by Trassati (Section II.5), at $q \ll 0$ it can be assumed that the water dipoles are all oriented and the difference in potentials at which corresponding charges are attained at the two interfaces in this potential region is proportional to the difference between the nondipole part of the potential at the two interfaces. Thus the difference between this value and Δb_0 gives the surface dipole potential drop, 0.33 V. This value characterizes the difference between the surface water structure in the two systems.

Kir'yanov et al.[50] note that qualitative agreement between the observed adsorption behavior for ions and the theoretical relationships can be obtained, but the determination of the particular parameters in the systems of equations are subject to considerable error due to the nature of the approximations in the models used in their calculation. Independent confirmation of these contributions is difficult to obtain but will be necessary before such an approach can be considered definitive.

V. DISCUSSION AND CONCLUSIONS

From the data in Section III, it is evident that the theoretical approaches in Section IV do not reflect the complexity of the experimental data. This disparity is particularly apparent in aqueous solutions, where much of the effort has been directed at explaining the hump in the capacity–potential function. In aqueous media this is clearly a complex characteristic of the system depending on the solvent and the ions. There seems little point in analyzing further the theoretical treatment in Section IV for aqueous systems. The situation in nonaqueous systems is clearer. The characteristics of the capacity–potential function in formamides at $q \ll 0$ are clearly controlled by the solvent and not by the ion properties and it is more sensible to apply the solvent theories in Section IV.5 to these rather than the aqueous systems.

Nancollas et al.[212] have extended and modified the Watts-Tobin theory. The influence of solvent dipole orientations θ and η

in the two positions at the interface (cf. equation 90b), are given by

$$C_{\text{solvent}} = \frac{N\mu^2(\sin\theta + \sin\eta)^2}{d^2kT}$$

$$\times \frac{\exp[(V - \phi)\mu(\sin\theta + \sin\eta)/dkT]}{\{1 + \exp[(V - \phi)\mu(\sin\theta + \sin\eta)/\alpha kT]\}^2} \quad (189)$$

This is identical with equation (90b) if $\sin\theta = \sin\eta = 1/\sqrt{3}$. However, using equation (189), it is not possible to produce an adequate fit to the experimental data. Nancollas invokes cation adsorption to improve the fit, although there seems little justification for cation adsorption from any of the solutions studied since the hump is evidently the result of solvent properties alone. The theory has already been criticized in Section IV.5 and has since been superceded by Levine's approach.[217] This was applied by Fawcett and Loutfy to the hump at $q \ll 0$ in N-methyl formamide.[247] This hump is very similar to the hump in formamide (cf. Section III.8). Using equation (126) and the experimental data at the maximum of the hump ($\lambda = 0$), the isotropic polarizability α for the component of the anisotropic polarizability perpendicular to the electrode, and a Courtauld molecular model for the molecule, conclusions were drawn regarding the orientation of the molecules at the interface. At the pzc, $\lambda = -1$, which shows that all the dipoles in the inner layer have their negative ends to the electrode. At the capacity peak the density of solvent molecules at the interface was estimated from equation (126). It corresponds to a molecular diameter of 6.4 Å, as compared with the Courtauld molecule value of ~ 4.0 Å. The dielectric constant of the inner layer was also calculated using the surface potential, equation (122). The calculated value, 3.5, is lower than the value calculated from the integral capacity, 11.5. The choice of N, C_e, and α is critical to these calculations and since it has already been shown that this theory is really only a first description of the system (Section IV.5), a fit over the entire range was not attempted. No data have yet been fitted to the extended Bockris theory.

Good evidence for the structure of water in the inner layer can be derived from the studies of Conway and coworkers on organic adsorption at the Hg/H$_2$O interface.[264,265,266] The interpretation of the "Esin and Markov" effect using equations of the type 131,132,133 was attempted. By studying pyrazine adsorption, a rigid adsorbate having itself no dipole moment, a finite slope of this coefficient must be associated at $q = 0$ with some specific residual orientation of the water molecule. The coefficient becomes zero at $q \sim -2\mu C \, cm^{-2}$ which agrees closely with the observations of the B.D.M. theory. The observed Esin and Markov coefficients are therefore associated with the replacement of a number of water dipoles by the pyrazine. The effect of temperature is consistent with the thermal disorder induced by a rise in temperature. A similar interpretation of the pyridine system[265] is confused by orientation of the pyridine molecule but is consistent with the behaviour observed for pyrazine.

Variation of the temperature allowed calculation of the entropy of adsorption of pyridine and the entropy associated with the displacement of water molecules, 2–4 being replaced for each species adsorbed.[264,269]

The properties of the solvent in the inner layer are obviously not yet adequately described, but the advances will probably come from extensions and modifications of theories of the Bockris and Levine type. There is one problem that has not really yet been approached. There is probably some degree of interaction of the water molecule with the mercury (Section II.5). This should modify the dipole in the inner layer and must be defined precisely if a complete treatment is to be devised. On the other side of the inner layer, the interaction of the inner layer molecule with the second layer of solvent molecules must be included. Before leaving the subject of the solvent, a few comments about the aqueous system are pertinent. The water hump in the capacity–potential relationship is shown in Fig. 13 and is smaller than is usually assumed. The contention of Bockris that the solvent contribution to the hump is negligible is not completely correct,[107] but the effect is very much smaller than the Levine theory predicts.[217] This may be solely a result of the simplified approach adopted by Levine. Two-dimensional ordering phenomena may also be important in aqueous systems.[197,213] Inclusion of dipole–dipole lateral interactions in the inner layer would also improve the Levine theory.

There has been much discussion about the fits of the theoretical approaches to the ionic systems. Earlier fits were discussed by MacDonald and Barlow.[223] The first fits to the BDM isotherm have been discussed by various authors.[155,223,224,251–253] The fits to the alkali halides data appear to be satisfactory, but with oxyanions and pseudo-oxyanions further modification to the theory is required to give a good fit. The theory requires a saturation coverage, seldom approached in the ionic system, and a large error may be introduced.[155] The errors in oxyanion systems may in part be due to the problems discussed in Section III.3. At this stage, fits to the experimental data are probably less important than attempting to identify the relative magnitude of the various contributions to the adsorption energy. Wroblowa and Müller,[249] Bonciocat et al.,[250] and Levine[253] have all attempted to separate these various contributions.

The principal point of contention is the imaging situation which is assumed for the ions at the interface. The BDM isotherm assumes a single imaging situation, while Levine has consistently employed multiple or infinite imaging.[107,225] The earlier evidence is shown in Fig. 16, where, using Grahames' data, the infinite imaging model seems to fit the system more satisfactorily. The consequences of the more recent attempts to describe this region in terms of isotherms are more certain. The BDM isotherm has already been shown in Section IV.6 to be compatible with the general form of the capacity–potential function. The humps and minima in the capacity/potential function are a direct consequence of the isotherm and agree with the experimental data for the halide and simple oxyanion systems.[250] The Esin and Markov coefficients have also been calculated from the BDM isotherm[250] and are derived from equation (158) by differentiation. The values are of the correct order of magnitude but are not constant over the experimental range. The values show that the experimental trend $Cl^- > Br^- > I^-$ is reversed when calculated theoretically. This is because the coefficient depends theoretically on $1/r_{ion}^2$ while experimentally the relationship is $\simeq f(r)$. It has been suggested that if some degree of compressibility in the adsorbed ion were introduced, this anomaly might be resolved.[249]

Recently Baugh and Parsons[254] fitted the BDM model of the inner layer to their data on the adsorption of PF_6^- from aqueous solution, using the coefficient

$$\left(\frac{\partial \ln a_1}{\partial q}\right)_{q^1} = \frac{F}{RT}\left(\frac{4\pi}{\varepsilon}\right)h \qquad (190)$$

where $h = 2r_{solvent} - r_{ion}$. If the usual radii are employed, it is found that $h < 0$, which has little significance. This result will always occur when the ion radius is greater than two solvent radii. By itself, this is insufficient evidence to disprove the isotherm since in this case the system studied involved an ion whose solution properties are highly anomalous. It is not improbable that the hydrophobic nature of the ion produces effects in the inner layer which are not accounted for in the basic theory. Wroblowa and Müller also discussed the discrepancies in oxyanion systems.[249] They attributed the anomalous behavior to asymmetric charge distributions on these ions.

It is possible to analyze the Levine multiple imaging isotherm in much the same way as the BDM isotherm was analyzed in Section IV.6. If the isotherm is written in the form

$$[\theta/(1 - 2\theta)^2] = \text{const} + \ln a_\pm + A'q - C\theta \quad (191)$$

for $p = 2$, the condition for an inflection point can be derived by differentiation for $d^2\theta/dq^2$ which can be shown to be consistent with

$$4\theta^2 + 4\theta - 1 = 0 \quad (192)$$

The only significant root is θ_1 and the value is independent of the properties of the ion, in contrast with experimental result. The inflection point in the q^1 versus q relationship is directly related to the maximum in the capacity–potential function and consequently the hump derived from the BDM isotherm is anion dependent, while the multiple imaging isotherm results in an anion-independent hump, in contrast with experiment. It has already been shown that the BDM isotherm is capable of predicting a minimum at $q > q_{\text{hump}}$ in qualitative agreement with experiment. No such conclusion is possible from the multiple imaging theory. The dependence of the hump on temperature also seems to be a direct result of the single imaging theory.

Although this evidence seems conclusive in favor of single imaging, the single imaging theory has been criticized at a number of levels.[253] For monovalent ions the multiple imaging model of Levine can be written as[253]

$$\ln \theta + g(\theta) = F(P, T) + \ln a_- + (e_0\psi_A/kT) \quad (193)$$

where $g(\theta)$ is an entropic term[244,245] accounting for ion size and ψ_A is the micropotential. ψ_A is calculated using g as the correction factor in the real imaging system when compared with the ideal imaging system with perfectly conducting walls on either side of the inner layer. This was used in Section IV.6. The dielectric constant was considered to be variable across the inner region, and weakly adsorbed ions are considered to be located between the IHP and OHP, their distribution being described by a continuous function in this region. ψ_A is calculated from the integral capacities of the inner region,

$$\psi_A = \psi_2 + (q/K_2') + (Kq^1 g^*/K_2^{12}) \tag{194}$$

where

$$g^* = 1 - [(K_2' - K)^2/K(C_{\text{diff}} + K)] \tag{195}$$

K and K_2' are the experimental inner layer capacities.[74,190,133] For a perfectly conducting OHP, $g^* = 1$ and if the ions are located on the IHP, $K_2' = K_2$. The term K_2' is involved because of the ion distribution across the outer region of the inner layer. In most cases g is ~ 1 and deviations on either side of this value are usually $< 10\%$.[230,231] Levine therefore assumes $g^* = 1$ in the isotherm. If suitable substitutions are made, the isotherm equation (193) becomes

$$\ln \theta + g(\theta) = F(P, T) + \frac{e_0 \psi_2}{kT} + \frac{e_0 q}{kTK_2'} - Cg^* \theta + \ln c$$

$$- \frac{e_0^2 \kappa}{2\varepsilon kT(1 + \kappa b)} \tag{196}$$

where

$$C = \frac{e_0}{kT} \frac{Q_{\text{sat}}}{K_2'^2} K = \frac{4\pi \gamma^2 e_0}{\varepsilon_1 kTd} Q_{\text{sat}} \tag{197}$$

The final equivalence in equation (197) is obtained when the dielectric constant of the inner layer is uniform and the ions are restricted to the IHP [cf. equation (190)]. The values of C calculated from Levine's earlier theory by Wroblowa and Müller[249] are considered to be too large by Levine.[253] Using the Flory–Huggins model with $p = 2$, values of C of the correct order of magnitude were calculated. Tests of the adsorption isotherm also gave satisfactory agreement with theory over the middle range of concentrations.[253] Since the Esin and Markov coefficients involve C, errors in this constant are important. Levine writes the Esin and Markov coefficient as

$$\left(\frac{dq_-}{dq}\right)_{a_\pm} = -\frac{(e_0/kTK_2')Q_{max}}{(1/\theta) + (dq/d\theta) + C} \tag{198}$$

in the absence of diffuse layer contributions and suggests that the approximately constant value of the coefficients in the presence of adsorption is related to the form of $g(\theta)$. For the Flory–Huggins model, equation (196) has maxima at $\theta = 0.5$, 0.21, and 0.12 for $p = 1, 2$, and 3, respectively [cf. equation (76)]. For the Helfand et al. model,[256] $p \sim 1.8$. For $p = 1$ the maximum occurs at high coverage ($\theta n = 0.5$), while experimentally the coefficients are constant at $\theta \ll 0.5$. Levine therefore argues that p should be at least 2. The rate at which the Esin and Markov coefficients increase with coverage were not discussed by Wroblowa and Müller. On Levine's model this requires a sophisticated treatment of the quantity Cg^* as a function of θ.

Wroblowa and Müller[249] use the description of $g(\theta)$ according to the model of Flory and Huggins and set $p = 1$ (Langmuir isotherm). In addition the term $-Cg^*\theta$ in equation (196) is replaced by $-B\theta^{3/2} + D\theta^{5/2}$ [equation (158)]. The BDM theory seems to give much better agreement with the experimental parameters, but it has been criticized. The use of $p = 1$ and the implication of a layer of infinite thickness of dielectric constant six are two of the objections raised by Levine.[253]

The various arguments must be considered in the light of the discussions in Section II, III, and IV.7. Neither of the theories considers the limit of $q^1 \to 0$ and consequently the more weakly adsorbed systems are anomalous according to these theories. In the presence of adsorption of halide ions the possibility of partial charge transfer to the metal must be considered and this will modify both approaches. The imaging is in all probability between single and infinite and will be modified by these partial charge-transfer effects.[162,163] The parameter g will not be unity in this case.

The current situation in the organic systems is equally complex. Recently the position has been clarified and the discussion will center around these rationalizations. The BGM theory[239] emphasizes the displacement of oriented solvent molecules from the electrode. The principal contribution to the dependence of the free energy of adsorption of the organic species on the electrode charge was supposed to result from lateral interactions among adsorbed water molecules in the two orientations. The validity of this assumption has been questioned.[257] It has been pointed out that the theory cannot account for the size of the adsorbate molecules on adsorption and that energy change of the ionic double layer on adsorption is ignored.[155,257] Gileadi has shown that the Bockris theory if applied in full is no more at variance with thermodynamics, as Damaskin claims,[257,258] than is the Frumkin isotherm.

The Frumkin isotherm can be expressed in the form[82]

$$B(E)C = f(\theta) \tag{199}$$

and this is thermodynamically equivalent to equation (24). This is a particular form of

$$B(E, \theta)C = f(\theta) \tag{200}$$

where B in equation (199) is not a function of potential. The Bockris isotherm is derived at constant charge and the equivalent relations to equations (199) and (200) are

$$G(q)c = f(\theta) \tag{201a}$$

$$G(q, \theta)c = f(\theta) \tag{201b}$$

Damaskin showed that from equation (201a) the dependence of the free energy of adsorption on q, i.e., $G(q)$ as $f(q)$, was[257]

$$G(q) = G_0 \exp\left\{-A^{-1}\left\{[(K_0 - C')/2K_0 \cdot C']q^2 \right.\right.$$
$$\left.\left. + \int_0^q \Delta\chi\,dq + qE_N^{\text{org}}\right\}\right\} \tag{202}$$

In this equation $\Delta\chi$ is the surface potential associated with the orientation of water dipoles at the interface.

The equivalent relationship on the Bockris theory[239] (Section IV.5) is[107,239]

$$G = G_0 \exp -mR[(\mu\mathscr{E} - REn)/kT] \tag{203}$$

where m is the number of water molecules replaced by each organic molecule. This theory can be generally formulated in the same way as the Frumkin theory,

$$G(\mathscr{E})c = f(\theta) \tag{204}$$

Discussion and Conclusions

Both equation (202) and equation (203) are based on restrictions of the general thermodynamic relationship. While equation (202) requires an independence of B of θ, $\mathscr{E}$ is a function of both q and θ since they are related through the dielectric constant. The model invoked by Bockris is therefore not incorrect but *only* different. Damaskin pointed out the error in their use of equation (202); Bockris assumed a value of six for the dielectric constant for both the organic species and the solvent layer.[259] This is probably not a serious error since the experimental values of the inner layer dielectric constant calculated from the inner layer capacity are not particularly dependent on the solvent molecule (cf. Section III). Because the Frumkin theory involves the base solution capacity, the contributions of the solvent dipoles are considered to be included in the model. Since evidently neither of these theories is at variance with thermodynamics, when they are applied it is essential that the models which are of necessity introduced [equations (199)–(204)] are applicable to the particular experimental system under study.

In a recent study Frumkin has tried to show that the neglect of polarity of the dipole is not justifiable.[260] These dipole components have already been included in the BGM[239] theory and although the treatment of the dipole in the inner layer is different in the two approaches (Frumkin favoring an associative interaction system for the organic species), both approaches seem capable of interpreting the observed experimental data in aliphatic systems. While the BGM model interprets the adsorption at low coverages without the interaction of the organic residues from a basically molecular standpoint, Frumkin uses the thermodynamic approach and an isotherm, the parameters of which are then subjected to theoretical molecular analysis.

One consequence of the Bockris theory[239] is that the molecules with the largest dipole moment per unit surface area should be preferentially adsorbed at large electrode charges. However, while amino acids have large dipole moments they seem to be adsorbed like any other organic species, probably with their dipole *parallel* to the surface.[177,261,262]

Although these criticisms of the Bockris theory are impossible to refute, they must be considered in the context of the approximations on which the theory is based (cf. Section IV.8).[263] Interactions of the organic species with the electrode may be important but they were excluded by BDM and were included by BGM.[107,239] The Bockris theory recognized that the situation in the presence of organic adsorption is complex and that in certain cases the water–electrode interactions will predominate, while in other cases the organic species may predominate, as Butler's model showed qualitatively.[169]

VI. RECENT ADVANCES NOT DIRECTLY APPLICABLE TO METAL–SOLUTION INTERFACES

In the context of this section, two recent publications are of particular interest. In a recent study Levine et al.[269] have attempted to formulate a fundamental theoretical description of the methanol–water system at an AgI interface. The theory is complex and in its present form is only applicable to the mercury–solution interface for $q = 0$. The model involved is similar to that used in the description of the mercury–aqueous solution interface (Section IV.5) applied in the context of statistical mechanics. The monolayer approximation is assumed and the two-orientation model for the dipoles is invoked.

The adsorbed state can be described by its Helmholtz free energy, which can be written as the sum of a number of terms,

$$F = E^e_{\text{int}} + E_v + E^*_{\text{int}} + E' - kT \ln g \tag{205}$$

In this expression E^e_{int} is the work of producing the induced dipoles against the internal forces within the adsorbed molecules ($= E^\alpha_{\text{int}} + E^i_{\text{int}}$, the sum of permanent and induced dipole interaction energies), E_v is the total interaction energy per unit area (including interactions at both interfaces and self-image, dipole, and quadrupole terms, van der Waals interactions, and hydrogen bonding), and E^*_{int} represents the sum of the van der Waals polar interaction energies between nearest neighbors in the monolayer, assumed to be independent of dipole orientation and the van der Waals energy of mixing. E' represents the interaction energy between the adsorbed layer and the mixture due to the dependence of the self-image dipole term on mixture composition, van der Waals energy, and quadrupole terms. The final term g is an entropy term representing the number of configurations available, on the assumed hexagonal lattice site configuration of the monolayer, to methanol molecules in a random mixing model. Levine et al., after substituting suitable expressions for the terms in equation (205), were able to show that strong lateral dipole–dipole interactions resulted in reasonable χ potentials without the need to introduce an inner layer dielectric constant. The residual energies are such that for both dipoles, the positive ends are toward the interface. The adsorption process at equilibrium is visualized using a replacement or exchange reaction scheme (cf. Section IV.5), and the theoretical individual isotherms for methanol and water were obtained. The effect of van der Waals interactions between molecules is included. The general characteristics of the experimental data are well reproduced. The problems of extending this theory to the mercury–solution interface are huge, due to the introduction of the extra potential variable.

An alternative approach which may prove fruitful in constructing a theoretical double-layer model may derive from recent studies on liquid water, using the molecular dynamics technique.[270,271] This approach uses a sample of water consisting of a fixed number of rigid molecules whose internal angles and bond lengths are invariant. The system evolves in time by the laws of classical dynamics under the influence of an effective pair potential that incorporates the principal structural effects of many-body interactions in real water. The advantage of this technique is that it gives both kinetic and equilibrium information on the system in terms of fundamental intermolecular parameters. Moderately good agreement with suitable experimental data was obtained by Rahman and Stillinger.[271] In principle, it is possible to extend this technique to the double-layer interphase by incorporating the correct interactions with the potential and metallic interface, but the problem of the structure of ionic solutions still requires solution before the type of system in which the electrochemist is interested can be described. A major problem with this type of study is the vast amount of computing time which it requires in order to give sensible solutions.

ACKNOWLEDGMENTS

The author wishes to thank Professor G. J. Hills for Section III.4 and Dr. R. Parsons for invaluable discussions of subject matter and interpretation. The use of the facilities of the University of Bristol is gratefully acknowledged.

REFERENCES

[1] G. Gouy, *J. Phys.* 9[4] (1910) 457; *Compt. Rend.* 149 (1910) 654.
[2] D. L. Chapman, *Phil. Mag.* 25[6] (1913) 475.
[3] D. C. Grahame, *J. Am. Chem. Soc.* 63 (1941) 1209; 68 (1946) 301.
[4] P. Delahay, *Double Layer and Electrode Kinetics*, Interscience, New York, 1965.
[5] D. M. Mohilner, in *Electroanalytical Chemistry*, Ed. by A. Bard, Vol. 1, p. 241, Dekker, New York, 1966.
[6] I. Prigogine, P. Mayer, and R. Defay, *J. Chim. Phys.* 50 (1953) 146.
[7] F. P. Buff and N. S. Goel, *J. Chem. Phys.* 51 (1969) 4983.
[8] F. P. Buff and N. S. Goel, *J. Chem. Phys.* 51 (1969) 5363.
[9] K. M. Joshi and R. Parsons, *Electrochim. Acta* 4 (1961) 129.
[10] R. Parsons and S. Trasatti, *Trans. Faraday Soc.* 65 (1969) 3314.
[11] R. Parsons, in *Modern Aspects of Electrochemistry*, Ed. by J. Bockris, Vol. 1, Butterworths, London, 1954.
[12] F. C. Anson, R. F. Martin, and C. Yornitzky, *J. Phys. Chem.* 73 (1969) 1835.
[13] J. C. Henniker, *Rev. Mod. Phys.* 21 (1949) 322.
[14] B. V. Derjaguin, *Disc. Faraday Soc.* 42 (1966) 109.
[15] J. Timmermans and H. Bodson, *Compt. Rend.* 204 (1937) 1804.
[16] W. Drost-Hansen, *Ind. Eng. Chem.* 57[3] (1965) 38; 57[4] (1965) 18.
[17a] G. Peschel and K. H. Aldfinger, *Naturwiss.* 54 (1967) 614.
[17b] B. A. Pethica, *Expt. Cell Res. Suppl.* 8 (1961) 123.
[18] J. Lyklema, *Disc. Faraday Soc.* 42 (1966) 81.
[19] T. P. Melia and W. P. Moffitt, *J. Coll. Sci.* 19 (1964) 433.
[20] L. S. Palmer, A. Cunliffe, and J. M. Hough, *Nature* 170 (1952) 796.
[21] N. V. Afanas'ev and N. S. Nestik, in *Research in Surface Forces*, Ed. by N. V. Derjaguin, p. 177, Consultants Bureau, New York, 1966.
[22] D. Michel, *Z. Naturforsch.* 22A (1967) 1751.
[22] S. Reball and H. Winkler, *Z. Naturforsch.* 19A (1964) 861.
[24] J. H. Pickett and L. B. Rogers, *Anal. Chem.* 39 (1967) 1892.
[25] C. T. Deeds and H. Van Olphen, in *Advances in Chemistry, No.* 33, 1961, p. 332.
[26] B. M. Anderson and P. F. Low, *Proc. Soil Sci. Am.* 22 (1958) 99.
[27] A. N. Frumkin, Z. A. Iofa, and R. A. Gerovicz, *Russ. J. Phys. Chem.* 30 (1956) 1455.
[28] N. H. Fletcher, *Phil. Mag.* 7 (1962) 255.
[29] H. F. Stillinger and A. Ben-Naim, *J. Chem. Phys.* 47 (1967) 4431.
[30] W. E. Claussen, *Science* 156 (1967) 1226.
[31] J. W. McBain, R. C. Bacon, and H. D. Bruce, *J. Chem. Phys.* 7 (1939) 818.
[32] I. Langmuir, *J. Am. Chem. Soc.* 39 (1917) 1897.
[33] W. D. Harkens and E. C. Gilbert, *J. Am. Chem. Soc.* 48 (1926) 604.
[34] C. Wagner, *Physik Z.* 25 (1924) 474.
[35] L. Onsager and N. N. T. Samaras, *J. Chem. Phys.* 2 (1934) 528.
[36] A. N. Frumkin, *Z. Physik. Chem.* 109 (1924) 34.
[37a] J. E. B. Randles, *Disc. Faraday Soc.* 24 (1957) 194.
[37b] J. E. B. Randles, *Advances in Electrochemistry and Electrochemical Engineering*, Ed. by P. Delahay, 1966, Vol. 3, p. 1.
[38] R. W. Gurney, *Ionic Processes in Solution*, McGraw-Hill, London, 1953.
[39] N. A. Hampson, *Chem. Soc. Specialist Reports* 2 (1972).
[40a] J. Clavilier and N. Van Huong, *J. Electroanal. Chem.* 41 (1973) 193.
[40b] A. Hamelin and J. P. Bellier, *J. Electroanal. Chem.* 41 (1973) 179.
[41] D. Leikis and E. S. Savastianov, *J. Electrochem. Soc.* 113 (1966) 1314.
[42] A. N. Frumkin and N. B. Grigor'ev, *Elektrokhimiya* 4 (1968) 533.

References 363

[43] A. M. Morozov, I. A. Bagotskaya, and E. A. Preis, *Elektrokhimiya* **5** (1969) 40.
[44] N. B. Grigorev, S. A. Fateev, and I. A. Bagotskaya, *Elektrokhimiya* **7** (1971) 1852.
[45] A. N. Frumkin, N. B. Grigor'ev, and I. A. Bagotskaya, *Electrokhimiya* **2** (1966) 329.
[46] Z. A. Oifa, E. I. Lyakhovetskaya, and K. Shaviov, *Dokl. Akad. Nauk SSSR* **84** (1952) 543.
[47] A. N. Frumkin, N. S. Polyanovskaya, and N. B. Grigorev, *Dokl. Akad. Nauk SSSR* **157** (1964) 455.
[48] A. N. Frumkin, N. B. Grigor'ev, and I. A. Bagotskaya, *Dokl. Akad. Nauk SSSR* **157** (1964) 957
[49] A. N. Frumkin, N. S. Polyanovskaya, N. B. Grigor'ev, and I. A. Bagotskaya, *Electrochim. Acta* **10** (1965) 793.
[50] V. A. Kiry'anov, V. S. Krylov, and N. B. Grigor'ev, *Elektrokhimiya* **4** (1968) 408.
[51] A. N. Frumkin, *Electrochimica Acta* **9** (1964) 465.
[52] A. N. Frumkin and A. Gorodetzkaya, *Z. Phys. Chem.* **136** (1928) 215.
[53] A. N. Frumkin, *Svensk. Kem. Tidskr.* **77** (1965) 300.
[54] S. D. Argade and E. Gileadi, in *Electrosorption*, Ed. by E. Gileadi, Plenum, New York, 1967.
[55] R. Vasenin, *Zh. Fiz. Khim.* **27** (1953) 878; **28** (1954) 1672.
[56] S. Trasatti, *J. Electroanal. Chem.* **33** (1971) 351.
[57] E. V. Osipova, N. A. Shurmovskaya, and R. KH. Burshtein, *Electrokhimiya* **5** (1969) 1139.
[58] K. Müller, *J. Res. Inst. Catal., Hokkaido Univ.*, **14** (1966) 224.
[59] A. N. Frumkin, *Ereb. Exact. Naturwiss.* **7** (1928) 235.
[60] J. E. B. Randles and K. W. Whiteley, *Trans. Faraday Soc.* **52** (1956) 1509.
[61] A. N. Frumkin, Z. A. Iofa, and M. A. Gerovich, *Zh. Fiz. Khim.* **30** (1956) 1455.
[62] R. Parsons and F. G. R. Zobel, *J. Electroanal. Chem.* **9** (1965) 333.
[63] S. Trasatti, *J. Electroanal. Chem.* **28** (1970) 257.
[64] N. B. Grigorev and O. N. Machavariani, *Elektrokhimiva* **6** (1970) 89.
[65] N. B. Grigorev and O. N. Machavariani, *Elektrokhimiya* **5** (1969) 87.
[66] N. B. Grigorev and I. A. Bagotskaya, *Elektrokhimiya* **2** (1966) 1449.
[67] L. Pauling, *The Nature of the Chemical Bond*, Cornell University Press, Ithaca, New York, 1960.
[68] A. N. Frumkin, *Electrochimica Acta* **2** (1960) 351.
[69] B. Case and R. Parsons, *Trans. Faraday Soc.* **63** (1967) 1224.
[70] J. E. B. Randles and D. J. Schiffrin, *J. Electroanal. Chem.* **10** (1965) 480.
[71] J. R. Macdonald and C. A. Barlow, *J. Chem. Phys.* **39** (1964) 412.
[72] J. R. Macdonald and C. A. Barlow, *J. Chem. Phys.* **44** (1966) 202.
[73] E. P. Gyftopoulos and J. D. Levine, *J. Appl. Phys.* **33** (1962) 67.
[74] D. C. Grahame, *Chem. Revs.* **41** (1947) 441.
[75] R. Parsons and M. A. V. Devanathan, *Trans. Faraday Soc.* **49** (1953) 404.
[76] J. W. Gibbs, *Collected Works*, Yale University Press, 1948, pp. 229–31.
[77] A. N. Frumkin, *Phil. Mag.* **40** (1920) 363.
[78] E. Dutkiewicz and R. Parsons, *J. Electroanal. Chem.* **11** (1966) 100.
[79] H. D. Hurwitz, *J. Electroanal. Chem.* **10** (1969) 35.
[80] R. Parsons, *Trans. Faraday Soc.* **55** (1959) 999.
[81] R. Parsons, in *Proc. 2nd Int. Congress of Surface Activity*, Butterworths, London, 1957, Vol. 3, p. 38.
[82] B. B. Damaskin, O. A. Petrii, and V. V. Batrakov, *Adsorption of Organic Compounds on Electrodes*, Plenum, New York, 1971.
[83] M. W. Breiter and P. Delahay, *J. Am. Chem. Soc.* **81** (1959) 2938.
[84] R. Parsons, *J. Electroanal. Chem.* **8** (1964) 93.

[85] A. N. Frumkin, R. V. Ivanova and B. B. Damaskin, *Dokl. Akad. Nauk SSSR* **157** (1964) 1202.
[86] A. N. Frumkin, B. B. Damaskin, and A. A. Survila, *J. Electroanal. Chem.* **16** (1968) 493.
[87] B. B. Damaskin, *Electrokhimiya* **6** (1970) 1135.
[88] A. N. Frumkin, B. B. Damaskin, and A. Chizhov, *J. Electroanal. Chem.* **28** (1970) 93.
[89] R. S. Hanson, D. J. Kelsh, and D. H. Grantham, *J. Phys. Chem.* **67** (1963) 2316.
[90] D. C. Grahame and R. Parsons, *J. Am. Chem. Soc.* **83** (1961) 1291.
[91] J. A. Harrison, J. E. B. Randles, and D. J. Schiffrin, *J. Electroanal. Chem.* **25** (1970) 197.
[92] G. Gouy, *Ann. Chim. Phys.* **29** (1903) 159.
[93] R. Parsons, *Can. J. Chem.* **37** (1959) 308.
[94] R. Parsons, in *IUPAC Resource Book, Colloid and Surface Chemistry*, Ed. by H. Van Olphen and K. Mysels.
[95] G. J. Hills and R. M. Reeves, *J. Electroanal. Chem.* **38** (1972) 1.
[96] R. Parsons and F. G. R. Zobel, *J. Electroanal. Chem.* **9** (1965) 333.
[97] B. B. Damaskin, A. N. Frumkin, V. F. Ivanov, N. I. Melekhova, and V. F. Khonina, *Elektrokhimiya* **4** (1968) 1336.
[98] R. I. Kaganovich, B. B. Damaskin, and I. M. Ganzhina, *Elektrokhimiya* **4** (1968) 867.
[99] G. J. Hills and R. Payne, *Trans. Faraday Soc.* **61** (1965) 316.
[100] G. J. Hills and R. Payne, *Trans. Faraday Soc.* **61** (1965) 326.
[101] S. Hsieh, Thesis, Southampton 1969.
[102] G. J. Hills and S. Hsieh, *Chem. Ing. Technik.* **44** (1972) 216.
[103] R. Zana and E. B. Yeager, *J. Phys. Chem.* **71** (1967) 521.
[104] B. E. Conway and L. H. Laliberté, in *Hydrogen Bonded Solvent Systems*, Ed. by A. K. Covington, Taylor and Francis, London, 1968; B. E. Conway, R. E. Verrall, and J. E. Desnoyers, *Trans. Faraday Soc.* **62** (1966) 2738; *J. Phys. Chem.* **75** (1971) 3031.
[105] D. J. Schiffrin, Thesis, Birmingham, 1962.
[106] J. A. Harrison, Thesis, Birmingham, 1960.
[107a] J. O'M. Bockris, M. A. V. Devanathan, and K. Müller, *Proc. Roy. Soc.* **A274** (1963) 55.
[107b] B. E. Conway, D. J. MacKinnon and B. V. Tilak, *Trans. Faraday Soc.* **66** (1970) 1203.
[108] A. Bewick and A. M. Tuxford, private communication; S. Gottesfeld and B. E. Conway, Extended Abstracts, No. 104, Electrochemical Society Meeting, Houston, Texas, May 1972.
[109] H. D. Hurwitz, private communication.
[110] B. B. Damaskin, N. V. Nikolaeva-Fedorovich, and A. N. Frumkin, *Dokl. Akad. Nauk SSSR*, **121** (1958) 129.
[111] R. Payne, *J. Electroanal. Chem.* **7** (1964) 343.
[112] R. D. Armstrong, W. P. Race, and H. R. Thirsk, *J. Electroanal. Chem.* **14** (1967) 143.
[113] D. C. Grahame, *J. Electrochem. Soc.* **98** (1951) 343.
[114] R. Payne, Thesis, London, 1962.
[115] R. Payne, in *Advances in Electrochemistry and Electrochemical Engineering*, Ed. by P. Delahay, Interscience, New York, Vol. 7, p. 1.
[116] A. Stockton and R. Parsons, *J. Electroanal. Chem.* **25** (1970) App 10.
[117] L. Gierst, E. Nicolas, and L. Tytgart-Vandenberghen, *Croat. Chem. Acta* **42** (1970) 117.

References

[118] S. Minc and J. Jastrzebska, *J. Electrochem. Soc.* **107** (1960) 135.
[119] D. C. Grahame, *J. Am. Chem. Soc.* **79** (1957) 2093.
[120] D. C. Grahame, *J. Am. Chem. Soc.* **76** (1954) 4815.
[121] G. J. Hills and R. M. Reeves, *J. Electroanal. Chem.* **31** (1971) 269.
[122] R. Payne, *J. Electrochem. Soc.* **113** (1966) 999.
[123] N. F. Mott and R. J. Watts-Tobin, *Electrochim. Acta* **4** (1961) 79.
[124] D. C. Grahame, ONR Report No. 14, 1954.
[125] A. W. M. Verkroost, M. Sluyters-Rehbach, and J. H. Sluyters, *J. Electroanal. Chem.* **24** (1970) 1.
[126] D. J. Schiffrin, *Trans. Faraday Soc.* **67** (1971) 3318.
[127] G. J. Hills and R. M. Reeves, *J. Electroanal. Chem.* **42** (1973) 355.
[128] L. M. Baugh and R. Parsons, *J. Electroanal. Chem.* **41** (1973) 311.
[129] D. P. Aleksandrova and D. J. Leikis, *Elektrokhimiya* **1** (1965) 241.
[130] V. S. Ivanov, B. B. Damaskin, A. N. Frumkin, A. A. Ivaskshenko, and N. I. Peshkova, *Elektrokhimiya* **1** (1965) 279.
[131] V.Sh. Palanker, A. M. Skundin, and V. S. Bagotskii, *Elektrokhimiya* **2** (1966) 640.
[132] B. G. Dekker, Thesis, Utrecht, 1970, p. 69.
[133] D. C. Grahame, *J. Am. Chem. Soc.* **80** (1958) 4201.
[134] D. C. Grahame, *Z. Electrochem.* **62** (1958) 264.
[135] S. Minc, J. Jastrzebska, and M. Grezostowska, *J. Electrochem. Soc.* **108** (1961) 1160.
[136] E. Dutkiewicz and R. Parsons, *J. Electroanal. Chem.* **1** (1966) 196.
[137] B. B. Damaskin, R. V. Ivanova, and A. A. Survila, *Elektrokhimiya* **1** (1965) 767.
[138] B. B. Damaskin and Y. M. Povarov, *Dokl. Akad. Nauk SSSR* **160** (1961) 394.
[139] V. D. Bezuglyi and L. A. Korshikov, *Elektrokhimiya* **3** (1967) 390.
[140] V. D. Bezuglyi and L. A. Korshikov, *Elektrokhimiya* **1** (1965) 1422.
[141] R. Payne, *J. Am. Chem. Soc.* **89** (1968) 489.
[142] J. Lawrence and R. Parsons, *Trans. Faraday Soc.* **64** (1968) 751.
[143] R. Payne, *J. Phys. Chem.* **71** (1967) 1548.
[144] T. N. Anderson, J. L. Anderson, and H. Eyring, *J. Phys. Chem.* **73** (1969) 3562.
[145] D. J. Barclay, *J. Electroanal. Chem.* **28** (1970) 443.
[146] D. J. Barclay, *J. Electroanal. Chem.* **19** (1968) 318.
[147] G. Klopman, *J. Am. Chem. Soc.* **90** (1968) 223.
[148] D. C. Grahame, *Z. Elektrochem.* **59** (1955) 740.
[149] J. D. Garnish and R. Parsons, *Trans. Faraday Soc.* **63** (1967) 1754.
[150] I. M. Ganshina, B. B. Damaskin, and R. V. Ivanova, *Elektrokhimiya* **6** (1970) 709.
[151] V. D. Bezuglyi and L. A. Korshikov, *Elektrokhimiya* **1** (1965) 1422; **3** (1967) 390.
[152] I. M. Ganshina, B. B. Damaskin, R. I. Kaganovich, and R. V. Ivanova, *Elektrokhimiya* **7** (1970) 363.
[153] G. J. Hills and R. M. Reeves, *J. Electroanal. Chem.* **41** (1973) 213.
[154] R. W. Gurney, *Ionic Processes in Solution*, McGraw-Hill, London, 1953, Chapter 16.
[155] R. Parsons, *Rev. Pure Appl. Chem.* **18** (1968) 91.
[156] J. N. Andersen and J. O'M. Bockris, *Electrochim. Acta* **8** (1965) 347.
[157] C. Kemball, *Proc. Roy. Soc.* **A190** (1947) 177.
[158] W. Anderson and R. Parsons, in *Proc. 2nd Int. Congress of Surface Activity*, Butterworths, London, 1957, Vol. 3, p. 45.
[159] D. D. Bodé, *J. Phys. Chem.* **76** (1972) 2915.
[160] D. D. Bodé, in *Adv. Chem. Phys.* No. 21, 1971, p. 362.
[161] B. B. Damaskin, *Elektrokhimiya* **5** (1967) 771.
[162] K. J. Vetter and J. W. Schultze, *Ber. Bunsenges. Physik. Chem.* **76** (1972) 920.
[163] K. J. Vetter and J. W. Schultze, *Ber. Bunsenges. Physik. Chem.* **76** (1972) 927.

[164] J. W. Schultze and K. J. Vetter, *J. Electroanal.* **44** (1973) 63.
[165] A. N. Frumkin and B. B. Damaskin, *Modern Aspects of Electrochemistry*, Ed. by J. O'M. Brockris, Butterworths, London, 1964, Vol. 3, p. 149.
[166] B. B. Damaskin, O. A. Petrii, and V. V. Batrakov, *Adsorption of Organic Compounds at Electrodes*, Plenum, New York, 1971.
[167] G. Gouy, *Ann. Chim. Phys.* **29** (1903) 145.
[168] A. N. Frumkin and J. Williams, *Proc. Nat. Acad. Sci. Wash.* **15** (1929) 400.
[169] J. A. V. Butler, *Proc. Roy. Soc.* **A122** (1929) 399.
[170] G. Gouy, *Ann. Chim. Phys.* **8** (1906) 291; **9** (1906) 75.
[171] A. N. Frumkin, *Colloid Symp.* **A7** (1930) 89.
[172a] B. E. Conway, J. O'M. Bockris, and B. Lovrecek, in *Proc. 6th CITCE*, 1955, p. 207.
[172b] M. A. Gerovich and N. S. Polyanovskaya, *Nauchn. Dokl. Vysshei Shkoly. Khim. i Khim., Tekhnol.* **4** (1958) 651.
[172c] E. Blomgren and J. O'M. Bockris, *J. Phys. Chem.* **63** (1959) 1475.
[173a] B. E. Conway and R. G. Barradas, *Electrochim. Acta* **5** (1961) 319, 349.
[173b] A. N. Frumkin, R. I. Kaganovich, and S. Popova, *Dokl. Akad. Nauk SSSR* **141** (1961) 670.
[174] R. I. Kaganovich, B. B. Damaskin and M. A. Suranova, *Elektrokhimiya* **7** (1971) 1158.
[175] N. A. Borovaya and B. B. Damaskin, *Elektrokhimiya* **7** (1971) 571.
[176] R. I. Kaganovich, V. M. Gerovich, and O. Yu. Gusakova, *Electrokhimiya* **3** (1967) 946.
[177] L. M. Baugh and R. Parsons, *Croat Chem. Acta* **45** (1973) 127.
[178] J. M. Parry and R. Parsons, *Trans. Faraday Soc.* **59** (1963) 241.
[179] A. Murtazaev and A. Gorodetskaya, *Acta Physicochem. URSS* **4** (1936) 75.
[180] N. B. Grigorev and I. A. Bagotskaya, *Elektrokhimiya* **2** (1966) 1449.
[181] G. Gouy, *Ann. Phys. Paris* **7** (1917) 163.
[182] O. A. Esin and B. F. Markov, *Acta Physicochem. URSS* **10** (1939) 353.
[183] A. N. Frumkin, *Uspekhi Khim.* **4** (1935) 987.
[184] O. A. Esin and V. M. Shikov, *Zh. Phys. Chem. URSS*, **17** (1943) 236.
[185] B. V. Ershler, *Zh. Phys. Chem.* **20** (1946) 679.
[186] R. Payne, *J. Chem. Phys.* **42** (1965) 3371.
[187] V. G. Levich, V. A. Kir'yanov, and V. S. Krylov, *Dokl. Akad. Nauk SSSR* **135** (1960) 1425.
[188] J. R. Macdonald and C. A. Barlow, in *Proc. 1st Aust. Conf. Electrochem.*, Pergamon, 1963, p. 199.
[189] J. R. MacDonald, *J. Chem. Phys.* **22** (1954) 1857.
[190] D. C. Grahame, *J. Chem. Phys.* **18** (1950) 903.
[191] F. Malsh, *Physik. Z.* **29** (1928) 770.
[192] R. Page and N. I. Adams, *Principles of Electricity*, Van Nostrand, New York, 1949, p. 50.
[193] E. L. Mackor, *Rec. Trav. Chim.* **70** (1951) 763.
[194] J. O'M. Bockris and E. C. Potter, *J. Chem. Phys.* **20** (1952) 614.
[195] D. C. Grahame, *J. Am. Chem. Soc.* **79** (1957) 2093.
[196] D. C. Grahame, *J. Chem. Phys.* **25** (1956) 364.
[197a] H. P. Dhar, B. E. Conway, and K. M. Joshi, *Electrochim. Acta* **18** (1973) 789.
[197b] B. B. Damaskin, *Elektrokhimiya* **1** (1965) 1258.
[198] M. A. V. Devanathan, *Trans. Faraday Soc.* **50** (1954) 373.
[199] R. Payne, *J. Phys. Chem.* **70** (1966) 204.
[200] A. N. Frumkin, B. B. Damaskin, and Y. A. Chizmadzhev, *Elektrokhimiya* **2** (1966) 875.

References

[201] M. A. V. Devanathan and S. G. Canagaratna, *Electrochim. Acta* **8** (1963) 77.
[202] R. J. Watts-Tobin, *Phil. Mag.* **6** (1961) 133.
[203] L. Pauling, *Proc. Roy. Soc.* **A114** (1927) 181.
[204] C. Kemball, *Proc. Roy. Soc.* **A90** (1947) 117.
[205] T. Smith, *Adv. Colloid Interf. Sci.* **3** (1972) 161.
[206] J. T. Law, Ph.D. Thesis, Royal College of Science, London, 1951.
[207] A. D. Buckingham, *Disc. Faraday Soc.* **24** (1957) 151.
[208] A. Duncan and J. A. Pople, *Trans. Faraday Soc.* **49** (1953) 217.
[209] J. Verhoeven and A. Dymanus, *J. Chem. Phys.* **52** (1970) 3224.
[210] N. F. Mott and R. J. Watts-Tobin, *Electrochim. Acta* **4** (1961) 79.
[211] N. F. Mott, R. Parsons, and R. J. Watts-Tobin, *Phil. Mag.* **7** (1962) 483.
[212] G. H. Nancollas, D. S. Reid, and C. A. Vincent, *J. Phys. Chem.* **70** (1966) 3300.
[213] B. B. Damaskin, *Elektrokhimiya* **2** (1966) 828.
[214] R. Lerkkh and B. B. Damaskin, *Zh. Fiz. Khim.* **38** (1964) 1154; **39** (1965) 211.
[215] J. R. MacDonald and C. A. Barlow, *J. Chem. Phys.* **36** (1962) 3062.
[216] J. Topping, *Proc. Roy. Soc.* **A114** (1927) 67.
[217] S. Levine, G. M. Bell, and A. L. Smith, *J. Phys. Chem.* **73** (1969) 3534.
[218] R. H. Fowler and E. A. Guggenheim, *Statistical Thermodynamics*, Cambridge, 1949.
[219] J. R. Macdonald and C. A. Barlow, *Surface Sci.* **4** (1966) 381.
[220] M. Polanyi, *Trans. Faraday Soc.* **28** (1932) 316.
[221] R. A. Perotti and G. D. Halsey, *J. Phys. Chem.* **63** (1959) 680.
[222] J. O'M. Bockris, M. Green, and D. A. Swinkels, *J. Electrochem. Soc.* **111** (1966) 743.
[223] J. R. MacDonald and C. A. Barlow, *Advances in Electrochemistry and Electrochemical Engineering*, Ed. by P. Delahay, Interscience, New York, 1967, Vol. 6, p. 1.
[224] S. Levine, J. Mingins, and G. M. Bell, *J. Electroanal. Chem.* **13** (1967) 280.
[225] G. M. Bell, J. Mingins, and S. Levine, *Trans. Faraday Soc.* **62** (1966) 949.
[226] J. R. MacDonald and C. A. Barlow, *Can. J. Chem.* **43** (1965) 2985.
[227] S. Levine, J. Mingins, and G. M. Bell, *Can. J. Cam.* **43** (1965) 2834.
[228] F. P. Buff and F. H. Stillinger, *J. Chem. Phys.* **39** (1963) 1911.
[229] H. D. Hurwitz, *J. Chem. Phys.* **48** (1968) 1541.
[230a] S. Levine, K. Robinson, G. M. Bell, and J. Mingins, *J. Electroanal. Chem.* **38** (1972) 253.
[230b] K. Robinson and S. Levine, *J. Electroanal. Chem.* **47** (1973) 395.
[231] S. Levine and K. Robinson, *J. Electroanal. Chem.* **41** (1973) 159.
[232] R. Parsons, *Bull. Nat. Inst. Sci. India* **29** (1961) 53.
[233] R. Parry and R. Parsons, *Trans. Faraday Soc.* **59** (1963) 241.
[234] I. Langmuir, *J. Am. Chem. Soc.* **54** (1932) 1252, 2798.
[235] E. Schwartz, B. B. Damaskin, and A. N. Frumkin, *Zh. Fiz. Khim.* **36** (1962) 2419.
[236] M. A. V. Devanathan and B. V. K. S. R. A. Tilak, *Proc. Roy. Soc.* **A290** (1966) 527.
[237] G. J. Hills, *J. Phys. Chem.* **73** (1969) 3591.
[238] A. K. N. Reddy and S. Sathyanarayana, XV Int. Conf. on Chem., Brussels, 1972.
[239] J. O'M. Bockris, E. Gileadi, and K. Muller, *Electrochim. Acta* **12** (1967) 1301.
[240] I. Langmuir, *J. Am. Chem. Soc.* **39** (1917) 1848.
[241] V. A. Kir'yanov, V. S. Krylov, and B. B. Damaskin, *Elektrokhimiya* **6** (1970) 533.
[242] V. A. Kir'yanov, V. S. Krylov, B. B. Damaskin, and A. V. Chizhov, *Elektrokhimiya* **6** (1970) 1020.
[243] V. A. Kir'yanov, V. S. Krylov, B. B. Damaskin, and A. V. Chizhov, *Elektrokhimiya* **6** (1970) 1518.
[244] P. J. Flory, *J. Chem. Phys.* **10** (1942) 51.

[245] M. L. Huggins, *J. Phys. Chem.* **46** (1942) 151; *Ann. New York Acad. Sci.* **43** (1942) 1.
[246] V. A. Kir'yanov and V. S. Krylov, *Elektrokhimiya* **6** (1970) 412.
[247] W. R. Fawcett and R. O. Loutfy, *J. Electroanal. Chem.* **39** (1972) 185.
[248] H. Wroblowa, Z. Kovac, and J. O'M. Bockris, *Trans. Faraday Soc.* **61** (1965) 1523.
[249] H. Wroblowa and K. Müller, *J. Phys. Chem.* **73** (1969) 3528.
[250] N. Bonciocat, J. O'M. Bockris, and R. K. Sen, Submitted to *J. Electroanal. Chem.*
[251] D. J. Schiffrin, *J. Electroanal. Chem.* **23** (1969) 168.
[252] J. Lawrence, R. Parsons, and R. Payne, *J. Electroanal. Chem.* **16** (1968) 193.
[253] S. Levine, *J. Colloid Interf. Sci.* **37** (1971) 619.
[254] L. M. Baugh and R. Parsons, *J. Electroanal. Chem.* **40** (1972) 407.
[255] S. Levine, G. M. Bell, and D. Calvert, *Can. J. Chem.* **40** (1962) 518.
[256] E. Helfand, H. L. Frisch, and J. J. Lebowitz, *J. Chem. Phys.* **34** (1961) 1037.
[257] B. B. Damaskin, *J. Electroanal. Chem.* **23** (1969) 431.
[258] E. Gileadi, *J. Electroanal. Chem.* **30** (1971) 123.
[259] B. B. Damaskin, *J. Electroanal. Chem.* **30** (1971) 129.
[260] B. B. Damaskin and A. N. Frumkin, *J. Electroanal. Chem.* **34** (1972) 191.
[261] B. B. Damaskin, S. L. Dyatkina, and N. A. Borovaya, *Elektrokhimiya* **7** (1971) 712.
[262] N. A. Borovaya and B. B. Damaskin, *Elektrokhimiya* **7** (1971) 571.
[263] J. O'M. Bockris, *J. Electroanal. Chem.* **34** (1972) 201.
[264] B. E. Conway and L. G. M. Gordon, *J. Phys. Chem.* **73** (1969) 3609
[265] B. E. Conway and H. P. Dhar, *Croatica Chem. Acta.* **45** (1973) 109
[266] B. E. Conway, H. P. Dhar and S. Gottesfeld, *J. Coll. Interf. Sci.* **43** (1973) 303.
[267] B. E. Conway and D. J. Mackinnon, *Trans. Faraday Soc.* (in press).
[268] H. P. Dhar, B. E. Conway, and K. M. Joshi, *Electrochim. Acta* **18** (1973) 189.
[269] S. Levine, A. L. Smith, and E. Matijevic, *J. Coll. and Interf. Sci.* **31** (1969) 409.
[270] B. J. Alder and T. E. Wainwright, *J. Chem. Phys.* **33** (1960) 1439.
[271] A. Rahman and F. H. Stillinger, *J. Chem. Phys.* **55** (1971) 3336.

5

Electrocatalysis

A. J. Appleby*

Laboratoire d'Electrolyse C.N.R.S., Bellevue, France

I. INTRODUCTION

This article will attempt to discuss those factors which influence the overall rates of electrochemical reactions, a subject that has been treated several times elsewhere.[1-4,83] It is hoped that the question will be approached from somewhat different points of view from those expressed previously. In particular, concentration will be on the fundamental aspects of the rate equation for electron transfer as a function of the substrate, together with their experimental consequences, which will involve a discussion and review of the mechanism of the elementary act of electron transfer. Particular emphasis will be placed on the electrocatalytic case, i.e., that involving adsorbed reaction intermediates. This has been the subject of much dispute in the recent literature.[5] The reasoning will be applied to a discussion of those reactions that have been most studied with respect to the dependence of rate on the nature of the substrate: namely hydrogen evolution, oxygen reduction and evolution, and, to a lesser extent, hydrocarbon oxidation. Emphasis will be particularly on the oxygen electrode reaction, which has been reviewed elsewhere.[83,163,189]

II. ELECTRON TRANSFER AT THE METAL–SOLUTION INTERFACE

1. General

Electron transfer between a metal electrode and species in solution (or a species crossing the double layer, in the case where either

*Present address: Laboratoires de Marcoussis (C.G.E.), 91-Marcoussis, France.

reactant or product is adsorbed) is a radiationless process. In such cases the transfer condition is that the electron energy level in the metal must be equal to the energy level in the discharged species. During the electron transfer the electron crosses the gap between the electrode and solution levels (or vice versa) by a tunnel-type mechanism.

The literature contains two models that describe this process. In the first model ("thermal theory"), which dates from a paper by Gurney (1931),[6] the electrolyte is considered from the point of view of a condensed gas: The important interactions that are activated are considered to be those involving closest neighbors (i.e., strong forces). In the second model, which dates from papers by Libby (1952)[7] and Weiss (1954),[8] the electrolyte is considered from the viewpoint of a pseudocrystalline solid, and the activated interactions involve changes of weak, long-range electrostatic forces ("electrostatic" or "polaron" theory).

2. Thermal Theory

(i) Quantum Mechanical (Activated Electron Transfer) Theory

The thermal activation model is the older of the two concepts, and was from the outset applied to cases involving an atom transfer reaction (as is implied in all electrocatalysis). It arose from a quantum mechanical consideration of radiationless electron transfer by Gurney (1931)[6] as applied to a model for the hydrogen evolution reaction on mercury, for which the rate-determining step was considered to be

$$H_{aq}^+ + e^- \to H_{aq} \tag{1}$$

The fundamental hypotheses[6] were the following:

1. The product of the rate-determining step (H) is not adsorbed on the electrode.
2. The H^+ ion in solution is excited within its solvent cage to an energy level sufficient to receive an electron from the metal. A one-dimensional model of the potential energy–configuration diagram may be represented by a function of the Morse type.
3. The electron transfer takes place by Gamow tunneling (although the transfer distance was not considered in detail).

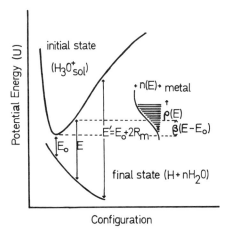

Figure 1. Model of proton discharge to give unadsorbed H (after Gurney[6]).

4. Because the tunnel transition time is very short compared with molecular movements, the Born–Oppenheimer approximation may be applied to all heavy particles.

A schematic view of the potential–distance curves is shown in Fig. 1 for the system H_2O–H^+ (initial state), and H plus H_2O (repulsive) (final state). For the population of H_2O–H^+ molecules, $N(E)$, having energies in the range E and $E + dE$, where E is the energy difference between the reactants and the products in the same nuclear configuration (Born–Oppenheimer principle, see Fig. 1) we can write for the rate of neutralization an expression of the form

$$dV = v\kappa(E, x)\rho(E)N(E)n(E)\,dE \qquad (2)$$

where v is a frequency factor (collision factor), $\kappa(E, x)$ is the electron transmission coefficient (which varies with the transfer distance x), $\rho(E)$ is the density of state of electrons at the electrode at energy E, and $n(E)$ is the Fermi–Dirac distribution of electronic levels in the electrode.

For a radiationless electron transfer each E value must equal $\psi + \varphi e$, where ψ is the electron energy level, e is the electronic charge, and φ is the Galvani potential difference between the metal level and the reaction zone. Supposing the potential energy–

configuration curves can be considered to be more or less straight lines,* we can write (see Fig. 1)

$$N(E) = N_0 \exp(\beta E_0/kT)\exp(-\beta E/kT) \quad (3)$$

where E_0 is the energy difference between the reactants in their ground state and the products in the same nuclear configuration, and β is some fraction between zero and unity, which is a weak function of E. N_0 is the number of H^+-H_2O molecules in the ground state in the reaction zone.

Integration of (1), using (2), gives the total current $I_{x(av)}$ at the interface, making the simplifying assumption that all the neutralization occurs at some mean distance x [$x(av)$]. Thus

$$I_{x(av)} = eN_0 v \exp(\beta E_0/kT) \int_0^\infty \kappa(E)\rho(E)n(E)\exp(-\beta E/kT)\,dE \quad (4)$$

$\kappa(E)$ and $\rho(E)$ are weak functions of E compared with the $\exp(-\beta E/kT)$ term: $\kappa(E)$ varies with $\exp -E^{1/2}$ for electron tunnel transitions, and $\rho(E)$ varies with $E^{1/2}$, assuming a simple electron gas model. To a first approximation, therefore, these terms need not be included in the integral. Bearing in mind that the metal levels are inverted with respect to those in the solution, we obtain:

$$I_{\text{total}} = neN_0\kappa^*\rho^* v \left(\exp\frac{\beta E_0}{kT}\right)\int_0^\infty \frac{\exp(-\beta E/kT)}{1 + \exp[(E_F - E)/kT]}\,dE \quad (5)$$

where E_F is the value of E corresponding to the Fermi level of the electrode, and $n(E)$ has been replaced by the Fermi–Dirac distribution. κ^*, ρ^* are mean values within the range considered.

A simple, and sufficiently accurate, integration can be performed by replacing the Fermi–Dirac function by a Boltzmann function (an approximation which is valid between E_0 and an energy very close to the Fermi level energy E_F), then changing the upper limit to E_F instead of infinity. It should be noted that differentiation of the integrand indicates a current maximum close to the Fermi level of the electrode (exactly at the Fermi level if $\beta = \frac{1}{2}$). A more complete calculation shows that the majority of the current

---

*See Gurney.[6] This idea is supported by experimental evidence for the hydrogen evolution reaction.[9]

comes from levels within $\pm kT$ of the Fermi level. The final expression obtained for the total current is

$$I_{\text{cath}} = eN_0\kappa^*\rho^*v[kT/(1-\beta)]\exp[\beta(E_0 - E_F)/kT] \qquad (6)$$

A precisely similar calculation for the current in the anodic direction gives the corresponding expression:

$$I_{\text{anod}} = eN_0'\kappa^*\rho^*v(kT/\beta)\exp[-(1-\beta)(E' - E_F)/kT] \qquad (7)$$

where E' is the value of E (assuming a constant β value) that corresponds to the minimum energy state of the reaction products $(H + H_2O)$, and N_0' is the number of H atoms in the ground state in the reaction zone.

Writing $E' = E_0 + 2R_m$, where R_m is the mean energy difference between the lowest energy state of the products and their state in the same nuclear configuration as the ground state of the reactants, and the ground state of the reactants and their state in the same configuration as the lowest energy state of the products (see Fig. 1), and equating the rates at equilibrium, we obtain the following expression for the equilibrium reaction rate for the process $H_{aq}^+ + e^- \rightarrow H_{aq}$:

$$I_{\text{equ}} = eN_0^{1-\beta}N_0'^{\beta}\kappa^*\rho^*vkT\beta^{-\beta}(1-\beta)^{-(1-\beta)}$$
$$\times \exp[-\beta(1-\beta)2R_m/kT] \qquad (8)$$

Excursions from equilibrium in anodic and cathodic directions are given by the multiplier $\{[\exp[(1-\beta)eV/kT] - \exp(-\beta eV/kT)]\}$, where V is the algebraic difference in potential across the metal–solution boundary between the applied conditions and equilibrium for the primary step. Equation (8), it should be noted, contains no parameters (beyond ρ^*, which represents an integrated density of states) that depends on the electronic energy of the electrode (e.g., the Fermi level or the work function). This may easily be demonstrated by other, quasithermodynamic arguments[10,11] which suppose that the electrons are transferred to or from the Fermi level. As we have seen, this is a valid assumption for a metal electrode.

The basic assumptions of Gurney's charge transfer model were modified by Butler (1936)[12] to give an account of the cases where the reaction product (H) is adsorbed on the electrode surface. A

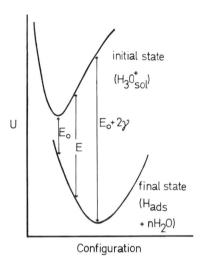

Figure 2. Model of proton discharge to give adsorbed H (after Butler[12]).

general diagram of the energy–configuration curves for this model is shown in Fig. 2.

Using an analogous argument to that presented above for Gurney's model, it is easy to show that the corresponding expression for the equilibrium rate of the process

$$H_{aq}^+ + e^- \to H_{ads} \tag{9}$$

is

$$I_{equ} \sim eN_0^{1-\beta} N_0'^\beta \kappa^* \rho^* \nu k T \exp[-\beta(1-\beta)2\gamma/kT] \tag{10}$$

where γ has the same meaning as R_m in equation (8), and represents the mean "rearrangement energy" for the system (i.e., the mean potential energy to change reactant or product in its ground state to the product or reactant in the same nuclear configuration). It should be noted that Butler regarded the Morse-type energy curve for the reactants as a simple potential–distance curve for the chemical part of the $H^+–H_2O$ bond, with the solvation energy of the complex affecting the ground state. As we will see later, this is an erroneous oversimplification. The potential–configuration curve in Fig. 2 for the reaction products is a summation of that for H adsorbed

Electron Transfer at the Metal–Solution Interface

on the electrode in absence of H_2O, plus the curves for the interaction of H and H_2O (repulsive) and H_2O and the electrode (neglected in the original).

Since the current in the forward direction is given by an equation analogous to equation (6), and since E_F, as we have seen, does not occur in the equilibrium rate (or in the rate expressed at any other potential on an arbitrary scale), then it is apparent that the rate over a series of metals will be essentially governed by changes in E_0, i.e., the energy difference between the ground state of the reactants and the energy of the products in the same nuclear configuration. Clearly, this depends on the ground state of the products (i.e., essentially on $-\Delta H_H^\circ$, where ΔH_H° is the standard heat of adsorption of the product, H), the shape of the curve, and the difference between the ground states of reactants and products along the configuration axis. If one supposes that the neutralization distance is always constant (representing the thickness of the Helmholtz double layer) and independent of the electrode substrate, and further that the general shape of the potential–configuration curve is independent of the electrode material (i.e., independent of ΔH_H°), we see that for any electrode (subscript m), at constant potential,

$$I_m = q\kappa_m^* \rho_m^* \exp(-\beta \, \Delta H_{H,m}^\circ / RT) \qquad (11)$$

where q is a constant.

A similar result may be obtained from the use of equation (10) for the rate at equilibrium for process (9), with the assumption that the product of the rate-determining step (H_{ads}) is in equilibrium with the final product in the overall process

$$H^+ + e^- \to H_{ads} \qquad \text{(rate-determining)} \qquad (12)$$

$$H_{ads} \rightleftharpoons \tfrac{1}{2} H_2(\text{gas}) \qquad (13)$$

(see normalized energy diagram in Fig. 2, with ionization potential difference between H^+ and H and H_2 removed).

We can therefore write, assuming Langmuir adsorption,

$$c'/N(1 - \theta) = [pH_2]^{1/2} \exp(-\Delta G_H^\circ / RT) \qquad (14)$$

where c' is the total concentration of H on the surface of the electrode (fractional coverage θ) and N is the total number of sites per unit area. Remembering that metal sites are reactants in process (9), their concentration should appear as $(1 - \theta)^{1-\beta}$ in equation (10). Use of the appropriate partition functions allows us to write $N_0 = cf_i^{-1}$, $N_0' = c'f_{ads}^{-1}$, and $\exp(-\Delta G_H^\circ/RT) = f_f^{-1/2}f_{ads}\exp(-U_H^0/kT)$, where f_f is the partition function of the final state (H_2), and U_H^0 is the potential energy of adsorption of H. Again, we can assimilate this quantity to ΔH_H^0, and we obtain an equation of the same form as equation (11), using (10) and (14), at equilibrium for the overall process [(12) followed by (13)]:

$$I_{exchange} = ec^{1-\beta}[pH_2]^{\beta/2}N^\beta(1 - \theta)f_i^{-(1-\beta)}f_f^{-\beta/2}\kappa^*\rho^*vkT$$
$$\times \exp[-\beta(1 - \beta)2\gamma/kT\exp(-U_H^0/kT) \qquad (15)$$

Strictly, equation (15) [unlike (11)] only applies at equilibrium; we will later show that it may also be expected to be valid under net current flow.

Butler's theory thus introduces the concept of electrocatalysis, i.e., the reaction rate of a multiple-step process depending on the adsorption properties of the electrode for reaction intermediates.

(ii) Developments of the Gurney–Butler Theory

The charge-transfer theory of Gurney has been developed in a very complete manner by Gerischer[13,14,58] for the case of a redox system at metal and semiconductor–solution interfaces. For proton transfer the theory of Butler[12] (making the simplifying assumptions that κ^*, the electron transmission coefficient or tunneling factor, multiplied by ρ^*, the integral density of states, could be placed equal to unity) has been considered by Parsons and Bockris[15] and Conway and Bockris,[16-18] and, taking into account the possibilities of proton tunneling, by Conway,[19] Bockris and Matthews,[20] and Bockris et al.[21]

(iii) Transition State Theory

The model of Horiuti and Polany[22] for the proton transfer reaction, proposed at about the same time as that of Butler,[12] uses the concept of stretching of the proton–water molecule bond to form an activated complex.[23,24]

It is of great importance to point out that in Gurney's theory (and, by extension, in Butler's) the transition from H^+ to H_{ads} takes place at a unique point in space and time. A transition state model supposes that a *gradual* transition (in space and time) takes place, i.e., the reactant H^+ changes slowly to the product H along the reaction coordinate. It is further assumed that the electron arrives spontaneously from the Fermi level of the electrode (on quasi-thermodynamic grounds, as discussed above).

Thus, the basic hypotheses of the transition state model are: (1) The electron is considered to be transferred from the Fermi level (for a metallic electrode; the situation in the case of a semiconductor electrode will be discussed later). (2) There are always sufficient electrons at the Fermi level of the metallic electrode to maintain the reaction (i.e., the electron activity is taken to be unity). (3) The proton transfer along the reaction coordinate is classical. A frequency factor of kT/h,[25] and a proton transmission coefficient,[25] thus appear in the rate equation.

It is easy to show[26] that the expression for the standard exchange current for the overall process [reactions (12) and (13)] using Horiuti and Polyani's model is of the form

$$I_{exchange} = ec^{(1-\beta)}[pH_2]^{\beta/2} N^\beta (1-\theta)(kT/h)\exp(-\Delta G^\ddagger/RT)$$
$$\times \exp(-\beta \Delta G_H^\circ/RT) \qquad (16)$$

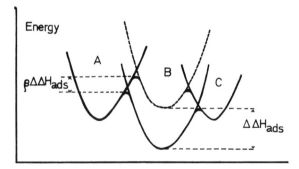

Figure 3. Transition state model of proton discharge (Horiuti and Polanyi[22]) showing effect of changing heat of adsorption on activation energy. A, initial state (H_3O^+); B, adsorbed intermediate (H); C, final state $(\frac{1}{2}H_2)$. Thickened parts of curves at intersections show actual path followed by system.

where $\Delta G^{\ddagger}$ is a standard free energy of activation (expressed at $\Delta G_H° = 0$) that is independent of the electrode material, and where the $\exp(-\beta \Delta G_H°)$ term results from the assumption of an equilibrium of the same type as in equation (14). Away from the equilibrium situation, the rate equation will also contain a term in $\exp(-\beta \Delta G_H°/RT)$, as "vertical" changes in the free energy of adsorption of the H radical will cause a change in the height of the energy barrier by $\beta \Delta G_H°$ (as shown in Fig. 3). In making such a derivation, it was assumed that the shapes of the free energy–configuration curves, as well as their relative positions along the reaction coordinate, do not change with vertical displacement (i.e., with $\Delta G_H°$). We will see later that this condition does not necessarily apply, due to variation of the entropy part of $\Delta G^{\ddagger}$ with $\Delta G_H°$.

A transition state theory approach was used in the majority of discussions on electrode reactions during the 1950's and 1960's (see, for example, Parsons[26] and Bockris[188]).

3. Electrostatic Theory

This idea grew out of the work of Libby (1952)[7] and Weiss (1954)[8] applied to homogeneous redox processes. In this theory the electrolyte is considered essentially as a polar crystal lattice (see Mott and Gurney,[27] Landau[28] and Pekar[29]). The stabilization energy E of a charge ze trapped within a perfectly conducting sphere of radius r within a polar crystal lattice may be expressed by the Born equation:

$$E = (z^2 e^2/2r)[1 - (1/\varepsilon)] \qquad (17)$$

where ε is the static dielectric constant of the medium. Libby considered that this term represented the solvation energy of the ion, and that it could change microscopically on account of thermal fluctuations of the dielectric. The charge-transfer complex was considered to possess an increased radius r, the net energy difference being the energy of activation. Weiss, on the other hand, considered that the act of charge transfer occurred at optical frequencies. This would be expected to result in a net quantity of energy stored in the solvent lattice for a length of time corresponding to the Debye relaxation time of the molecules. For a charge ze this energy is given by the difference between the two Born terms at optical and static frequencies,

$$E = \frac{z^2 e^2}{2r}\left(\frac{1}{n^2} - \frac{1}{\varepsilon}\right) \qquad (18)$$

where n is the refractive index of the solvent (see Landau,[28] Platzmann and Franck,[30] Pekar,[29] and Mott and Gurney[27]). Weiss attributed the reaction activation energy to this energy difference, and developed rate equations with preexponential factors calculated from the Landau–Zener[31,32] equation, for the case of nonadiabatic electron transfer processes with intersecting energy terms showing little splitting, and using the Gamow equation for those cases where tunneling of the electron was considered to be important.

Another approach, based on electrostatic repulsion and electron tunneling, was taken by Marcus et al. (1954),[33] and later with a repulsion term depending on a dielectric constant varying with field,[34] by Laidler.[35,36]

Marcus (1956)[37] showed that the energy difference in equation (18) represented the total energy difference on charge transfer (the energy required to change reactants to products in the same nuclear configuration). He considered that the free energy of activation λ involved the electrostatic work required to bring the reactants together ($\Delta\omega$) plus the appropriate continuum expression for reorganization at optical frequencies. In the early Marcus papers[38,39] the ion was regarded as a perfectly conducting sphere of unchanging radius, and only energy changes in the continuum were considered. The energy–configuration diagram for reactants and products according to the early Marcus theory is shown in Fig. 4. Plots are presented on a normalized basis, i.e., the ionization energy has been removed, so that the situation, as illustrated, represents equilibrium.

A simple electrostatic demonstration shows that in the case of an electrode reaction (involving one electron only)[39]

$$\lambda = \frac{e^2}{2}\left(\frac{1}{a} - \frac{1}{R}\right)\left(\frac{1}{n^2} - \frac{1}{\varepsilon}\right) + \Delta\omega \qquad (19)$$

where R is twice the ion–electrode distance and a is the radius of the ion in the activated condition. Marcus considered that the reaction took place with a high probability (i.e., adiabatically) when the ion solvation spheres (homogeneous case) or solvation sphere and electrode Helmholtz double layer (heterogeneous case) for suitably activated ions were in contact.

Hush[40] presented a generally similar theory to the original one of Marcus but containing a revised concept of the reorganizational

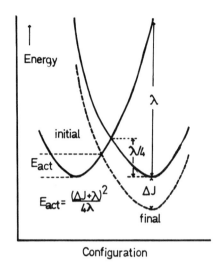

Figure 4. Parabolic energy—configuration surfaces for outer sphere reactions (Marcus[44]).

energy, λ, including that of the primary solvation shell. The intermediate states leading up to charge transfer (i.e., to the intersection of the energy curves) were regarded as having a quasiequilibrium character, unlike those of Marcus.[37-39] Final expressions derived using the two approaches are, however, similar. Hush and Marcus both consider their theories to be adiabatic, on grounds that are not clearly specified. Other variations on the above models, that account better for adiabaticity, have been proposed, for example, that of bridge-assisted charge transfer (Schmidt and Mark,[41,42] based on the ideas of Halpern and Orgel[43]).

The implications of the theory of Marcus for heterogeneous electron transfer will be discussed in some detail here, because of the influence that these concepts have had on later work on electrode reactions. Because of the assumption of a similar harmonic energy surface for reactants and products that depends only on the properties of the continuum, rather than on the properties of the individual ions involved in the reaction (the surfaces being expressed in configuration space), we can write, using a simple geometric argument,

$$F^{\ddagger} = (\Delta J + \lambda)^2/4\lambda \qquad (20)$$

where $F^{\ddagger}$ is the activation free energy of the process and ΔJ is the displacement of the system from the state corresponding to equilibrium. J may represent $e\eta$, where η is the overpotential. In addition, for a reaction in several stages ΔJ represents the free energy difference between the initial or final state and the ground state of reaction intermediates, corresponding to the ΔG_H term in equation (15). The Brønsted slope $(\partial F^{\ddagger}/\partial \Delta J)$ is thus given by the expression

$$\partial F^{\ddagger}/\partial \Delta J = \tfrac{1}{2} + (\Delta J/2\lambda) \qquad (21)$$

in the Marcus theory. This is also the value of the transfer coefficient.

Since 1961 Marcus has somewhat changed the original concept[44] of λ. It is apparent that for the majority of ions the greater part of the solvation energy resides in the *inner* solvation sheath, where the concept of a dielectric constant is meaningless. In such cases the solvation sheath must rearrange independently of the dielectric continuum. On this basis Marcus introduced a further λ term (λ_s, the subscript indicating solvation sheath), thus:

$$\lambda_{\text{total}} = \lambda_{\text{continuum}} + \lambda_s \qquad (22)$$

Marcus did not consider the method of activation by means of the electrostatic continuum, but it is interesting to note that the pre-exponential term of his rate expression, a simple form of which is given by

$$I = eNz \exp[-(\Delta J + \lambda)^2/4\lambda kT] \qquad (23)$$

where z is the collision number, implies the existence of classical levels of movement (vibration or translation) on the reacting ion. On this basis it is difficult to see why classical energy levels should be related only to the dielectric continuum. Hence, the electrostatic optical frequency rearrangement term λ is not an energy of activation indicative of the height of the charge transfer barrier to be surmounted, but an energy difference appearing rather *after* the elementary act of charge transfer. In typical cases the Born solvation energy only represents about 25% of the total solvation energy of an ion. If one supposes (typically) that activation energies are of the order of 10% of bond (or solvation) energies, then a Born term variation of 50% is needed to effect charge transfer. This is scarcely conceivable: The majority of the contribution to the activation energy must come via classical thermal levels of the discharging ion (vibrational and hindered translational levels).

In any case, the use of the Born–Landau optical frequency polarization equation introduces further problems. It may be expected to apply only when the energy of the trapped electron is very different from that of the other electrons in the crystal lattice.[45] This is not necessarily the case in a typical electrolytic solution. If the above condition is not obeyed, the normal Born equation for the excess energy difference [equation (17)] would be expected to give a more reasonable estimate of the energy change on transfer of charge.[45]

Dogonadze and Levich and their collaborators[46–54] have, in a long series of papers, attempted to develop a method for the calculation of the rates of redox reactions using a method that may be regarded as more exact than that of the original Marcus theory, but essentially using the same basic model as the latter. In their theory the only contribution to the energy of activation, at least in a formalistic manner, derives from the continuum. At the same time the implicit concept of thermal activation (via the collision preexponential term) of the original Marcus theory is abandoned in order to unify the activation mechanism and the energy difference represented by the Landau–Born expression. This approach, while intellectually satisfying, is probably not physically reasonable. Dogonadze and Levich use essentially polaron theory[55–57] (for energy transfer in a perfect crystal lattice) for their activation model. They consider that activation is due to stationary polarization waves thermally induced within the continuum, whose frequency is ω_0, where $1/\omega_0$ is of the order of the Debye relaxation time. If each solvent molecule is capable of transferring one quantum $\hbar\omega_0$, about 10^3 molecules are required to activate the discharging ion.[5] While this may be possible in a solid crystalline lattice, it seems improbable that this number of molecules can vibrate in phase in the required manner in a liquid such as water, which does not possess a discrete long-range structure: Only cybotactic groups, of the order of 25–40 molecules, can be said to maintain the properties of a crystalline lattice to some degree.[9] As in the early Marcus work, Dogonadze and Levich have not considered the possibility of activation of the solvation sheath, on the grounds that the frequencies of vibration occurring there are nonclassical. In addition, because of the Born–

Oppenheimer approximation, energy changes due to bond length changes as a result of charge transfer are neglected. (In very recent work the possibility of activation of the solvation shell has in fact been considered.[59]) The work of Levich and Dogonadze differs in two very important respects from that of Marcus, namely they regard the λ term (E_s in their terminology) as a heat, rather than a free energy term, on the grounds that it is the difference between two free energy terms whose entropy contributions are similar, and is in fact experimentally only slightly temperature dependent,[53] and second, they regard electrode reactions as not necessarily adiabatic with the electrons (at unit activity) all arriving from the Fermi level.

For the rate of an electrode reaction they write an equation similar in form to equation (4):

$$i = e \int_0^\infty C(x)\,dx \int_0^\infty W(x, E)n(E)\rho(E)\,dE \qquad (24)$$

where C is the total reactant concentration per unit area, and $W(x, E)$ is the transfer probability, which contains $N(E)$ and $\kappa(x, E)$. $W(x, E)$ was calculated assuming harmonic electronic energy terms, as in the theory of Marcus.[37-39,44] The electron transfer probability $\kappa(x, E)$ was estimated using time-dependent perturbation theory, analogous to the Landau–Zener[31,32] expressions of Weiss.[8] Tunneling of Gamow type (fixed barrier tunneling for a plane electron wave) was not considered on the grounds that a time-dependent barrier is involved, and tunneling is between a plane wave (in the electrode case) and a spherical wave around the ion. No final expressions for the electron transmission coefficient were obtained (except in the adiabatic case, for which it is unity) because of the difficulty of evaluating the exchange integrals.

The final expression of the Levich and Dogonadze model for electrode reactions contains an exponential term of the same form as that in equation (23) (a consequence of the assumption of similar parabolic energy terms for reactants and products) and a frequency factor essentially depending on ω_0, with the activation-energy-dependent κ value and a density of states term ρ^*.

More recently[60,61] the same concepts have been applied to an electrocatalytic reaction, involving bond breaking, atom transfer, and adsorption—the proton transfer reaction. This time, transition probabilities have been determined by the application of two successive Born–Oppenheimer approximations to the system: first to the solvent molecules, assuming the time constant for proton and electron motion is rapid compared with that of heavy nuclei, and second, to the solvent molecules and the proton, assuming electron movement is sufficiently rapid. The model assumes that the proton makes a short, quantized movement to its new position on the surface of the electrode metal of about 0.5 Å, measured as the distance between the proton ground states in the H_3O^+ (or $H_4O_9^+$) ion and that at the electrode surface. This proton transfer, like that of the electron, is considered to be of the tunnel type (Landau–Zener), and occurs in a radiationless transfer process. From this point of view it bears some resemblance to proton conduction in electrolytes.[62]

In order to simplify the calculation, a number of approximations are made. All electrostatic work terms[36] are ignored, and, as indicated in the present discussion of Gurney's theory, it is assumed that all electron–proton exchanges take place at some constant distance from the electrode x. If, in addition, one supposes an adiabatic proton transfer, the final expression for the current passing at the interface is

$$i_{(n,n')} = eC\delta\rho^*\kappa^*\omega_0 kT \exp(-E_{act}/kT) \qquad (25)$$

where E_{act} is defined (as $F^{\ddagger}$) by equation (20), in which the term in J contains (in addition to the reactant–product ground-state energy difference) the difference between the reactant and product vibration energies in the vibrational states n and n', respectively. In addition, J contains the reaction overpotential. The subscripts n, n' correspond to transitions from the nth vibrational level of the initial state to the n'th level of the final state. δ is the reaction zone thickness. As indicated above for Gurney's model, the electrons are transferred on the whole from levels that are within $\pm kT$ of the Fermi level.

The major problems involved in the Levich–Dogonadze model (apart from the application of polaronic concepts to a solvent such as water) are essentially related to the problem of agreement with experiment, in particular for Tafel plots for hydrogen discharge on mercury.[9] Such plots show a straight-line Tafel relationship for almost 1.5 V (12 decades in all if the early results of Bowden and Grew[63] are taken into account). This is inconsistent with the simplified assumption of a parabolic energy expression[37,54] (see Ref. 59).

In addition, the model does not account for the movement of the proton out of its solvation sheath and subsequently through the Helmholtz double layer, a distance of the order of 3 Å.[20,64] The model does not give any account of activation of the accepted classical energy levels of H_3O^+ ions,[65,66] and in addition, it predicts H–T separation factors that are not in good agreement with experiments[20,21] unless a nonadiabatic proton transfer is assumed.[100] However, even if one assumes that proton transfer is adiabatic, the model still predicts a preexponential factor in the transfer rate equation that is about two to three orders of magnitude less than that in transition state[25] or collision theories.[44] Since the H_3O^+ (or $H_9O_4^+$) ion is known to possess classical energy levels (cf. its heat capacity behavior[9]), such theories would seem to be more appropriate, and thus the Levich–Dogonadze model is not too probable for the proton transfer case (see Ref. 9). While it is clearly capable of improvement, along the lines of introducing classical activated levels, other than in the electrostatic continuum, it still suffers from two major defects: (1) its total inapplicability to particles much heavier than protons, which cannot tunnel out of their presumably nonclassical solvation sheaths, and (2) the fact that it assumes that reactants become products at a unique point in space and time. It would therefore seem that another approach is indicated, which depends on the participation of classical energy levels in nonequilibrium configurations of reactant and product molecules. Such an essentially transition-state approach has been recently assumed by Marcus for the case of atom transfer reactions,[67,68] based on Johnston's BEBO model,[69] which was first applied to the hydrogen evolution reaction case by Conway et al.[68]

4. Improved Transition State Theories for Electrode Reactions

In all classical accounts of proton transfer in the gas phase, and in extensions of these ideas to the liquid phase, nonequilibrium bond configurations in the reactants can lead to nonequilibrium classical vibrations of the proton, with a consequent smooth classical transition of the proton along the reaction coordinate from reactants to products. If such a situation occurs, then the concept of the proton making a quantum jump from nonclassical reactant levels to nonclassical product levels, as postulated by Dogonadze and

Levich,[5,59,60] is clearly not necessary. The nonequilibrium proton transfer condition, under classical conditions, would be mainly accounted for by the resonance sharing of an extended proton–water molecule bond, and a similarly extended hydrogen atom–metal electrode bond in a transition state. Such a state is not improbable; The proton, together with its tightly (nonclassically, under equilibrium conditions) bound solvation sheath, can be activated *as an entity* (compare Levich[54]) by thermal bombardment to an energy level corresponding to that for radiationless electron transfer. This is indicated by the specific heat of the H_3O^+ (or $H_9O_4^+$) ion[65] and by the classical "hindered translational" levels in its infrared spectrum.[66] The lifetime of such an activated state will necessarily be several orders of magnitude longer than the time constant for radiationless electron transfer. It is therefore possible to set up a condition of electron resonance between one or more of the metal energy levels and the proton, with the consequent formation of a partial bond (between H_2O, H, and the metal) which is dependent on time and on the configuration of the overall system. The cooperative effect of the two partial bonds sets up a classical proton vibration along the reaction coordinate. We may visualize therefore a transition state whose H^+–H energy difference [and the electron energy level in the metal corresponding to it, i.e., E in equation (1)] remains constant through the transition state, assuming that the resonating bond involves the same metal level during the transition state lifetime.

The above concept bears some resemblance to Gurney's theory,[6] but assumes that the reactants change gradually to the products of reaction (somewhat in the manner of Hush's model for redox processes[40]). The configuration of the transition state will be expected to change relatively slowly compared with the electron configurational changes in the system. Similarly, movement of nuclei much heavier than the proton may be neglected during the movement of the proton from reactants to products.* The Born–Oppenheimer approximation may therefore be applied twice to the system (cf. Dogonadze and Levich[60]). The condition for radiationless electron transfer from the electrode to the proton is that the donor level in the metal should have the same total energy as that of

---
*That is, water molecule movements may be ignored. This will not be the case for heavy particle, e.g., Cl^- ion, transfer.

the electron level in the neutralized proton. This energy will clearly not be the same as that of the empty level of the nonneutralized proton, except under infinitely slow electron transfer conditions,* since part of the dipole energy of the surrounding solvent that can respond to optical frequency charge transfer will be changed on the passage of the electron. This energy will be essentially the noninertial part of the nearest-neighbor dipole and electrostatic continuum interaction terms in the solvation energy. Thus a residual energy will be present in the surrounding solvent, equivalent to the inertial part of the ion–permanent dipole interaction for nearest-neighbor dipoles (the polarizability part operates at optical frequencies) together with the optical Born term for the electrostatic continuum and electrostatic work terms [λ in equation (17)].

It is therefore clear that the above model makes use of elements of both the thermal and the electrostatic theories.

The above approach thus simultaneously explains how a classical movement of the proton is possible along the reaction coordinate, despite the quantum vibrations ($v \gg kT/h$) of the proton chemical bond in the initial state, and the H bond to metal in the final state of the rate-determining step. It also explains how the proton can be transferred with a transmission coefficient of the order of unity through a distance of the order of 2–3 Å, i.e., through the Helmholtz double layer. The last condition is necessary to explain in a satisfactory manner the experimental H–T separation factors for this process under conditions where proton tunneling is important.[20]

The concept is basically the same as that in Gurney's theory. For the population of molecules accepting electrons by a radiationless mechanism from levels between E and $E + dE$, we can write an equation for neutralization frequency analogous to equation (2):

$$dV = \kappa(E)\tau(E)N(E)\rho(E)v(E)\,dE \qquad (26)$$

$\tau(E)$ is the nuclear transmission coefficient, and $v(E)$ is the frequency at which molecules pass from the reactant state to the product state (under conditions where arrival of molecules in the reaction zone is not rate determining). It is assumed that $\tau(E)$ contains the effect of nuclear tunneling. For simplicity, it has been assumed that neutralization takes place at some average distance (corresponding

---

* The word adiabatic has been avoided, cf. Laidler.[36]

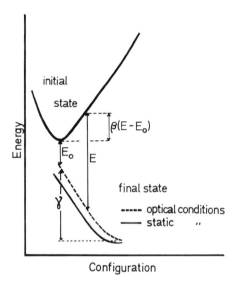

Figure 5. Proton transfer model under conditions of optical frequency transfer of charge. Zero point energies neglected.

to the length of the extended H_2O–H–metal bonds). The other symbols have the same meanings as before.

A diagram of the potential energy surfaces for reactants (H^+ in its solvation sheath) and products (H adsorbed on the metal surface in the presence of water) as a function of configuration is shown in Fig. 5. These curves contain the work necessary to bring the reactants together into a favorable position for electron transmission and proton movement to occur, together with the energy necessary to arrange the equilibrium ground state of the products to a configuration corresponding to the ground state of the reactants (i.e., a rearrangement energy). For the forward process two sets of curves are shown: The continuous line corresponds to the case where electrostatic equilibrium has been attained in the electrolyte and the solvation sheath. The nonequilibrium curve, corresponding to the case of optical electron transfer, is shown superimposed.

As in the case of the discussion of Gurney's model, and as suggested by experiment,[9] we can assume that β is relatively independent of E. It seems reasonable to suppose that the curves will in fact show a limiting potential value as the configuration diverges from the equilibrium value, in a curve that is somewhat of Morse form. The latter has been used as a model by a number of authors.[15–22] On the basis of a constant β value, we may therefore use equation (3). Hence we can write, after performing an integration over all metal levels, and with the same assumptions as in equation (6), including $(1 - \theta)$ for metal sites:

$$I_{cath} = e\kappa^*\rho^*\tau^*v^*(1 - \theta)[kT/(1 - \beta)]\exp[\beta(E_0 - E_F)/kT \quad (27)$$

where E_0 corresponds this time to optical conditions (see Fig. 5), and where the preexponential terms with asterisks represent mean values. Although there is not one unique activated state, the most probable one is situated close to the Fermi level of the electrode (for metallic electrodes). Assuming the applicability of the Gibbs distribution function, we can write

$$N_0\,H = cf_i^{-1}f_{act} \quad (28)$$

where $c(=C\delta)$ is the total reactant concentration in the reaction zone, and f_i and f_{act} are the partition functions of the initial and activated states, respectively. This expression is equivalent to the usual transition state assumption of thermodynamic quasiequilibrium between initial and activated states.[25] The partition function of the activated state can be split up in the usual way[25] into a part along the coordinate of reaction, together with a part at right angles to it ($f^\ddagger$). Along the reaction coordinate this represents the value for a loose vibration (kT/hv^*). Thus

$$f_{act} = (kT/h)f^\ddagger \quad (29)$$

where kT/h is the universal frequency factor.

It is easy to show that the corresponding rate in the anodic direction is [cf. equation (7)]

$$I_{anod} = e\kappa^*\rho^*\tau^*c'f_{ads}^{-1}f^\ddagger\frac{kT}{h}\frac{kT}{\beta}\exp[-(1-\beta)(E' - E_F)kT] \quad (30)$$

where E' has the same definition as before, though this time it refers to a transfer under optical conditions. c' is the concentration of product (H) atoms in the reaction zone, and f_{ads} is the vibrational–rotational partition function of the final state (adsorbed H plus water). By equating rates at equilibrium for the surface redox process $H_{aq}^+ + e^- \to H_{ads} + nH_2O$, we obtain for the net equilibrium reaction rate (expressed as a current), using equations (27)–(30),

$$I_{equ} = ec^{1-\beta}c'^{\beta}\kappa^*\rho^*\tau^*(1-\theta)^{1-\beta}f_i^{-(1-\beta)}f_{ads}^{-\beta}f^{\ddagger}(kT/h)kT \quad (31)$$
$$\times \beta^{-\beta}(1-\beta)^{-(1-\beta)}\exp[-\beta(1-\beta)2\gamma/kT]$$

where γ represents the mean reorganizational energy between states [analogous to γ in equation (10)], and contains the potential energy differences involved in making or breaking chemical bonds, plus the inertial solvent terms (ion–dipole and continuum). The latter are strictly free energies,[36,44] but probably vary with temperature only insofar as liquid structure varies $[(1/n^2) - (1/\varepsilon)$ is approximately temperature independent,[53] as previously noted]. A calculation of γ, or alternatively, of the activation energy of an electrocatalytic reaction under conditions of equilibrium for the rate-determining step (reactants and products in their standard states) from first principles is not possible, unless a simple semiempirical approach (for example, BEBO[69]) is used. Such a method ignores nonequilibrium optical polarization. It should be noted that all the above models make the simple assumption that microscopic equilibrium applies between the reactant and activated states, which is neither self-evident nor rigorously supposable. Such an assumption can only be justified by arguments based on quantum statistical methods, as used in other applications by Van Hove[70] and Kubo,[71] and in electrochemistry by de Hemptinne.[72]

Finally, it is of interest to point out that the preexponential factor in equation (31) differs by more than two orders of magnitude from that in the adiabatic proton transfer equation of Dogonadze, Levich, and co-workers.[60] The difference will be even greater in the case of the nonadiabatic proton transfer theory proposed later[100] to account for experimental values of the H–T separation factor. This suggests that the mechanism of Dogonadze and Levich is intrinsically less probable than one involving a transition state, given that the condition exists for setting up a resonant-bond activated complex.

III. EFFECT OF ADSORPTION ON THE RATE OF REACTION

We have already seen that if the product of the rate-determining step is adsorbed on the electrode, as in the example cited above (H_{ads}), then, at equilibrium for the overall process [(12) plus (13), with (12) rate determining], the exchange current will be given by equation (31), with c' substituted using the relation in (14). Since $f_{ads}^{-\beta} f^{\ddagger}$ should be metal independent, we can write the exchange current rate in the form

$$I_{equ} = ec^{1-\beta}[pH_2]^{\beta/2}\kappa^*\rho^*\tau^*(1-\theta)N\frac{kT}{h}$$

$$\times \exp-\frac{\Delta G^{\ddagger}}{RT}\exp-\beta\frac{\Delta G_H^{\circ}}{RT} \quad (32)$$

which is identical to (16). It is reasonable to suppose that a similar equation will apply out of equilibrium, based on a similar idea to that discussed already [equation (16)]: Translations of energy curves, for reaction intermediates in the vertical plane, without changing the general shape of the curves, will cause a proportional shift at the intersection point, i.e., the activated complex (see Fig. 3). The multiplier [equation (14)] only applies for Langmuir adsorption (i.e., at limitingly low or at limitingly high coverage). In intermediate cases some other isotherm, involving a coverage-dependent heat or free energy of adsorption, will apply. Often it is noted experimentally that the change in heat of adsorption with coverage is approximately constant, giving rise to the simplified Temkin isotherm[73] in the intermediate-coverage region (for review, see Ref. 74). The linear relationship between heat of adsorption and coverage has been attributed to particle–particle interactions[75–77] as well as on the bond-strength change caused by the effect of oriented dipoles at the surface, via the work function.[78] In electrolytic solutions adsorption necessarily involves displacement of adsorbed water molecules, which are bound by dispersion forces and have heats of adsorption approaching 20 kcal mole^{-1}.[79] Heats of adsorption in electrochemistry are therefore the difference between the heat of adsorption in vacuum and the heat of adsorption of the displaced water dipoles under the appropriate potential conditions. A theory of competitive adsorption, which applies particularly well to comparatively weakly adsorbing organic molecules,[64a] has been developed on this basis by Bockris et al.[64,80] Their isotherm contains a number of terms involving dipole–dipole, ion–ion, and dipole–ion interactions. Such an isotherm will be valid unless dissociation, and associated strong chemisorption, occurs. Under such conditions a Temkin isotherm can be expected to be a good approximation to the observed behavior in the medium-coverage range, and it can be expected that the same value of the Temkin term (change in heat of adsorption as coverage goes from limitingly low to limitingly high values) will apply to all species of similar chemibonded character that are simultaneously adsorbed.[74]

For Langmuir adsorption at low coverage the $(1 - \theta)$ term in equation (32) may be ignored. The same is true under Temkin (medium coverage, $0.1 < \theta < 0.9$) conditions, since the isotherm contains no preexponential θ terms. However, for high coverages, the $(1 - \theta)$ term cannot be ignored. As a consequence, the overall rate equation under equilibrium conditions (i.e., for the exchange current density) for reaction $(12) + (13)$ becomes (at constant H^+, pH_2)

$$I_{equ} = ec^{1-\beta}[pH_2]^{\beta/2}\kappa^*\rho^*\tau^* \frac{kT}{h} \exp - \frac{\Delta G^{\ddagger}}{RT} \exp - \beta \frac{\Delta G_H^{\circ}}{RT} \quad (33)$$

(low coverage)

Under high coverage conditions we must rewrite the $(1 - \theta)$ term in equation (32), using equation (14), with $c'/N = \theta$, as

$$(1 - \theta) = \{1 + [pH_2]^{1/2} \exp(-\Delta G_H^{\circ'}/RT)\}^{-1}$$

$$\sim [pH_2]^{1/2} \exp(-\Delta G_H^{\circ'}/RT) \quad (34)$$

Hence, from equations (32) and (34)

$$I_{equ} = ec^{1-\beta}[pH_2]^{-(1-\beta)/2}\kappa^*\rho^*\tau^* N \frac{kT}{h} \exp - \frac{\Delta G^{\ddagger}}{RT}$$

$$\times \exp + \beta \frac{\Delta G_H^{\circ'}}{RT} \quad \text{(high coverage)} \quad (35)$$

In equations (33)–(35) ΔG_H° and $\Delta G_H^{\circ'}$ are the respective standard free energies of formation of the rate-determining-step product under conditions of low and high coverage. The overall log reaction rate for the process (12) + (13), as represented by equation (32), would therefore be expected to be an ascending and descending function of $\Delta G_H°$ and $\Delta G_H°'$, respectively. This introduces the idea of the *volcano plot* first considered in gas-phase catalysis by Balandin,[81] and introduced into electrochemistry by Conway and Bockris,[16] Parsons,[26] Gerischer,[82] and (experimentally) by Kuhn et al.[148]

A similar argument holds if the *reactant* in the rate-determining step is adsorbed, as in the hypothetical sequence

$$A_{sol} \rightleftharpoons A_{ads} \quad (36)$$

$$A_{ads} + e^- \to A_{sol}^- \quad \text{r.d.s.} \quad (37)$$

Using arguments similar to those given in equations (26)–(34), the rate of reaction (37) at equilibrium can be written in the form

$$I_{equ} = e\theta^{1-\beta}(1-\theta)^\beta [A^-]^\beta \kappa^* \rho^* \tau^* (kT/h) \exp(-\Delta G^\ddagger/RT) \quad (38)$$

where θ is the coverage of A on the electrode, $[A^-]$ is the concentration of A_{sol}^-, and $\Delta G^\ddagger$ is the free energy of activation under standard conditions. If we assume that the rate-determining step [reaction (37)] is slow compared with reaction (36), we can write a Langmuir isotherm of the form

$$\theta/(1-\theta) = [A_{sol}] \exp(-\Delta G_A°/RT) \quad (39)$$

where $\Delta G_A°$ is the free energy of adsorption of A relative to A_{sol}. Hence, from (38) and (39) the exchange current density for the overall process (36) plus (37) is of the form

$$I_{exchange} = e[A_{sol}]^{1-\beta}[A^-]^\beta \kappa^* \rho^* \tau^* \frac{kT}{h}$$

$$\times \exp - (1-\beta)\frac{\Delta G_A°}{RT} \quad (40)$$

for low coverage and

$$I_{exchange} = e[A_{sol}]^{-\beta}[A^-]^\beta \kappa^* \rho^* \tau^* \frac{kT}{h} \exp - \frac{\Delta G^\ddagger}{RT} \exp \beta \frac{\Delta G_A^{°'}}{RT} \quad (41)$$

for high coverage. In these equations $\Delta G_A°$ and $\Delta G_A^{°'}$ are the standard free energies of adsorption of A (with respect to A_{sol}) under high and low coverage conditions, respectively. Equation (41) is exactly equivalent to looking at the barrier in the reaction from the bottom of the potential energy well for the reactants, and considering full coverage of the surface (see Conway and Bockris[16]), rather than from the outside, and substituting $(1 - \theta)$ from (39). For a general rate-determining step*

$$X_{sol}^+ + Y_{ads} + e^- \to \sum P_{ads} \qquad (42)$$

it is not difficult to show that the anodic and cathodic currents at constant potential V (thus eliminating the electrochemical reaction orders) are given by expressions of the form

$$I_{cath} = ne \prod R(1 - \theta_T)^z \kappa^* \rho^* \tau^* \frac{kT}{h} \exp\left[-\frac{\Delta G^\ddagger}{RT} - \frac{(1-\beta)\Delta G_Y^\circ}{RT}\right.$$
$$\left. - \frac{\beta \sum \Delta G_p^\circ}{RT}\right] \exp -\frac{\alpha FV}{RT} \qquad (43)$$

$$I_{anod} = ne \prod P(1 - \theta_T)^z \kappa^* \rho^* \tau^* \frac{kT}{h} \exp\left[-\frac{\Delta G^\ddagger}{RT} - \frac{(1-\beta)\Delta G_Y^\circ}{RT}\right.$$
$$\left. - \frac{\beta \sum \Delta G_p^\circ}{RT}\right] \exp \frac{(1-\alpha)FV}{RT} \qquad (44)$$

where θ_T is the total coverage on the electrode surface, ΔG_Y° and $\sum \Delta G_p^\circ$ are the standard free energies of adsorption of Y and products P with respect to the initial and final states of the overall process, respectively (at $V = 0$), V is the potential on an arbitrary scale, α is the transfer coefficient (see below), n is the number of electrons transferred in the overall process, and z is the reaction order for sites in the rate-determining step. $\prod R$, $\prod P$ are the products of reactant and product activities.

A similar equation may be written for the general case of the combination of two adsorbed products to give a series of adsorbed and unadsorbed products. "Unadsorbed" products are stipulated, since it is necessary to transfer charge completely across the double layer in the rate-determining step, so that the reacting particle sees the whole of the metal–solution potential difference. It is improbable

---

*It may always be assumed that only one electron can be transferred in the rate-determining step, since otherwise the reaction will involve too great an activation energy. This is discussed by Vorotyntsev and Kuznetsov.[84]

Effect of Adsorption on the Rate of Reaction 395

that a transfer of charge between two adsorbed species, with no complete charge transfer across the double layer, can ever be rate-determining. A discussion of this type of reaction will be given later in reference to the oxygen electrode.

Equations (43) and (44), like equations (33), (35), (40), and (41) represent a volcano-type relationship if log rate is plotted against free energy of adsorption of species adsorbed. In general, free energy of adsorption involves an entropy term that should not be very dependent on the substrate, since it represents the loss of translational degrees of freedom. As will be shown later for the oxygen electrode reaction, changes in vibrational–rotational degrees of freedom can probably be generally neglected. The ΔG°_{ads} terms in equations (43) and (44) over a range of electrode materials can therefore be roughly assimilated in the changes of the ΔH°_{ads} terms. In addition, the values of the ΔH°_{ads} terms may be expected to change by similar amounts for each particular reactant or product as one goes from one electrode material to another, in proportion to the degree of bonding (i.e., percentage of single-bond character) of the substance to the electrode. Thus, in all cases, log i at constant potential will be expected to be on a simple basis a rising function of ΔH°_{ads} for any particular adsorbate, at low total coverage of adsorbed material, whether it participates in the reaction or not, until a point is reached where the total coverage becomes high so that the reaction is then inhibited because of lack of sites for adsorption on the metal surface, when log i will fall with rising ΔH°_{ads}.

For complex reactions the overall scheme may be worked out in individual cases (cf. Parsons,[26] Gerischer,[82] Conway and Bockris,[16] and Kuhn et al.[148])

While the concept of the volcano plot is useful, it should be treated with care as a basis for analysis of an electrocatalytic process. The simple exposition given above, for example, ignores the possibility that the rate-determining step under equilibrium conditions may change with ΔH_{ads}. In addition, the rates of reactions for substrates which show small ΔH_{ads} ($\theta \to 0$), or large ΔH_{ads} ($\theta \to 1$), may be so slow as to be only experimentally accessible at high overpotential, i.e., under conditions that are far from equilibrium. If electrons are transferred in fast steps in the overall process, then the total coverages under presumed equilibrium conditions will be potential dependent, and thus the transfer coefficients (α) as measured experimentally from the Tafel slopes will differ from β. Under Langmuir conditions of low coverage, its value will be $\alpha = n'/v + \beta$, where n' is the number of electrons transferred before the rate-determining step, and v is the stoichiometric number (i.e., the number of times the rate-determining step occurs in the overall process.[85–87] Under high coverage conditions the transfer coefficient must be calculated by taking into account the potential dependence of the $(1 - \theta)^z$ term in (44), together with $\alpha = n'/v + \beta$. Calculations of transfer coefficients of various suggested rate-determining steps for the oxygen evolution process based on this method have been given by Bockris.[88] Under Temkin conditions (medium coverage) the effective transfer coefficients may be calculated by eliminating the coverage-dependent parts of the ΔG°_{ads} (or ΔH°_{ads}) terms in equations (43) and (44) by means of the respective isotherms (see Gileadi and Conway,[74] and Damjanovic and Brusic[89]). Because of the potential dependence of coverage in processes involving successive charge-transfer steps and the relative change in the rates of individual steps with potential, the rate-determining steps will often be a function of both ΔH°_{ads} and overpotential. A good example is given again by the hydrogen evolution reaction. At low coverages (small bond strength or $-\Delta H_H^\circ$) of H the process

$$H^+ + e^- \to H_{ads} \quad \text{(discharge)} \qquad (45)$$

is rate determining, whereas at high coverage (large bond strength) the corresponding process is (see Conway and Bockris[16])

$$H^+ + e^- + H_{ads} \to H_2 \quad \text{(ion plus atom)} \qquad (46)$$

The following reaction must also be considered:

$$H_{ads} \to \tfrac{1}{2}H_2(gas) \quad \text{(combination)} \qquad (47)$$

In the general case the behavior of a reaction sequence as a function of changing ΔH_{ads} and overpotential can only be described by writing the rate equation for the individual steps in each path, using the steady-state assumption (cf. Bockris[88]). However, a shorthand method exists which can give an account of the overall behavior of a system as a function of ΔH_{ads} and overpotential. The method involves the construction of volcano plots for all the steps in each possible path, assuming that each step in each path is in turn rate-determining, with the remaining steps in supposed equilibrium.

The path followed will be the one whose slowest step indicates the greatest rate under the given conditions, and the rate-determining step will be of course this step. This method can be used if the rates of the individual steps are in fact known under some given conditions of coverage, so that the positions of the volcanos can be located relative to each other. An illustration of the fairly successful use of this method will be given later in a discussion of the hydrogen evolution reaction.

Such an analysis, however, ignores a further factor: the contribution of nuclear tunneling.[20,21,139] For strong heats of adsorption the charge transfer barrier will be expected to be considerably narrower in a transition state mechanism, allowing subbarrier leakage. This evidently increases the effective nuclear transition probability τ^* in equations (43) and (44). Such an effect may be important for light atom (proton) transfer. Other effects may also complicate the overall picture: for example, the presence of adsorbed films (of oxygen, or surface-active materials) that block reaction sites,* and also change the electronic properties of the electrode surface,[90] and possibly the electron transmission coefficient. Two other effects must be included, namely possible changes in entropy of the activated state with ΔH_{ads},[3] and the effect of the diffuse double layer.

These matters are discussed separately in more detail below.

IV. FACTORS OTHER THAN ΔH°_{ads} AFFECTING REACTION RATES

1. Nuclear Transmission Coefficient

This quantity, in the absence of nuclear tunneling, may be taken to be approximately unity for a transition state mechanism. For light particles, especially protons, it may exceed unity on account of subbarrier penetration for low and narrow barriers. This situation has been treated in detail by Bockris and Matthews[20] and Christov.[139] Under normal conditions of overall reaction rate the effect of tunneling on the rate is comparatively small (much less than one order of magnitude for protons[20] and negligible for heavier particles).

---

*That is, change of θ_T, especially in cases where the active catalytic area is less than the geometric area of the electrode for reactions requiring special sites.

2. Electron Transmission Coefficient

It is reasonable to expect considerable splitting of the energy terms in the case of electrode reactions involving adsorption. Quasi-adiabatic behavior and κ^* values of the order of unity will therefore be expected generally.

3. Effect of the Diffuse Double Layer

The act of electron transfer will take place preferentially at the outer Helmholtz plane. For charged reactants it is clear that the concentrations in equations (43) and (44) will not be the same at the reaction site as in the bulk solution because of the effect of the diffuse double layer (Frumkin effect[91]). It has been shown[92] that the concentration change is a function of electrode material, since it depends on the potential of zero charge of the electrode, and indirectly, on the electronic work function. Under normal conditions of solution concentration the change of exchange current for a process involving highly charged anions or cations may be considerable—perhaps up to an order of magnitude for tripositive ions.[92] Even so, it will be very much a second-order-of-magnitude effect for reactions involving adsorption. For protons its effect will be small, e.g., a factor of two or three over a range of materials.

4. Effect of the Electronic Structure of the Electrode Material

The density of states term ρ^* in equations (43) and (44) is a weighting factor for each electron level which should appear in the final integrated equation for the overall reaction rate as the total number of states in the energy level region that provides (or accepts) the majority of electrons involved in the process. This may be taken to be about $\pm kT$ around the Fermi level for metallic electrodes. This integral number of states depends on the narrowness of the electron band, and may be taken to be roughly proportional to the density of states and the number of electrode atoms per cm^2 at the Fermi level.[93] For transition metals, where the d band is the major electron donor and acceptor, the effective density of states term may be as much as an order of magnitude higher than in metals where the s band predominates.[94] Wide-gap semiconductors may show even higher effective densities of states in the narrow conduction band.[93,95] However, this will always be more than offset by the effect of the forbidden gap in the material. For semiconductors the integration of the rate equation (26) must be carried out only as far as the conduction or valence band edges, rather than to the Fermi level, as is the case for a metal (cf. Gerischer[14,58] and Levich[54]). A simple calculation shows that in such cases the current density equations (43) and (44) should be multiplied by a factor equal to $\exp[-\beta|E - E_F|/kT]$, where $E - E_F$ is the energy difference between the band edge at the surface under equilibrium conditions for the reaction and the electron energy at the Fermi level. For semiconductors with surface states $E - E_F$ will represent the energy difference between the surface state and the Fermi level. Tafel slope values will, in general, differ from those on metals, because of the potential drop within the semiconductor. A general treatment of this problem has been given.[96] In the case of metal oxide type semiconductors, where conduction is via a valence-change hopping mechanism, a further multiplier of the order of $\exp(-\gamma E_c/kT)$ will appear in the rate equation, where γ is some fraction and E_c is the activation energy for conduction in the material.[97] In general, it would therefore not appear that reaction rates on thick semiconducting electrodes will exceed those on metals under the same heat of adsorption conditions. This may, however, not be true at very thin semiconducting films on other electrodes, where electrons can be transferred by tunneling (for films of up to a few monolayers) and use can be made of the superior adsorption properties of the film. Some examples of this effect will be discussed later.

5. Effect of $\Delta H°_{ads}$ on the Entropy of Activation

In reactions involving adsorption or desorption of a charged particle with accompanying electron transfer it is reasonable to suppose that the total entropy change across the system during the act of transfer will not vary to a great degree with substrate, since it involves (in the absence of considerable vibrational–rotational entropy of the adsorbate) only a change in restricted translational degrees of freedom. However, the entropy involved in destroying or forming the solvation sheath of the charged particle may be considerable. On a simple basis those reactions involving adsorbed products with a small $-\Delta H°_{ads}$ may be expected to have an activated state which resembles the products, and those with a large $-\Delta H°_{ads}$,

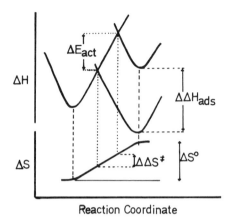

Figure 6. Schematic plot to illustrate compensation effect. Entropy of activation changes with degree of reorganization in transition configuration, implying a change in shape of free energy surfaces according to their relative positions. Zero point energies neglected.

an activated state resembling the reactants (see Fig. 6). The effective charge on the transition state, as seen by the solvent (cf. Hush[40]) will vary along the reaction coordinate, and hence with the $\Delta H°_{ads}$ value. The tightness of binding, hence the vibrational entropy, of the solvation sheath will similarly vary. In the case of proton transfer the solvation sheath must vary enormously—from a tightly bound shell around the proton to disordered water molecules after adsorption of the product (H_{ads}). Thus the entropy of the activated solvation sheath may be expected to partly or wholly compensate the effect of ΔH_{ads} in the reaction in a "compensation effect" (cf. Bond[98]). Evidence for such compensation exists in the electrochemical literature[99,100] and will be discussed in more detail later. Essentially, the product $f^{-\beta}_{ads} f^{\ddagger}$ in equation (31) is then not substrate independent, but varies with $\Delta G°_{ads}$ (or $\Delta H°_{ads}$), i.e., the shape of the free-energy curves for reactants and products varies with $\Delta G°_{ads}$ (see Fig. 6).

V. EXPERIMENTAL RATE CORRELATIONS

1. General

In comparing reaction rates under experimental conditions on different electrode materials, it is important to first ascertain that two necessary conditions are met:[2]

(i) The process in question must be the same on all the electrodes under study. That is, the rate-determining step must be the same in each case, and the reaction path, at least as far as the rate-determining step, must also be the same.

(ii) The process must be compared over a series of electrodes of different composition under identical conditions, so that rates are only affected by interactions between reaction intermediates, the solvent, and the electrode material. In particular, great care must be taken to avoid impurities in the solution which affect rates either by making the electrode surfaces unavailable for reaction or, in less extreme cases, by changing the effective activity coefficients of intermediates, including activated complexes, at the surface. In a similar way, the surfaces on the electrodes must be carefully prepared and free from adsorbed impurities and chemisorbed films. Methods of preparing solutions, reagents, and electrode surfaces have been frequently described in the literature (see Ref. 187). On finely divided materials, for example, metal black electrodes of the type used in fuel cells, poisoning of electrode surfaces by impurities is less important, due to the very large electrode area/solution volume ratio. If electrodes of this type are used, the necessity for intensive solution purification is reduced, but reaction rates at such electrodes are difficult to measure accurately due to diffusion limitations (see, however, Ref. 156), and surface areas, hence real current densities, are difficult to measure. Rates have therefore normally been measured under high-purity conditions on smooth electrodes, using steady-state techniques so that intermediate concentrations will approach their equilibrium values. Transient techniques have been widely used to determine coverages of adsorbed intermediates by coulometry, and in some cases other techniques, for example, ellipsometry, reflectance methods, electron spin resonance, and infrared absorption, have been used to determine the concentrations and nature of the adsorbed species during the course of the reaction. Although there is a great deal of information in the literature on electrode reaction rates as a function of substrate, much of it is not of great value because the necessary precautions have not always been taken to ensure surface reproducibility and solution purity. This is especially true of work in the earlier literature.

The most broadly studied processes involving absorbed intermediates have been: (a) hydrogen evolution in acid solution;

(b) oxygen evolution on noble metal oxides in acid solution; (c) certain hydrocarbon reactions, for example, the oxidation of ethylene.

2. The Hydrogen Evolution Process

It is generally considered that the mechanism of the hydrogen evolution reaction in acid solution on the majority of materials that have been studied involves the sequence of reactions shown in equations (45)–(47).

Attempts to correlate the overpotential of the hydrogen evolution process on different materials date back several decades. Bonhoeffer (1924)[101] noted a connection between the rate of the catalytic combination of hydrogen atoms and the hydrogen overvoltage on several metals. Correlations between exchange current and a number of metal parameters, for example, interatomic distance,[102–105] metal surface energy,[106,107] cohesion energy,[108] melting point,[104] and compressibility[109] have been shown to occur. Bockris and his co-workers[110–112] showed that the thermionic work function of the metal and its hydrogen overvoltage were empirically related. Horiuti and Polanyi,[22] as discussed earlier, showed that the activation energy of the discharge reaction (45) should be affected by the potential energy of the product of that reaction (i.e., adsorbed hydrogen on the metal surface). They reasoned that the intersection point of the potential energy surfaces would be changed by a fraction of the vertical displacement of the potential energy curve for the reaction product, i.e., that the Brønsted rule should apply. If no entropy changes are involved, then the logarithm of the reaction rate should therefore be proportional to increasing heat of adsorption. The first attempt to establish such a relationship was carried out by Ruetschi and Delahay,[113,114] who used data available in the literature[110,115–120] on the rate of the hydrogen evolution reaction on a number of metals. They attempted to calculate the change in activation energy (hence change in reaction rate, assuming a constant entropy of adsorption) by using the equation proposed by Pauling for the calculation of the strength of a partially polarized one-centered covalent bond.[121] This equation takes into account an approximation for the orbital overlap term together with the lattice energy terms in chemisorption. It is reasonable to suppose that the other terms (for example, those due to the image effect,[122] the field at the interface, to particle–particle interactions, and magnetic coupling) may be regarded as being approximately constant over a series of similar substrates, so that the estimation may be regarded as an approximate method of determining the change in the strength of the adsorption bond as a function of material. In its basic form, Pauling's equation[121] may be written

$$D_{M-X} = \tfrac{1}{2}(D_{M-M} + D_{X-X}) + 23.06(\chi_M - \chi_X)^2 \qquad (48)$$

where D_{M-X} is the bond dissociation energy of the M–X bond, D_{M-M} and D_{X-X} are the dissociation energies of the M–M and X–X bonds (all in kcal mole^{-1} units), and χ_M and χ_X are the respective electronegativities of M and X. It must be stressed that this formula is only approximate and there is considerable uncertainty in the electronegativity term. Strictly, it only applies to isolated molecules of each species, but the possibility of making use of the formula for chemisorption processes was first proposed by Eley.[123] In such an application it is necessary to make an approximate estimation of the D_{M-M} parameter (i.e., the bond strength of the metal substrate) in a direction perpendicular to the metal surface. If it can be assumed that the properties of the metal are the same in the bulk and at the surface, and may be directionally averaged (i.e., the material is isotropic), then a rough estimate of this quantity for a metal with a coordination number of 12 will be given by $L_s/6$, where L_s is the latent heat of sublimation of the metal at the temperature of the experiment. This approximation depends on the assumption that the interatomic forces in the metal have a sufficiently short range for the effect of second nearest neighbors to be ignored. An assumption of this type may be criticized on the grounds that although many metallic properties are similar at the surface and in the bulk (for example, interatomic distance), it seems hardly realistic to suppose that electron properties are the same in the bulk metal and at the surface, where differentiation of the common metallic orbital into separate atomic orbitals must occur. These points must be borne in mind when attempting to apply the Pauling equation to chemisorption, and have been discussed in detail by Bond.[124]

Ruetschi and Delahay's calculation was criticized,[16] since it ignored the electronegativity terms (on the grounds that the small moments of M–H bonds imply that electronegativity differences are small, in view of the uncertainty of the absolute electronegativity

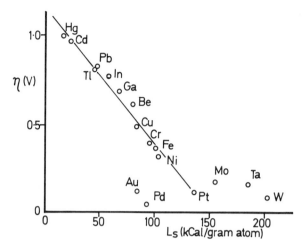

Figure 7. Hydrogen overpotential—electrode latent heat of sublimation correlation (Rüetschi and Delahay[113]).

values that were available). In addition, they made the unjustified assumption that the discharge reaction could be regarded as rate-determining on the metals they considered. However, despite these limitations, their data show an excellent correlation between hydrogen overvoltage and heat of adsorption based only on the Pauling D_{M-M} term (i.e., on the latent heat of sublimation of the metal). Their correlation is shown in Fig. 7. Since L_s is closely related to hardness, melting point, compressibility, and surface energy, the relationship that these authors obtain is in good agreement with other correlations based on the above parameters.[104,106–109]*

Conway and Bockris,[16] developing earlier ideas of Bockris and co-workers,[110,112] showed that an approximately linear relationship appears to exist between log i_0 for the hydrogen evolution reaction in acid solution and the electronic work function of many metals. A second (inverse) relationship is apparent for a few metals (Tl, Hg, Pb) that show very low activity for hydrogen evolution (i_0 in the range 10^{-11}–10^{-13} A cm^{-2} at room temperature). The same authors

*As Bond[124] has pointed out, a good deal of care must be used in interpreting such correlations, in the sense that they must not be used at face value, i.e., a correlation based on, say, compressibility only indicates that the same factors that influence catalysis also influence compressibility.

point out that because of the empirical linear relationships that exist between work function and L_s, and between work function and χ_M, which hold for many metals, it is possible to rewrite Pauling's equation in the approximate form

$$D_{M-X} = A\Phi^2 + B\Phi + C \tag{49}$$

where Φ is the electronic work function, and A, B, and C are constant for a particular adsorbate. In consequence of this relationship, and since $\log i_0$ and Φ are linearly dependent (at least for a considerable number of metals), then a quadratic relationship connects $\log i_0$ and calculated (or experimental) values of D_{M-H}.[126,127] The plot of i_0 (or overpotential) against D_{M-H} that Conway and Bockris obtain has like the $\log i_0$ versus Φ plot, two distinctly linear regions. These have a positive slope (for overvoltage) for the transition metals and group Ib metals, together with (among those metals studied) Al, Ga, and Be, and a negative slope for Hg, Tl, and Pb. All the latter are metals showing very high overvoltage. Conway and Bockris[16] attribute this difference of slope to the fact that on the latter three metals, which have probably much lower D_{M-H} values than were calculated, due to their electronic structures, and show low adsorbed hydrogen coverage under the conditions of experiment, the discharge process (45) is the rate-determining step, whereas on the remaining metals, which show higher hydrogen coverages, the ion plus atom process (46) is rate determining. They thus accounted for the overpotential versus D_{M-H} slopes by arguing that increasing D_{M-H} should cause an increase in rate, from the Brønsted concept of Horiuti and Polanyi,[22] in the case of the discharge reaction, whereas for the ion plus atom reaction, where the adsorbed species occurs on the reactant side in the rate-determining step, the opposite should be true.

The same authors[16] also showed that relationships exist between Φ and the percentage d character (i.e., the percentage d-orbital contribution to intermetallic dsp hybdrid-bond orbitals), which itself is a function of atomic number and interatomic distance.[125,163] This relationship serves to correlate earlier observations of rate dependence on the latter quantity.[102–105] Conway and Bockris[16] show that a linear % d character versus $\log i_0$ relationship for the hydrogen evolution process is observed for the transition metals.

3. Volcano Plots for the Hydrogen Evolution Reaction

Using equations (43) and (44), we can write the forward rate equations at constant potential for the discharge and ion plus atom reactions (45) and (46) (for the special case of a metallic electrode) in the form

$$i = nF\frac{kT}{h}[\text{H}^+](1 - \theta_T)\exp\left[-\frac{\Delta G^\ddagger}{RT} - \beta\frac{\Delta G_\text{H}}{RT}\right]$$

$$\times \exp\left[-\beta\frac{FV}{RT}\right] \quad \text{(discharge)} \tag{50}$$

$$i = nF\frac{kT}{h}[\text{H}^+]^2(1 - \theta_T)\exp\left[-\frac{\Delta G'^\ddagger}{RT} - (1-\beta)\frac{\Delta G_\text{H}}{RT}\right]$$

$$\times \exp\left[-(1+\beta)\frac{FV}{RT}\right] \quad \text{(ion + atom)} \tag{51}$$

In the above equations it is assumed that the discharge step is in quasiequilibrium when the ion plus atom step is rate determining. $n\,(=2)$ is the number of electrons transferred in the overall process (each unit rate-determining step occurring once). ΔG_H is the free energy of adsorption of atomic hydrogen on the electrode surface, which appears in the product state of the discharge process, and in the reactant stage of the ion plus atom process, so that the appropriate Brønsted (i.e., symmetry) factors are β and $1 - \beta$, respectively. One electron is transferred before each unit of the rate-determining step in the ion plus atom process, hence the transfer coefficient for this process is $(1 + \beta)$. The coverage of available metal sites appears in the reactant stage of the rate-determining step in the discharge process and in the product stage of the ion plus atom process, in both cases with a reaction order of unity. V is the potential of the electrode on an arbitrary scale, and the $\Delta G^\ddagger$ and $\Delta G'^\ddagger$ terms are the substrate-independent free energies of activation of the rate-determining step at $V = 0$, and $\Delta G_\text{H} = 0$, and are presumed to contain κ^*, τ^*, and ρ^*.

In order to obtain a complete description of the current density at a metallic electrode as a function of ΔG_H at constant V, it is necessary to establish the form of the relationship between θ_T (the total fractional coverage of material adsorbed on immobile sites) and the concentrations of other materials present in the system. If we assume that no impurities are present, θ_T may be equated to the total coverage with adsorbed hydrogen atoms. Under conditions of considerable overpotential, we can assume tentatively that the ion plus atom reaction (46) is in quasiequilibrium when the discharge reaction is rate determining. Under such conditions, and assuming Langmuir adsorption, we have

$$\theta/(1 - \theta) = [H^+]^{-1} pH_2 \exp(-\Delta G_H/RT) \exp(FV/RT) \qquad (52)$$

or

$$1 - \theta = \{1 + [H^+]^{-1} pH_2 \exp(-\Delta G_H/RT) \exp(FV/RT)\}^{-1} \qquad (53)$$

The equilibrium condition for the discharge process (45) under Langmuir conditions is

$$\theta/(1 - \theta) = [H^+] \exp(-\Delta G_H/RT) \exp(-FV/RT) \qquad (54)$$

or

$$1 - \theta = \{1 + [H^+] \exp(-\Delta G_H/RT) \exp(-FV/RT)\}^{-1} \qquad (55)$$

If expressions (53) and (55) are combined with equations (50) and (51), respectively, typical relationships of the volcano type result.

When the discharge process is rate determining we have

$$i = k_1[H^+] \exp\{-\beta(\Delta G_H/RT) - \beta(FV/RT)\}, \qquad \theta \ll \tfrac{1}{2} \quad (56)$$

$$= k_1[H^+]^2 pH_2^{-1} \exp[(1 - \beta)(\Delta G_H/RT)] \exp[-(1 + \beta)FV/RT],$$

$$\theta \gg \tfrac{1}{2} \quad (57)$$

Similarly, when the ion plus atom process is rate determining

$$i = k_2[H^+]^2 \exp\{-(1 - \beta')(\Delta G_H/RT) - (1 + \beta')(FV\;RT)\},$$

$$\theta \ll \tfrac{1}{2} \quad (58)$$

$$i = k_2[H^+] \exp[\beta'(\Delta G_H/RT)] \exp[-\beta'(FV/RT)], \qquad \theta \gg \tfrac{1}{2} \quad (59)$$

where

$$k_1 = 2F(kT/h) \exp(-\Delta G^\ddagger/RT)$$

and

$$k_2 = 2F(kT/h) \exp(-\Delta G'^\ddagger/RT).$$

By using the same reasoning, the rate equations with the combination reaction (47) in equilibrium, and with the discharge and ion plus atom reactions rate determining may be obtained. When the discharge reaction is rate-determining at $\theta \ll \tfrac{1}{2}$ the expression is identical with (56); similarly, when the ion plus atom reaction is rate determining at $\theta \gg \frac{1}{2}$ the rate is identical with (59). The remaining cases are given by:

Discharge r.d.s., $\theta \gg \frac{1}{2}$:

$$i = k_1[H^+]pH_2^{-1/2} \exp[(1 - \beta)(\Delta G_H/RT)] \exp(-\beta FV/RT) \quad (60)$$

Ion plus atom r.d.s., $\theta \ll \frac{1}{2}$:

$$i = k_2 pH_2^{1/2}[H^+] \exp[-(1 - \beta)(\Delta G_H/RT)] \exp(-\beta FV/RT) \quad (61)$$

Finally, using the same analysis on the combination reaction as a rate-determining step we obtain

$$i = k_3[H^+]^2 \exp[-2(1 - \beta'')\Delta G_H/RT] \exp(-2FV/RT),$$

$$\theta \ll \tfrac{1}{2}, \text{ case 1} \quad (62)$$

$$= k_3 \exp(2\beta \Delta G/RT), \quad \theta \gg \tfrac{1}{2}, \text{ case 2} \quad (63)$$

$$= k_3[H^+]^{-2} pH_2^2 \exp[-2(1 - \beta'')\Delta G_H/RT] \exp(2FV/RT),$$

$$\theta \ll \tfrac{1}{2}, \text{ case 3} \quad (64)$$

case 1 refers to equilibrium discharge, case 2 to equilibrium discharge or ion plus atom, and case 3 to equilibrium ion plus atom; where k_3 is the rate of the combination reaction (in current density units) when the adsorbate is in its standard state (i.e., at $\Delta G_H = 0$), and where β'' is the Brönsted coefficient for the process.

In each case, all the above reactions will have a maximum rate for a particular value of V when ΔG_H is such that $\theta = \tfrac{1}{2}$. If $V = 0$ is taken to be the reversible potential for the process at $p_{H_2} = 1$ atm and $[H^+] = 1$ (i.e., at E_0), then the rates of the discharge, ion plus atom, and conbination reactions become k_1, k_2, and k_3, respectively at $V = 0$, when $\theta = \tfrac{1}{2}$. As Parsons[26] has pointed out, this situation applies approximately at the reversible potential in acid solutions on platinum. Using experimental data, it is therefore possible to estimate the values of k_1, k_2, and k_3, which are the highest possible exchange currents for the three processes in an aqueous solvent at room temperature, under Langmuir conditions of adsorption, and with hydrogen ions and hydrogen gas in their standard states.

According to Parsons' analysis,[26] the k values are about $10^{-1}, 10^{-4}$, and 10^{-2} cm^{-2}, respectively (see Bockris and Azzam[118]).

Using the above k values for the three reactions, and if the β values are known, we can calculate the net rate of the hydrogen evolution reaction as a function of V and ΔG_H. The path followed in each case will be that giving the highest rate, i.e., the pair of reactions involved will be those whose slower step is faster than the slower step in any other possible pair. The equations obtained above will only strictly apply at limitingly high or limitingly low coverage, where the Langmuir isotherm may be expected to be obeyed. In situations where θ lies in the range $0.1 < \theta < 0.9$ the Temkin isotherm will in most cases be more applicable.

For metals with large positive ΔG_H values, θ will still be small under the experimental conditions where Tafel plots for hydrogen evolution occur (i.e., at large negative V values). Equations (56), (58), (61), (62), and (64) will therefore be valid. Assuming $\beta = \beta' = \beta'' = \frac{1}{2}$ in the interests of simplicity, it can be seen that for all mechanisms involving the discharge or ion plus atom reactions as the rate-determining step the rate will depend at constant V on $\exp(-\frac{1}{2}\Delta G_H/RT)$. However, the reaction with the greatest rate at high negative values of V will in general be the ion plus atom process following the discharge process, because of its low Tafel slope ($\sim 2.3 \times 2RT/3F$). This will only be true, however, at overpotentials greater than about 200 mV because of the three orders of magnitude difference between k_1 and k_2. At constant V, mechanisms involving the combination reaction as rate-determining step will have rates depending on $\exp(-\Delta G_H/RT)$. Thus, for large positive ΔG_H, their rates will not be significant, in spite of the favorable Tafel slope ($\sim 2.3 RT/2F$) for the combination reaction following an equilibrium discharge process. In consequence, mechanisms involving the combination reaction, either in equilibrium or rate determining, may be discounted, and the most probable mechanism will therefore be the simple discharge process (rate determining) followed by a quasiequilibrium ion plus atom process. A similar analysis for ΔG_H large and negative (i.e., $\theta \gg \frac{1}{2}$) shows that the most probable reaction mechanism would be rate-determining ion plus atom, preceded by a quasiequilibrium discharge reaction [equation (59)]. Parsons[26] gives diagrams of the Tafel slopes that might be expected in accordance with the above reasoning, and indicates the potential regions where different mechanisms will be rate determining as a function of ΔG_H. In the same paper Parsons (see also Gerischer[82]) derives the volcano relationships for the discharge and ion plus atom rate-determining steps, but assumes that adsorbed hydrogen is in equilibrium with gaseous hydrogen via the combination reaction. This situation (as he points out) will only be true near equilibrium, but in one discussion of hydrogen electrode catalysis[129] it has been claimed that this simple model will account for Bockris and Conway's data[16,126,127] which generally refer to conditions far removed from equilibrium. This clearly cannot be the case. However, as has been shown above, Bockris and Conway's results[16,126,127] are perfectly consistent with the existence of two different mechanisms, both of which give volcano (i.e., ascending and descending log rate ΔG_H) expressions with a maximum rate when ΔG_H for the rate-determining step is equal to zero. However, only the ascending branch (ΔG_H positive, $\theta \ll \frac{1}{2}$) of one mechanism (discharge r.d.s.; ion plus atom equilibrium) is experimentally accessible. Similarly, only the descending branch (ΔG_H negative, $\theta \gg \frac{1}{2}$) of the other mechanism (ion plus atom r.d.s. preceded by discharge equilibrium) is observed. For metals that have ΔG_H close to zero, a third possible volcano expression will arise where, due to their respective k values, the combination reaction will be rate determining with the discharge reaction in quasiequilibrium. The experimental log i (or overpotential) versus ΔG_H plot for hydrogen evolution should therefore consist of the appropriate regions of the volcano plots for the three separate mechanisms superimposed. Mechanisms that apply for hydrogen evolution on different metals have been determined by a variety of methods, including reaction orders, Tafel parameters [giving the number of electrons transferred before* the rate-determining step—see equations (43), (44), and what follows], stoichiometric number determinations,† and isotope

---

*This is only true in simple cases. Results are much more difficult to interpret when the $(1 - \theta)$ term becomes significant in equations (43 and 44), i.e., at high coverage, where θ is potential dependent. As $\theta \to 1$ the ion plus atom r.d.s. following the discharge process in quasiequilibrium thus has a Tafel slope of $RT/\beta F$ [equation 59)], in good agreement with experiment.[16] Under Temkin conditions results can also be very difficult to interpret.

†This quantity may be defined as n'/n, where n is the number of electrons transferred in the overall process for each unit of the rate-determining step [see equations (43) and (44)], and n' is the number of electrons involved overall when the process is written stoichiometrically in the normal way. It may be determined either from the relative slopes of the anodic and cathodic Tafel plots in cases where the same mechanism occurs at considerable anodic and cathodic overpotentials,[137] or, alternatively, from the polarization resistance at equilibrium in cases where this can be measured accurately.[138]

separation coefficients.[4] In general, the various techniques available give results that are in agreement with regard to the mechanism followed, but occasionally conflicting evidence has arisen. For example, mechanistic data seem to indicate that the ion plus atom process is rate determining on Fe electrodes, whereas separation coefficient measurements favor the combination reaction.[131] However, it seems to be generally agreed that the discharge reaction is rate determining on Pb, Hg, and Tl electrodes in the Tafel region[16,126,127,132] (i.e., at high overpotentials), whereas for the group 1b and transition metals (except for some of the platinoid metals at low overpotentials) the ion plus atom reaction is the rate-determining step.[16,136] Near equilibrium the combination reaction is rate determining on Pt and possibly other group VIII noble metals,[132-134] in absence of diffusion of H_2.[130] It will be seen that

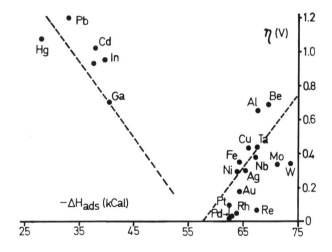

Figure 8. Volcano plots for hydrogen evolution in acid solution. Overpotential data from Conway and Bockris[16] and Kita.[149] Heat of adsorption of H calculated from Pauling equation using L_s values from Bond[98] and electronegativities from Gordy and Thomas.[267] For the sp metals, heat of adsorption is assumed to be 22.5 kcal less than Pauling value (mean between Pauling and gas-phase values). On *sp* metals, discharge is rate determining, followed by fast ion plus atom. On the d metals, except at low overpotentials, ion plus atom is rate determining, followed by fast discharge. Slopes of lines are 23.06 kcal/V.

this is in very good agreement with the above predictions using the volcano expressions for the different groups of metals. The transition metals adsorb hydrogen strongly (i.e., $\Delta G_H < 0$ in the potential range of the experiments), whereas for the platinoid metals ΔG_H appears to be close to zero near equilibrium, and becomes more negative at increasing overpotentials, thus favoring the ion plus atom process as the rate-determining step. Such results are in good agreement with Bockris and Conway's calculations of bond strength[16] based on the Pauling–Eley equation.[121,123] For the remaining group of metals (Pb, Hg, and Tl), bond strength values calculated from the Pauling–Eley equation are likely to be some 40–50 kcal mole^{-1} too great, as Conway and Bockris point out,[16] due to the electronic structures of the metals. When this correction is made the general agreement with the predictions of the volcano expressions is very good, as shown in Fig. 8, which also contains results for other metals reported by Kita.[149] The plot for metals that show the ion plus atom process intersects the axis corresponding to zero overpotential (at a current density of 10^{-3} A cm^{-2}) at a D_{M-H} value calculated from the Pauling–Eley equation of about 58 kcal mole^{-1}. The slope of the line that has been inserted is that calculated from equation (59), assuming that changes in $-\Delta G_H$ may be equated to changes in bond strength. Since, according to Parsons,[26] the rate of the ion plus atom reaction at $V = 0$ and $\Delta G_H = 0$ should be about 10^{-4} A cm^{-2}, then, to a rough approximation, 10^{-3} A cm^{-2} corrresponds to an overpotential of $-4.6RT/F$, or -0.12 V, for a metal with $\Delta G_H = 0$. This corresponds to a D_{M-H} value of about 61 kcal mole^{-1}, which should therefore be equated with $\Delta G_H = 0$. By contrast, the discharge reaction at $V = 0$ and $\Delta G_H = 0$ should have a rate equal to about 10^{-1} A cm^{-2}, so that a line of the same numerical slope, but of opposite sign [from equation (56), at constant i] should pass through the overpotential versus D_{M-H} coordinates of metals that show the discharge rate-determining step, and should pass through an overpotential of about $+(2 \times 4.6RT/F)$, or $+0.24$ V, for a metal with $\Delta G_H = 0$ (i.e., $D_{M-H} = 61$ kcal). The D_{M-H} values for Pb, Hg, Cd, Zn, In, Ga, and Tl in Fig. 8 correspond to those calculated from the Pauling equation minus 45 kcal mole^{-1},[16] and the D_{M-H} versus overpotential slopes are $+23$ and -23 kcal V^{-1} for the discharge and ion plus atom reactions, corresponding to the differentials of equations (56) and (59) at constant i. Although extrapolation in Fig. 8 has been carried out over a considerable overpotential range, the agreement with the above predictions is too good to be fortuitous, although some scatter does exist for metals (e.g., Al and Be) which are covered with oxide under the conditions of the experiments. Re, Mo, and W are anomalous, but are also oxide covered and can form hydrogen bronzes.[152] No data are shown plotted for metals on which the combination reaction may be rate determining (i.e., some of the platinum group at low overpotentials), but Pt, Rh, and Pd show overvoltages of ~ 50 mV or less at 10^{-3} A cm^{-2}, and all have estimated D_{M-H} values in the 62–64 kcal mole^{-1} range, i.e., close to $\Delta G = 0$. This estimated ΔG value is also in accordance with the hydrogen coverage values in the medium coverage range that occur on these metals.[26]

Although the overall picture is therefore satisfactory, certain more detailed problems remain. One of these is quite obvious—the relation between the approximate calculated heat of adsorption values to experimental data at low coverage. A second, and related, point is the effect of coverage dependence on the heat of adsorption values. Finally, there is the problem of both entropy of adsorption and of activation, which may possibly vary, at least in some cases, with heat of adsorption. If this should be the case, then the assumption that heat changes will be equivalent to free energy changes is clearly not justified.

The first point has been discussed at length by Conway and Bockris,[16] who provide a list of spectroscopic and chemical data for the D_{M-H} bond strengths of a number of metals. The agreement between calculated and experimental D_{M-H} values is generally quite good, provided a correction is made in the Pauling calculation for the electronic structures of Pb, Hg, and Tl, and provided the values as a whole are corrected for the "image energy," which is of the order of 15 kcal mole^{-1} for an ideal metal. When these corrections are made the agreement between calculated and experimental data is often within 2–3 kcal mole^{-1}.

The effect of changing heat of adsorption with coverage should be most apparent on the descending side of the volcano, i.e., with those metals that have the ion plus atom reaction as rate-determining step. Conway and Bockris[16] give a log i_0 versus heat of adsorption plot for these metals, using experimental, rather than calculated, values for heat of adsorption of hydrogen. This plot is shown in

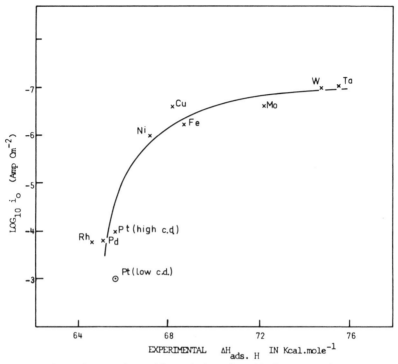

Figure 9. Log i_0—experimental heat of adsorption of hydrogen correlation for H_2 electrode (Conway and Bockris[16]).

Fig. 9. According to equation (59), the slope of the $\log_{10} i_0$ versus ΔG_H plot should be $-\beta/2.3RT$ (about 0.36 with $\beta = \frac{1}{2}$). It is clear from Fig. 9 that the slope of the experimental plot is considerably greater than this in going from Pt (at low overvoltage) to Ni, and considerably less on going from Ni to Ta.

It is reasonable to suppose that considerable discrepancies would be expected in this area of the volcano plot, depending on the length of the Temkin region (in the cases where it occurs) on each metal. Unfortunately, there are insufficient data to be able to test this relationship, but it seems possible that it may explain the large rate change on going from Pt to Ni. The problem becomes more complicated by the fact that, at least for Pt and Ir, there appear to be at least two different types of adsorbed hydrogen with differing bond strengths on the surface, i.e., the metal may have to be regarded as a heterogeneous surface.[99] If the proportions of these two types of adsorbate differ on different metals, this could cause further discrepancy between calculated and experimental rates.

Finally, it is perhaps of some interest to examine briefly the maximum rates of the discharge, combination, and ion plus atom reactions at 25°C under standard conditions. According to the charge transfer theory discussed in Section II, adiabatic proton transfer under standard conditions should have a rate that is determined by the reorganizational energy of the system. All three reactions may be regarded, in a broad sense, as proton transfers. It is clear that in the discharge process the total reorganizational energy involved is equal to the net solvent energy difference between a proton in solution and one in the adsorbed state. For the combination reaction it is the difference between two adsorbed protons and atomic hydrogen. Hence, the reorganizational energy in the ion plus atom reaction[46] should be equal to the sum of the corresponding energies in the remaining two steps, regarding all processes as cross-reactions.[44] The logarithm of the rate of the combination reaction should therefore be the mean of the logarithms of the discharge and ion plus atom steps under standard adsorbate conditions. This is in very good agreement with Parsons' figures[26] for the relative rates of the three reactions, if $i_{\text{discharge}} = 10^{-1}$, $i_{\text{ion plus atom}} = 10^{-4}$, and $i_{\text{combination}}$ is in the range 10^{-2}–10^{-3}, with the rates expressed in A cm^{-2} units.

4. Heats of Absorption and Frequency Factors for Hydrogen Evolution

It is clear that much of the doubt regarding the relative effect of the heat of adsorption of hydrogen could be removed if reliable data on activation energies at constant potential for the hydrogen evolution reaction were available. Using equations (56) and (59), we see that (with $\beta = \frac{1}{2}$) the change in activation energy in going from metal to metal at constant potential should be equal to $+\frac{1}{2}\Delta\Delta H$ on the rising side of the volcano ($\Delta G_H > 0$) and $-\frac{1}{2}\Delta\Delta H$ on the falling side of the volcano ($\Delta G_H < 0$), where $\Delta\Delta H$ is the change in heat of adsorption with substrate. At present very little data are available on rates as a function of temperature. Glasstone et al.[25]

quote some early values that indicate a general similarity of preexponential terms. Later values, obtained under conditions of high purity by Parsons[135] for the combination and discharge reactions on platinum, indicate that the activation energy is in both cases about 5.2 kcal mole^{-1} at the reversible potential. In the same work Parsons[135] attempts to theoretically derive the preexponential terms for the combination, ion plus atom, and discharge reactions. The model used is based on classical absolute rate theory. For the discharge reaction Parsons finds about 2×10^6 A cm^{-2} ($\theta \to 0$) and 2×10^7 A cm^{-2} ($\theta \to 1$); for the combination reaction 3×10^2 A cm^{-2} ($\theta \to 0$) and 2×10^7 A cm^{-2} ($\theta \to 1$); and for the ion plus atom reaction, assuming a mobile activated complex, 40 A cm^{-2} ($\theta \to 0$) and 10^4 A cm^{-2} ($\theta \to 1$).

Perhaps the most interesting aspect of Parsons' calculations is the proposition that the adsorbate may be mobile. This does represent one approach by which an entropy of adsorption that depends on the strength of the adsorption bond[99] might be explained. Another possible approach is via the entropy of vibration of the chemisorbed species, provided that the vibrational frequencies (either stretching, i.e., movement perpendicular to the adsorbing surface, or bending, parallel to the surface) are less than kT/h. These possibilities will be discussed in further detail later, with reference to the oxygen electrode. In the case of hydrogen adsorbed on metal surfaces in electrolytic solutions there is little doubt about the fact that the assumption that entropy changes may be neglected is in many cases an oversimplification. There is evidence that a compensation effect between ΔS_H and ΔH_H (i.e., the entropies and heats of adsorption) exists, implying that on surfaces where D_{M-H} is high ΔS_H is low, implying a high degree of order in the adsorbed film and solvent layer. Similarly, on surfaces with low D_{M-H} increasing disorder (translational or vibrational) and hence an increased ΔS_H result. This implies that free energy changes will in general be less than those assumed from heat of adsorption values with no entropy correction. This point is very well illustrated by Breiter's[99] experimental correlation of ΔH_H and ΔS_H on platinum in different electrolytes, where D_{M-H} is modified by differing degrees of anion adsorption. In addition, the configurational entropy of the solvation sheath of the activated complex may depend on ΔH_H, as discussed in Section III.

In order to relate heat and entropy factors to M–H adsorption Conway et al.[140] measured hydrogen evolution activation energies on a series of copper–nickel alloys. They found that in each case the activation energy appeared to be temperature-dependent, but the value on copper was less than that on nickel (in agreement with other workers[141,142]). The values for the alloys were nonlinearly dependent on the atom percentages of the components. However, at constant overpotential the difference in the heats of activation on copper and nickel was shown to be reasonably consistent with the experimental values of D_{M-H} on the two metals, assuming that equation (59) applies (ion plus atom is the rate determining step at high coverage). Thus, the fact that i_0 on nickel is higher than the value on copper indicates that the Arrhenius pre-exponential factors* are very different on the two metals, and shows the operation of a "compensation effect," to which Conway et al.[140] directed attention. Similar effects have been noted by Gossner et al.[140a]

Recently Bockris et al.[143] have examined the hydrogen electrode reaction on a number of pure metal and alloy systems, though unfortunately rates were not examined as a function of temperature. The systems were chosen to minimize the change in one particular parameter, for instance, d-orbital vacancies or internuclear distance. The results they obtained again show the dependence of rate constant on percent d-character and d-orbital vacancies, with a particularly good correlation based on the former. In addition, i_0 appears to be dependent on internuclear distance (which depends on L_s and electronegativity). Essentially, this work confirms that for similar systems reaction rate depends on those factors that control the strength of the M–H bond.

The general picture that emerges is an approximate dependence of log i at constant V on the heat of adsorption of atomic hydrogen in the directions predicted by theoretical considerations, but experimental evidence on the effect of the preexponential factor is not at present available. It would appear that future research should be directed toward obtaining data as a function of temperature, and measuring free energies and heats of adsorption, preferably in an electrochemical environment.

---

*Like $\Delta H^\pm$, $\Delta S^\pm$ for an electrode process cannot be measured absolutely. *Differences* for various metals can, however, be meaningfully examined.[140]

5. Electrocatalytic Studies in Other Systems

Ruetschi and Delahay[144] examined the Hickling and Hill[145] figures for oxygen evolution overpotential at constant current density in alkaline solution. They considered the data in terms of the strength of the M–OH bond, using arguments similar to those used previously for the hydrogen evolution reaction. The rate-determining step for the oxygen evolution reaction on a number of metals in acid solution, and probably also in alkaline solution, appears to be a water-discharge process of the type[87,146]

$$S + H_2O \rightarrow S{-}OH + H^+ + e^- \quad (65)$$

in which S is an adsorption site on the surface of the electrode, which is invariably covered with oxide at the high overpotentials encountered. Oxide films on Pt reach a thickness of several angstroms on Pt under oxygen evolution conditions. Under these circumstances, it is reasonable to consider the S–OH bond strength as approximating to the bond strength of the metal hydroxide. Ruetschi and Delahay[144] established that a good correlation between bond strength and overpotential appears to exist (Fig. 10).

Dahms and Bockris[147] and Kuhn et al.[148] have examined an organic electrochemical reaction (ethylene oxidation) in acid solution on a series of noble metals and alloys. The latter workers

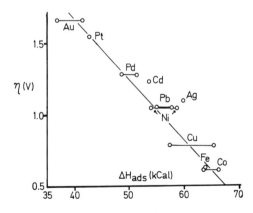

Figure 10. Correlation between oxygen evolution over-potential and M–OH bond strength (Ruetschi and Delahay[114]). Slope of line is 23.06 kcal/V.

Experimental Rate Correlations

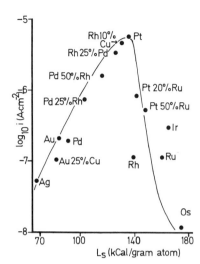

Figure 11. Latent heat of sublimation − log i plot at constant potential for ethylene oxidation (Kuhn, Wroblowa and Bockris[148]). L_s values for alloys considered proportional to composition (expressed at atom %).

established that a volcano-type relationship existed when log i at constant overpotential was plotted against d-orbital vacancies or L_s (cf. Kita,[149] for H_2 evolution). They show that rates increase by a factor of about 100 on going from Au to Pt, and show a subsequent fall by a factor of about 1000 on going from Pt through Ir, Rh, and Ru to Os. Their L_s versus log i plot is shown in Fig. 11.

The same authors[148] show that ethylene oxidation has a small negative order for ethylene on platinum, with a small positive order on the other substrates considered. By considering Tafel parameters and the relative reaction rates in normal and deuterated electrolyte, they conclude that the most probable reaction sequence is of the form

$$C_2H_4 \rightleftharpoons (C_2H_4)_{ads} \tag{66}$$

$$H_2O \rightarrow (OH)_{ads} + H^+ + e^- \tag{67}$$

$$(C_2H_4)_{ads} + (OH)_{ads} \rightarrow (-CH_2-O-CH_3)_{ads} \tag{68}$$

The adsorbed ethylene molecule is perhaps dissociated on the surface of the metal (but see Ref. 190). The rate-determining step is either the water discharge reaction (67), which applies only in the case of platinum, or the reaction of adsorbed OH radicals on the electrode surface with the adsorbed organic molecules (68). The latter process appears to apply to all the other substrates studied. This concept gives a satisfactory explanation of the negative reaction order observed on platinum, as increasing coverage of ethylene decreases the number of sites available for reaction (67). Wojtowicz and Conway[148a] showed, however, how other conditions, especially Temkin adsorption, could give rise to negative order without the assumptions of Ref. 148. Thus, with the exception of platinum, it would appear that the most satisfactory explanation of the rate variation with substrate in a volcano form is given by the assumption that the rate of the chemical step (68) first rises with increasing free energy of adsorption of the reactants and products in the rate-determining step, and finally falls after having gone through a maximum as the number of available metal sites, i.e., the $(1 - \theta)$ term in equation (44), falls with increasing free energy of adsorption.

There is a considerable amount of information in the fuel-cell literature concerning the rate of organic oxidation reactions (for example, of methanol, hydrocarbons, and carbon monoxide) as a function of the substrate. Unfortunately, little if any of this information is of value in establishing relationships between reaction rate and the adsorption properties of the substrates, since rates have generally been measured on porous electrodes under conditions that were almost entirely under diffusion control. (See, however, Ref. 156 for a method of obtaining kinetic data on porous electrodes.) However, some of the more striking examples of improved rates have been achieved by the use of noble metal alloys, for example, Pt–Rh or Pt–Ru,[150] or even mixtures of such alloys with other materials, for example WO_3.[151] Insufficient data are available to establish why such materials show improved reaction rates, although for WO_3 mixtures the formation of hydrogen tungsten bronzes as reaction intermediates has been suggested.[152] If the reactions do not proceed by a redox couple mechanism (i.e., a series of chemical steps in which the organic material reduces the electrode material to a lower valence state, followed by an electrochemical step, which would normally not be rate determining, in which the electrode material is oxidized to its original valence state), then the reactions seem to imply a mixed interface situation of the type discussed by Parsons.[268] As Parsons points out, a reaction proceeding only at interfaces, (because of their favorable free energy of adsorption conditions) would not be expected to show a high preexponential term, since the probability of finding a suitable reaction site will be small.[268] However, recent work on hydrogen evolution at growing electro-deposited ruthenium films on a mercury substrate suggests that the current may be concentrated at crystal edges.[153] It is not clear, however, if this effect is intrinsic, i.e., due to the fact that higher free energies of adsorption may exist in such circumstances, since atoms at the edge of a growing film will have fewer closest neighbors, or extrinsic, i.e., due to improved nucleation of hydrogen or enhanced (radial) diffusion at the film edges.[154] Other studies have shown that for hydrogen evolution on noble metal amalgams electrocatalysis is not associated with individual (i.e., dissolved) noble metal atoms, but appears to be a genuine bulk property of the noble metal lattice.[155] Further information is therefore clearly required before the interpretation of these and related "interfacial" cases of catalysis can be satisfactorily explained.

VI. ELECTROCATALYSIS AND THE OXYGEN ELECTRODE

1. Introduction

It is quite natural to compare the hydrogen and oxygen electrodes, both from a technological viewpoint, as regards their relative performance when used in energy conversion devices, and also from the theoretical viewpoint, as the hydrogen electrode has been by far the most widely studied of the electrochemical processes that can be termed "electrocatalytic," i.e., those involving a rate-determining step with adsorbed intermediates. Certain very important contrasts are apparent between the two reactions[157]:

(i) Because the standard potential of the oxygen electrode, i.e., for the overall process

$$O_2 + 4H^+ + 4e^- \rightarrow 2H_2O \qquad (69)$$

is 1.23 V on the standard hydrogen scale, it falls above the standard potentials of almost all of the solid elements. The choice of suitable substrates for the electrode material in this reaction is therefore small. Among the metallic elements, all materials except perhaps gold are thermodynamically unstable with respect to water in the potential range 0.9–1.2 V SHE. In acid solution, therefore, those metals that do not spontaneously dissolve (i.e., do not form passivating oxides) will at least be covered by oxide films or adsorbed oxygen derived from the oxidation of water, except for gold, below about 1.1 V. Only the group VIII noble metals, together with gold, can be considered among the metals as possible substrates, although in many cases their surfaces show high coverages of adsorbed oxygen at potentials about 0.8 V or above. Other possible stable substrates, for example, interstitial compounds, are likely to have totally oxidized surfaces. The only remaining possibilities as catalysts in acid solution are stable semiconductors, for example, insoluble oxides and certain organic materials. In alkaline solution the situation is more favorable, since the oxygen electrode potential is about 800–850 mV more negative due to the pH difference. In addition, the choice of oxides is much wider, since the majority of metal oxides are insoluble in alkali. Further, it must be remembered that substances that are thermodynamically unstable often show so slow a transformation that they can be considered stable from the practical point of view.

(ii) Interpretation of mechanisms is often difficult because of changes in the surface state of the catalyst. In the well-documented case of platinum, at potentials above about 1.0 V versus the hydrogen electrode in the same solution,* true phase oxides are present on the surface, i.e., Pt atoms are pulled out of their original lattice positions.[157,163] Consequently, the oxygen evolution reaction always takes place on a platinum oxide substrate.[87,158] Because the oxide, once formed, is highly irreversible,[159–161,165] reduction of oxygen at potentials below 1.0 V may be on either an oxide-covered surface or on a surface that is nominally oxide free.[166] However, electrodes

---

*Following a convention used in earlier work,[158] potentials with respect to an isothermal hydrogen electrode in the same solution will be referred to as HRE (hydrogen reference electrode) potentials. It should be noted that when used as a working tool the quoted potentials of such electrodes (at 1 atm H_2 nominal) are rarely corrected for electrolyte vapor pressure.

with no previous history of oxidation have fractional coverages of adsorbed oxygen at potentials above about 0.8 V HRE.[89,164] This oxygen adsorbate, derived from the oxidation of water molecules,[89,164] is essentially reversible.[159,167] The oxygen electrode mechanism on the phase oxide-covered surface[87,146,158] is quite different in both mechanistic and kinetic character from that on the reduced surface,[89,168,169] which is only accessible in the cathodic direction. This situation is quite distinct from the corresponding one for the hydrogen electrode, where no separate surface phases occur at normal overpotentials, and where a surface of the same type, involving equilibrium coverage of hydrogen atoms, exists in both the anodic and cathodic directions. The initial stages of Pt surface oxidation are reversible, as shown by cyclic voltammetry and modulated reflectance in the work of Kozlowska et al.[169a] and Gottesfeld and Conway.[169b]

(iii) A very large number of reaction intermediates, hence possible mechanisms and rate-determining steps, may occur for the oxygen electrode. This is again in complete contrast to the hydrogen electrode, where the only intermediate appears to be adsorbed hydrogen atoms (H_2^+ has also been suggested[162] as a possible intermediate, but there is little evidence for its occurrence other than as an activated complex). Some of these intermediates (for example, H_2O_2, HO_2^-, and, in some solvents, O_2^-) are sufficiently stable under some conditions to be the end products of the reaction. If this situation occurs, then the potential of the working oxygen electrode will be far below that predicted for the overall four-electron reduction process [reaction (69)]. For example, the standard potential for the overall process

$$O_2 + 2H^+ + 2e^- \to H_2O_2 \quad (70)$$

is only 0.68[170] (or 0.71[171]) V SHE. In energy conversion devices it is imperative that the highest possible thermodynamic efficiency is attained for economically feasible operation, hence the reduction of oxygen to water must take place as the overall process.

(iv) Like the hydrogen electrode, the oxygen electrode is seriously affected by impurities[146,169]; consequently, in order to obtain reproducible data, it is necessary to take special precautions to remove contaminants. In the case of the hydrogen electrode the most likely trace impurities to affect rates markedly will be heavy metal ions, whose electrodeposits show high overpotentials (e.g., Pb, As), as well as such obvious possibilities as organic greases. With oxygen the most obvious possibilities are oxidizable substances, for example, traces of organic materials and hydrogen and possibly hydrogen peroxide. Extended preelectrolysis[187] will usually suffice to remove impurities from solutions for hydrogen electrode studies, but in oxygen electrode work this technique should be used with considerable care, especially if platinum electrodes are used, since under certain conditions anodic preelectrolysis in the working electrode compartment can seriously contaminate the electrolyte with Pt and introduce considerable errors in electrocatalytic studies. Similarly, platinum counterelectrodes are best avoided.[172] Gold (which has a low oxygen electrode activity) is a satisfactory substitute.[172] In cases where it can be easily decomposed by strong heating (for example, in concentrated phosphoric acid), hydrogen peroxide treatment provides a useful way of removing oxidizable matter from the electrolyte,[158] Catalytic pyrodistillation of water appears to be the most satisfactory method for removal of organic contaminants.[187a] As for the hydrogen electrode, purification of the electrolyte when large-area black electrodes are used (e.g., fuel cell electrodes) is much less critical.[156]

(v) The most striking contrast between the behavior of the noble metals when functioning as substrates for oxygen and hydrogen electrodes is the very large difference in rate between the two systems. Typically, exchange current values for the noble metals are in the 10^{-10}–10^{-11} A cm^{-2} range in normal alkaline solution at room temperature, with a relatively small range of variation between the group VIII noble metals Pt,[89] Pd,[173] silver,[174] and gold.[173,175] In dilute acid Pt is the most active metal,[89] closely followed by Pd[176] (i_0 values about 10^{-10} and 10^{-11} A cm^{-2} respectively), with gold about $2\frac{1}{2}$ orders of magnitude worse.[176] These are many orders of magnitude less than the hydrogen evolution exchange currents on the same metals, and must be counted among the slowest group of electrode reactions that are known.*

(vi) In practical systems under normal conditions (e.g., fuel cells) the low rate of oxygen reduction process on platinum is

---

*Along with hydrogen evolution on the soft metals, such as Pb, Hg, and Tl,[16] and hydrocarbon oxidation processes.[148] We discount such oddities as N_2 evolution from azides.[188]

shown by the fact that the reversible potential is rarely observed unless special precautions are taken. In general, open-circuit potentials of 1 V or less are observed, which correspond to mixed potentials set up between cathodic oxygen reduction and spurious anodic processes, usually either oxidation of impurities[164] in the system or adsorbed oxygen formation on platinum.[177] In some cases a reversible O_2/H_2O_2 potential, representing the equilibrium in reaction (70), may be set up. This seems to be the case on Hg,[177] C,[178] and Au[163,175] electrodes, and possibly also on other substrates, in alkaline solution. In acid solution, potentials set up at platinum electrodes in oxygenated H_2O_2 solutions are independent of O_2 partial pressure, and are controlled by decomposition mechanisms.[179]

The significance of the low rest potential has been discussed by many authors,[189,163,164,177,180,181] and its theoretical importance has probably often been overemphasized. On platinum electrodes that are initially in a clean, phase-oxide-free state, rest potentials of about 1.0 V HRE are frequently observed at 1 atm oxygen partial pressure.[89,163] This appears to be so even in solutions in which the most stringent precautions with regard to purification have been taken. Such electrodes have the high log p_{O_2} dependence of RT/F after short times at rest,[89,164,180] which falls to lower values after long periods at the rest potential.[177,181] These rest potentials have been explained by means of various models, including potentials based on a peroxide radical system,[180] oxide mixed potentials,[177] and mixed potentials due to a limiting current for impurity oxidation.[164] In view of the very small peroxide radical concentrations required to maintain a rest potential of about 1 V HRE, and the pronounced irreversibility of the oxygen–hydrogen peroxide couple even at high hydrogen peroxide concentrations, the first model may be regarded as untenable for platinum in acid solution.[163] The impurity model gives a good account of the existence of the RT/F dependence on log p_{O_2},[164] but has difficulty in accounting for the lower value observed after long times,[177,180] and the lack of stirring dependence of the rest potentials.[177] In addition, it is hard to see how it can apply in the very purest solutions, in which currents down to 10^{-11} A cm^{-2} have been measured in the oxygen electrode reaction. Recently, a comprehensive mixed potential theory involving adsorbed oxygen film formation, and above 1.0 V, oxide film formation, as the anodic processes, has been given.[182] This appears to give a reasonably adequate account of the experimental data available in acid solutions for platinum electrodes. The details of the treatment are irrelevant to the present discussion, but they follow directly from the assumption of a Temkin isotherm operating for the adsorption of oxygen on platinum,[89,164,169] where the oxygen, in the form of O and OH radicals, is derived from water oxidation. This process is discussed in more detail below.

2. The Oxygen Electrode on Platinum Oxide Surfaces

It is logical to first consider electrocatalytic processes on oxide substrates, because such systems are in many cases less complex than the corresponding processes on oxide-free metals, and because, historically, they appear to have been the first systems to be studied. In early work emphasis was on the oxygen evolution process. Bowden (1929)[183] studied oxygen evolution (on platinum) in dilute sulfuric acid and determined the Tafel slope to be about $2.3 \times 2RT/F$. He also determined the temperature dependence of reaction rate. Hoar (1933)[184] made an extensive examination of oxygen evolution and reduction on platinum electrodes in a variety of electrolytes, and noted that it took a considerable time to attain a reproducible potential at constant anodic current in dilute sulfuric acid or dilute KOH solution. He attributed this effect to the gradual thickening of the oxide film on the metal surface. After a reproducible surface had been attained, he was able to confirm Bowden's[183] $2.3 \times 2RT/F$ Tafel slope in acid solution. Hoar[184] was able to show that anodic and cathodic Tafel slopes on platinum electrodes that had been subjected to previous heating in air intersected close to the calculated potential for the overall four-electron oxygen reduction process, reaction (69). Exchange current values were very low. Apart from the latter observation, the oxygen electrode was thus shown to exhibit normal behavior, even though it was not thermodynamically reversible at equilibrium.

Bockris and Huq[146] later showed that this irreversible behavior at equilibrium could be explained (for electrodes that had been previously well anodized) by the presence of depolarizing impurities in the system. They argue that the very low oxygen electrode exchange current is in general some orders of magnitude less than impurity limiting currents, and therefore mixed rest potentials, lower than the reversible oxygen potential, will inevitably occur in practice. In oxygenated dilute sulfuric acid that had previously been purified by extensive preelectrolysis they showed that well-anodized platinum electrodes took up rest potentials close to the thermodynamic value for reaction (69).[146] They were also able to show that the effect of oxygen partial pressure on this equilibrium was approximately that predicted by the Nernst equation for reaction (69). This conclusion has since been confirmed by other workers.[185,186] As in previous work, Bockris and Huq[146] were able to show that on well-anodized electrodes anodic and cathodic Tafel slopes were close to $2.3RT/F$ in dilute sulfuric acid.

Both Hoar[85] and Bockris and Huq,[146] were able to derive probable mechanisms for the anodic evolution and cathodic reduction processes in dilute acid and alkali, using stoichiometric number data derived from the Tafel slopes and (in the case of the latter workers) from the polarization resistances under reversible conditions. They started from the basic proposition that the mechanism of reaction and the rate-determining step was the same in the anodic and cathodic directions. This is reasonable in view of the intersection of both anodic and cathodic Tafel lines (even when extrapolated from high overpotentials) at the reversible potential for the overall process, which implies identity of rates at equilibrium, and hence, identity of rate-determining step. Bockris and Huq[146] showed that under steady-state conditions, and assuming Langmuir adsorption conditions for reaction intermediates, the reaction

$$S + H_2O \rightarrow S-OH + H^+ + e^- \tag{71}$$

was the most probable rate-determining step for Pt in acid solution. In alkaline solution Hoar[85] proposed a chemical rate-determining step in a mechanism involving a simultaneous two-electron transfer reaction to occur for the low observed anodic and cathodic Tafel slopes. In a reexamination of the available data, Riddiford[191] made the suggestion that the same mechanism may apply in both acid and alkaline solutions. He argued that the nonintegral values of the experimental stoichiometric numbers, and the change in mechanism under non-steady-state conditions, noted by Bockris and Huq[146] in acid solution, were best explained in terms of two separate charge transfer rate-determining steps, one controlling anodically, the other cathodically. Whereas Bockris and Huq suggested a mechanism of the type (in the anodic direction)

$$4(S + H_2O \rightarrow S\text{–}OH + H^+ + e^-) \tag{72}$$

$$4S\text{–}OH \rightleftharpoons 4S + O_2 + 2H_2O \tag{73}$$

Riddiford[191] proposed that the mechanism

$$2(S + H_2O \rightarrow S\text{–}OH + H^+ + e^-) \tag{74}$$

$$2(S\text{–}OH \rightarrow S\text{–}O + H^+ + e^-) \tag{75}$$

$$2S\text{–}O \rightleftharpoons O_2 \tag{76}$$

might instead determine the rate. Using the steady-state approximation for the concentration of adsorbed OH, the rate of such a mechanism can be written in the form

$$i = 2F \frac{k_1 k_2 \exp[(1 - \beta)FV/RT]}{k_{-1}[H^+] \exp(-FV/RT) + k_2} \quad \text{(anodic)} \tag{77}$$

$$= 2F \frac{k_{-1} k_{-2} \theta [H^+]^2 \exp[-(1 + \beta)FV/RT]}{k_{-1}[H^+] \exp(-FV/RT) + k_2} \quad \text{(cathodic)} \tag{78}$$

in which k_1, k_{-1}, k_2, and k_{-2} are the standard forward and backward rate constants of the reactions (74) and (75), respectively, at $V = 0$. θ is the coverage of adsorbed oxygen on the electrode surface, presumed to be in equilibrium with molecular oxygen. We can see that if, at potential V, $k_2 \simeq k_{-1}[H^+] \exp(-FV/RT)$, then reaction (74) will control anodically at potentials above V, while reaction (75) will control cathodically at potentials below V. This V value will clearly show a pH dependence of $-RT/F$, as does the equilibrium potential. Accepting that the critical V value is close to the reversible potential for the overall process, Riddiford's concept is therefore reasonable, and reaction order evidence for the participation of mixed mechanisms of this type in the oxygen electrode processes on oxidized platinum has been given. However, this view seems to be in conflct with considerations based on comparative rates on the noble metal oxides, as will be discussed in Sections VI.4 and VI.5. Although considerable progress has been made toward understanding the oxygen electrode mechanism on oxidized platinum in acid solution, it is very difficult to carry out definitive experiments that would clearly differentiate between alternative rate-controlling steps and reaction mechanisms, due to the instability of the oxide film at cathodic overpotentials.

In alkaline solution Riddiford's mechanism[191] is not strongly supported by the recent results of Damjanovic et al.,[87] at least if Langmuir conditions of adsorption are assumed. In principle, it is, however, possible to account for the Tafel slopes assuming that a Temkin isotherm holds for the adsorption of –OH in mechanism (72)–(76), provided that the coverage of other adsorbed species (principally adsorbed –O radicals) is small enough to be ignored.[158] This restriction may be unrealistic, and in addition it is difficult to account for the wide potential range in which the Temkin isotherm is presumed to operate (about 0.9 V). This point has been discussed, for other mechanisms, by Damjanovic et al.,[87] and on this basis the path is unlikely. In alkaline solution, the most probable path under Langmuir conditions is that originally proposed by Krasil'shchikov[192] for nickel electrodes, which may be written in the form

$$S + OH^- \rightarrow SOH + e^- \qquad (79)$$

$$SOH + OH^- \rightarrow SO^- + H_2O \qquad (80)$$

$$SO^- \rightarrow SO + e^- \qquad (81)$$

$$2SO \rightarrow O_2 \qquad (82)$$

The chemical step (80) is rate determining cathodically and to potentials up to about 1.6 V HRE anodically. At higher potentials the discharge step (79) becomes rate determining.

The thickness of the oxide film on the surface of platinum electrodes under anodic conditions has been examined by a large number of authors.[87,158-161,165] Thicknesses up to several monolayers have been shown to occur at high potentials (but see Ref. 161). Coulometric methods, and also ellipsometry,[160] have been used for film thickness determinations, and there is generally good agreement between the two techniques. It is generally accepted that the film consists of Pt oxide phases whose stoichiometry varies with potential.[161,193] X-ray diffraction experiments fail to show a pattern on either anodized platinum or anodized rhodium, which produces particularly thick films.[163] One concludes, therefore, that the films have crystallites that are smaller than 50 Å in diameter, or, alternatively, are essentially amorphous.[163]* Oxide films produced by prolonged anodization,[146,186] or by chemical treatment of platinum electrodes,[185,194] appear to have similar properties with regard to reversible behavior at the equilibrium potential for the four-electron oxygen electrode process.

It has been frequently observed that the rate of the oxygen evolution process falls with increasing anodization, or that the overpotential rises at constant current density on platinum electrodes.[146,158,184]

Activation energy values for the reaction in both acid and alkaline solutions have been obtained by a number of authors. Bowden's data,[183] corrected by Yoneda,[195] gives an activation energy of about 13–14 kcal mole^{-1} at the equilibrium potential in dilute (N/5) sulfuric acid. Damjanovic et al.[87] give a value of 11.4 kcal at potential of 1.5 V in N perchloric acid. This would appear to give a value of about 12.1 kcal for the exchange current values, assuming that their experimental symmetry factor value is temperature independent. The value of 13.1 kcal was recently obtained for the exchange currents in concentrated phosphoric acid.[158] Exchange currents in the latter electrolyte[158] were about one decade lower than those in N perchloric acid at room temperature.[87] In alkaline solution, data in the literature vary between 25.3[196] and 8.3[197] kcal in $N/10$ NaOH and N NaOH, respectively, though these refer to somewhat ill-defined experimental conditions. Yoneda[195] quotes a value of about 16 kcal at the reversible potential in $N/10$ NaOH. The most reliable data appears to be that of Damjanovic et al.,[87] who obtain a value of 15.5 kcal in N KOH at a constant potential of 1.5 V HRE. This would correspond to a value of ∼17 kcal for the exchange currents.

3. The Oxygen Electrode on Other Oxidized Metals

Damjanovic et al.[198] have carried out an extensive examination of the oxygen evolution and dissolution reactions on iridium and

---

*By contrast, films on nickel electrodes that have been anodized in alkaline solution have been shown to be crystalline.[163] However, the films are thicker than on Pt.

rhodium electrodes in both acid and alkaline solution, on the same lines as their previous work on platinum. They also examine the effect of alloying in platinum–rhodium electrodes. The results they obtain show the same type of Tafel slope pattern as that observed on platinum in alkaline (though not in acid) solution, with the difference that the cathodic slope is in all cases of the order of 2.3 ×

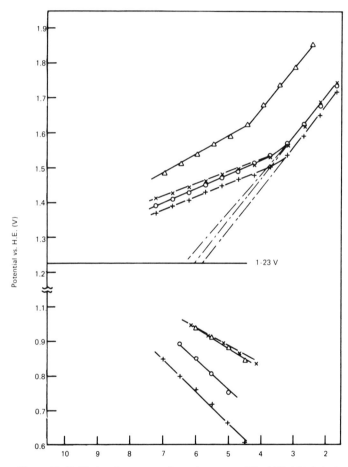

Figure 12. Tafel plots for oxygen electrode on preoxidized Rh (○), Ir (+) Pt (△) and Pt 40 at %Rh (×) in 1 V KOH (Damjanovic, Dey and Bockris[198]). Abscissa: units are $-\log_{10} i$.

$2RT/F$, which is about twice the cathodic slope for platinum electrodes in alkaline solution. Tafel plots for Pt, Rh, Ir, and Pt–Rh alloys are shown in Figs. 12 and 13. All the electrodes show a 2.3 × $2RT/F$ slope at high anodic potentials, which (assuming Langmuir adsorption conditions at low total coverage) is indicative of a primary charge-transfer reaction as rate-determining step. The only reaction that would satisfy the necessary conditions is a rate-determining water discharge process, as on platinum at all anodic

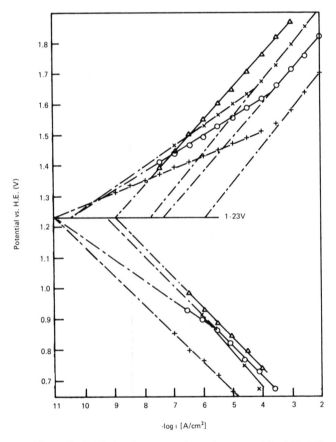

Figure 13. Tafel plots for oxygen electrode on preoxidized Rh (○), Ir (+), Pt (△) and Pt 40 at %Rh (x) in 1 V HClO$_4$ (Damjanovic, Dey and Bockris).[198] Abscissa: units are $-\log_{10} i$.

potentials in acid solution, and at high anodic potentials in alkali [reaction (72) or (74); in alkali, pH dependence on platinum points to OH^- ion discharge rather that water discharge, i.e., a rate-determining step of the same type as reaction (79)]. In the case of iridium the low $2.3 \times 2RT/3F$ Tafel slope that appears at low anodic overpotentials in both acid and alkaline solutions is explained by the fact that mechanism (74)–(76) (the "electrochemical oxide" path[88]) applies on this metal, and a change of rate-determining step from (74) to (75) takes place. Since the latter step follows a quasi-equilibrium charge transfer, its transfer coefficient should therefore be equal to $[1 + (1 - \beta)]$ [equation (44), et seq.], assuming Langmuir conditions of adsorption at low coverage, where β is the symmetry factor of step (75). The same reaction is presumed to be rate determining, as a primary charge transfer, in the cathodic direction. The same situation is presumed to hold on rhodium in alkaline solution.

In acid solution rhodium presents a situation that is not unlike platinum in alkaline solution, with the addition of the appearance of a $2.3 \times 2RT/F$ slope at high cathodic overpotential. Damjanovic et al.[198] therefore suggest a path of the Krasil'shchikov type,[192] which may be written in the anodic direction in acid solution in the form

$$S + H_2O \rightarrow SOH + H^+ + e^- \tag{83}$$

$$SOH \rightarrow SO^- + H^+ \tag{84}$$

$$SO^- \rightarrow SO + e^- \tag{85}$$

$$2SO \rightarrow 2S + O_2 \tag{86}$$

The primary charge transfer process (83) is presumed to be the rate-determining step at high anodic overpotentials, with the chemical step (84) rate determining at low anodic and cathodic over-potentials. Finally, at high cathodic overpotentials the primary (in the cathodic direction) charge transfer step, reaction (85) is presumed to be rate determining. The results suggest that Pt–Rh alloy electrodes behave as a mixture of patches of the respective Pt and Rh oxides.[198] Both Ir and Rh electrodes showed coverages of several monolayers of oxides in the potential range studied.[198] No activation energy data have been reported on Rh and Ir electrodes. Other oxygen evolution studies on rhodium[199] and iridium[200] electrodes have been reported by Hoare. His results deviate significantly from those of Damjanovic et al.[198] In $2N$ sulfuric acid he obtains an intermediate slope of about $2.3 \times 3RT/2F$ on iridium, and on rhodium his results show a dual slope plot, with a slope at low overpotential of about 100 mV/decade ($\sim 2.3 \times 3RT/2F$), rising to $2.3 \times 3RT/F$ at high overpotentials. These discrepancies probably reflect differences in surface preparation of the anodes.

Some studies on the oxygen evolution reaction on gold have been reported in acid solution. Barnartt[201] gave a Tafel slope of $2.3 \times 3RT/4F$ in N H_2SO_4, whereas Huq[202] indicates that the slope after prolonged anodizing is $2.3 \times 2RT/F$ in N $HClO_4$ solution. The latter slope would indicate that the water discharge process is also rate determining in this instance. On the other hand, Hoare's[199] results indicate that a Tafel slope of $2.3 \times 2RT/F$ refers not to the oxygen evolution reaction, but to Au_2O_3 formation, and he obtains a lower slope, of the same order as that reported by other authors,[201,203] in $2N$ H_2SO_4 at high overpotentials. On gold, the cathodic region is not accessible, due to the reduction of oxide, but the anodic process would merit further investigation.

Other materials that have been studied in acid solution for the oxygen evolution process include Pd,[203,206] oxides of lead (PbO_2),[204] and Fe (i.e., Fe_2O_3 or Fe_3O_4).[205] The latter material was investigated in the pH range 2.5–4.0. Low slopes ($2.3RT/2F$ and $2.3 \times 3RT/4F$, respectively) were recorded. PbO_2 gives slopes of $2.3 \times 2RT/F$, but in the case of β-PbO_2 a slope of $2.3 \times 3RT/4F$ was recorded.[204] The behavior of Pd appears to be similar to that of Pt,[203] although Hoare[206] reports a rapidly increasing Tafel slope at high current densities.

Much more data are available in alkaline solution. On gold, Hoar[184] reported a Tafel slope $2.3 \times 1.3RT/F$ in the potential range 1.6–1.7 V HRE in $N/10$ NaOH. Hickling and Hill[145] report a more complex behavior, with an initial high slope, a slope of about $2.3RT/F$ in the range 1.8–2.2 V HRE, followed by a rapid rise to a $2.3RT/F$ slope at potentials of about 2.7 V HRE and above in N NaOH. Hickling and Hill's results include a wide range of other materials, including Pt, Pd, Ag, Ni, Pb, Cd, Cu, Fe, and Co. In general, materials with low overvoltages (e.g., Co and Fe) show low Tafel slopes ($\sim 2.3RT/F$), whereas materials with relatively low activities show Tafel slopes at high overpotentials that are on the order of $2.3 \times 2RT/F$ or greater. In many cases considerable rapid rises in potential are noted between regions of different slope. It is not entirely clear from the original work as to whether the results these authors obtained represent steady-state data, i.e., the extent of, and nature of, any hysteresis effects is not indicated. More recently, Pd, Au, and alloys have been examined.[203] At high current densities on Pd (as on Pt), a $2.3 \times 2RT/F$ Tafel slope occurs.[87,203] On Pd–Au alloys and pure Au potential jumps were shown to occur between regions of different Tafel slope. It was suggested that the high ($2.3 \times 4RT/F$) Tafel slopes observed at high current densities could be accounted by a potential drop in the oxide film[203,207] (barrier-layer effect). The reaction mechanism on nickel oxide has been investigated in detail.[192,208,209] The overpotential characteristics depend on the initial anodization potential of the electrode.[208,209] On well-anodized electrodes the $2.3 \times 2RT/F$ slope appears at high current densities, with a low ($\sim 2.3RT/2F$) slope at low current densities.[209] On unoxidized nickel the slope is about $2.3 \times 2RT/3F$ at low current densities.[209] Krasilshchikov[209] proposes that a modified path of type (79)–(82) applies, in which the second charge transfer step is replaced by a modified step in which Ni_2O_3 is oxidized, with the participation of O^-, to Ni_2O_4. The path thereby becomes

$$S + OH^- \rightarrow S\text{--}OH + e^- \quad (87)$$

$$S\text{--}OH + OH^- \rightarrow S\text{--}O^- + H_2O \quad (88)$$

$$Ni_2O_3 + O^- \rightarrow Ni_2O_4 + e^- \quad (89)$$

$$Ni_2O_4 \rightarrow Ni_2O_3 + O \quad (94)$$

$$2O \rightarrow O_2 \quad (91)$$

In the high current density region, step (87) is rate determining, satisfying the observed Tafel slope and pH dependence. At low current densities it is considered that step (90) is rate determining on well-oxidized electrodes, with step (89) rate determining on unoxidized electrodes. A similar mechanism has been proposed on cobalt electrodes[209,210] in alkaline solution, with the difference that in this case step (87) is never rate determining (i.e., no $2.3 \times$

$2RT/F$ Tafel slope is observed under any conditions), and steps (88) and (89) are presumed to be rate-limiting on oxidized and unoxidized cobalt, respectively. For copper electrodes a similar scheme has been proposed,[211] with step (87) rate determining at high current densities, and Krasil'shchikov[209] suggests a similar mechanism for gold in acid solutions. Damjanovic[163] suggests an alternative possibility on nickel electrodes, where the simple "oxide path" variant of system (72) and (73)[88] may apply on anodized electrodes, in which the two stages after H_2O or OH^- discharge are

$$2S-OH \rightarrow 2S-O + H_2O \qquad (92)$$

$$2S-O \rightarrow 2S + O_2 \qquad (93)$$

Step (92) would then, from equation (44) (at low coverage), have a Tafel slope of $2.3RT/2F$. He suggests that the "electrochemical oxide"[88] mechanism (74)–(76) may apply on unoxidized electrodes, with the second step rate determining. However, Conway and Bourgault,[208] in a recent extensive examination of the nickel oxide anode,[208] find evidence for Temkin adsorption conditions (i.e., medium total coverage) by pseudocapacitance measurements, and hydrogen–deuterium kinetic isotope effects.[212] On this basis they conclude that the final desorption step

$$2SO \rightarrow 2S + O_2 \qquad (94)$$

is rate determining under Temkin conditions. They assume that the coverage of $-O$ and $-OH$ radicals increases with potential, with consequent fall in the free energies of adsorption. However, their pseudocapacitance analysis does indicate that Temkin conditions may also occur on other electrode systems, and that considerable care must be used in interpreting data in terms of only Langmuir conditions of adsorption.

4. Electrocatalysis of the Oxygen Evolution Reaction

At first sight Ruetschi and Delahay's[144] correlation of Hickling and Hill's[145] O_2 evolution results appears to be an excellent verification of the Brønsted rule in electrochemical systems. Their plot is shown in Fig. 10; their methods for obtaining the bond energies of the metal–OH bond have been discussed earlier (Section V.5). However, this analysis must be accepted with considerable caution. Hickling and Hill's results[145] differ in many respects from those of other workers, and were obtained using a clean electrode at each current density. Oxide film thicknesses were therefore not constant throughout the plots. In addition, the results were plotted for potentials where very thick visible oxide films occur on platinum, gold, and palladium electrodes.[145,203] However, it is possible that this fact will make the results on these metals more comparable with those on metals with high bond energies. If we accept this proposition and assume Langmuir conditions of adsorption, the rate of the limiting discharge reaction (72) may be put in the form [from equation (44)]

$$i = nFK(1 - \theta)\exp[-(1 - \beta)\Delta G/RT]\exp[(1 - \beta)FV/RT] \quad (95)$$

where ΔG is the free energy of adsorption of $-OH$ (at $V = 0$) and K is a substrate-independent constant.

If a path of the "oxide" type (72)–(73) holds, $(1 - \theta)$ must be replaced by $[1 + \exp(-\Delta G/RT)]^{-1}$, whereas if the "electrochemical oxide" path (74)–(76) is followed, the expression will be $[1 + \exp(-\Delta G/RT)\exp(-FV/RT)]^{-1}$. Hence we obtain

$$i = nFK \exp[-(1 - \beta)\Delta G/RT]\exp[(1 - \beta)FV/RT], \quad \theta \to 0 \quad (96)$$

$$i = nFK \exp(\beta \Delta G/RT)\exp[(1 - \beta)FV/RT], \quad \theta \to 1 \quad (97)$$

$$i = nFK \exp(\beta \Delta G/RT)\exp[(2 - \beta)FV/RT], \quad \theta \to 1 \quad (98)$$

where (97) is for the oxide path and (98) is for the electrochemical oxide path. Since, experimentally, rates increase with rising ΔG (assuming that changes in ΔG values may be equated with changes in bond energies), then it is reasonable to suppose that low coverage of OH is involved at the current density studied, i.e., equation (96) holds. Thus a Tafel slope of $2.3 \times 2RT/F$ would be expected, which is in accordance with the experimental observation for all substrates at high overpotential, with the possible exceptions of gold, iron, and cobalt. At constant $\log i$ equation (100) should show a V versus ΔG slope equal to 23, if ΔG is expressed in kcal and V in volts. Again, this is in accordance with experiment. In general,

Ruetschi and Delahay[144] were careful to ensure that the valence state to which the calculation refers would apply in the potential range in which the results were obtained.

The conclusions are of considerable interest, but it should be borne in mind that, in general, the oxide film formed anodically on metals will be semiconducting. If the film is only a few angstroms thick, the effect of gap width term (in the absence of a degenerate surface[96]) will not be significant, since electron tunneling from the surface of the metal will be possible. As thickness increases, the electronic properties of the film itself will become more apparent.[51] In semiconducting films with nondegenerate surfaces little potential change with overpotential will be expected in the Helmholtz double layer, as has been discussed in a previous section, but this will only be so if the film thickness is less than the Debye length.[96] Thus for *thick* films with nondegenerate surfaces the hole current will have an apparent Tafel slope of $2.3RT/F$ in a primary discharge anodic process, and the electron current will be independent of overpotential (for discussion see Refs. 14, 54, and 96), and a low overall current will be measured because of the gap-width term. Thin films with nondegenerate surfaces should show the possibility of tunneling from the metal, since tunneling probability depends on $\exp(-kx)$, where k is a level-dependent constant and x is film thickness. They will therefore show larger currents, of a progressively more metallic character, and will show a typical $2.3 \times 2RT/F$ primary discharge anodic Tafel slope (with $\beta = \frac{1}{2}$) since the potential drop in the space-charge region will be quite small. However, they will be expected to show currents at a given overpotential that are appreciably lower than those on a metal with the same adsorption properties, because of the effect of the tunneling probability term. This effect will not be expected on "metallized" films, i.e., those with degenerate surfaces.

Assuming platinum oxide films are semiconductors with nondegenerate surfaces, it is possible to explain the observation that the current for oxygen evolution falls with increasing anodization (i.e., increasing film thickness) because of the tunnel effect. Another explanation is possible, however, which requires a heterogeneous catalyst surface. If special sites are involved in the reaction, for example, grain boundaries, they may be expected to become fewer with increasing film growth. However, this seems to be contrary to X-ray evidence.[163]

5. Discussion of the Mechanism of the Oxygen Electrode on Oxidized Metals

All the mechanisms that have been discussed above involve the O–O combination reaction as the final step in the anodic direction, and it seems reasonable in view of reaction order and Tafel slope evidence that the same mechanism holds in the cathodic direction. Direct splitting of the O–O bond would appear to be unlikely at first sight, in view of the high bond strength of the oxygen molecule (117.2 kcal mole^{-1}). It may, however, come about, kinetically, by means of a series of proton transfers, provided that the bond strengths of the intermediates are sufficiently high. A possible reaction sequence, in which two adsorbed O radicals are formed in a 1:1 coupled mechanism, is given in Ref. 158. On oxide surfaces, especially on substrates which exhibit variable valence, comparatively high adsorption energies for O radicals would be expected.[213] It is important to note that there is little evidence (at least on Pt) for actual participation of the oxide ions in the oxide film in the reaction sequence. This conclusion is based on the interpretation[163,225] of results of ^{18}O tracer experiments conducted by Rozenthal and Veselovskii.[214] In the next section it will be shown that the O–O bond splitting process does not occur to any extent on phase-oxide-free platinum group metals. This seems to argue that heats of adsorption of O and OH radicals are appreciably greater on oxides than on bare metal surfaces. To some extent this concept is supported by the experimental activation energies at the reversible potential for oxygen reduction on oxidized platinum (~ 13 kcal mole^{-1} in 85% orthophosphoric acid[158]) and phase-oxide-free platinum (~ 14.3 kcal mole^{-1}) after correction for Temkin adsorption at 460 mV overpotential, i.e., about 19.6 kcal mole^{-1} at the reversible potential.[169] The preexponential factor for the reaction on oxidized platinum is much lower than that on phase-oxide-free platinum (i.e., the reaction has a much lower probability of taking place). This is probably due to a combination of the tunnel effect and the fact that special sites may be required in the reaction (i.e., pairs of sites with the required metal ion valence to ensure breaking of the O–O bond). Why the charge transfer mechanism that occurs with no bond breaking (as on phase-oxide-free noble metals) is improbable on oxide substrates with high heats of adsorption of reaction intermediates will become clear from the discussion in the next section, where it is shown that materials with high bond strengths do not favor this process.

If we now consider the data of Damjanovic et al.,[198] which were obtained on noble metals carrying very thin oxide films (about eight monolayers on iridium), we can see that the relative reaction rates in the anodic direction, in both acid and alkaline solution, are in accordance with predictions based on the bond strengths of adsorbed OH on the surfaces of the three metals, or on their hydroxide bond strengths. In agreement with Ruetschi and Delahay's correlation,[144] this indicates that the three metals lie on the rising (i.e., low coverage) side of a free energy of adsorption versus log reaction rate volcano. However, a difficulty arises on the cathodic side of the V versus log i plots. Because of the $2.3 \times 2RT/F$ Tafel slope on Rh and Ir in both acid and alkaline solutions in the cathodic direction, Damjanovic et al.[198] assume that a primary charge transfer reaction is rate determining. This reaction is a process of the form:

$$SO + e^- \rightarrow SO^- \qquad (99)$$

or

$$SO + H^+ + e^- \rightarrow SOH \qquad (100)$$

or

$$SO + H_2O + e^- \rightarrow SOH + OH^- \qquad (101)$$

depending on the pH and the path followed. Damjanovic et al.[198] suggest alternative possibilities (though on similar lines) for step (99). At low cathodic and low anodic potentials in acid on Rh, and over a somewhat wider potential range on Pt in alkali, the chemical combination of the product of reactions of the type (99) with H^+ or water molecules is considered to be rate determining. There are some difficulties in this interpretation. First, it is hard to see how a surface process involving charge between the surface of the metal and an adsorbed species situated well within the Helmholtz double layer [as in step (99)] can have an apparent symmetry factor of 0.5,[11,215] and, second, the order of activities appears to be reversed in the cathodic direction, which implies that the metals are on the descending side of a volcano (i.e., at high coverage).

The above difficulties would be resolved if we postulate that the anodic slope change on Ir and Rh is associated with the transition from high to low coverage, with or without an intervening Temkin region. This situation seems to be the case in alkaline solution for both metals, and for Ir in acid solution. If we consider the water or OH^- discharge process as rate determining in a path where OH coverage is potential dependent (for example, the Krasil'shchikov path), we can write, from equations (43) and (44) in the anodic direction for step (83) (in acid solution, with $[H_2O] = 1$)

$$i = nFk(1 - \theta)\exp[-(1 - \beta)\Delta G/RT]\exp[(1 - \beta)FV/RT] \quad (102)$$

and in the cathodic direction

$$i = nFk[H^+]^2 pO_2^{1/2}(1 - \theta)\exp[-(1 - \beta)\Delta G/RT]$$
$$\times \exp[-(1 + \beta)FV/RT] \quad (103)$$

where k is metal independent, and in which the equilibrium (via Krasil'shchikov's path) is assumed

$$2S + O_2 \rightleftharpoons 2SO \quad (104)$$

$$SO + e^- \rightleftharpoons SO^- \quad (105)$$

$$SO^- + H^+ \rightleftharpoons SOH \quad (106)$$

and, under Langmuir conditions,

$$1 - \theta = [1 + pO_2^{1/2}[H^+]\exp(-\Delta G/RT)\exp(-FV/RT]^{-1} \quad (107)$$

where ΔG may be taken to be the standard free energy of adsorption of OH at the standard equilibrium potential, if this is used as the reference potential ($V = 0$). The anodic process (with $\beta = \frac{1}{2}$) will therefore have a Tafel slope of $2.3 \times 2RT/F$ for $\theta \to 0$, and a Tafel slope of $2.3 \times 3RT/2F$ when $pO_2^{1/2}[H^+]\exp(-\Delta G/RT) \times \exp(-FV/RT) > 1$ (i.e., at $\theta \to 1$). The pH dependence of the latter process will be the same as if the second discharge step in the electrochemical oxide path is considered to be rate determining. In the cathodic direction the slope will be $2.3 \times 3RT/2F$ ($\theta \to 0$) and $2.3 \times 2RT/F$ ($\theta \to 1$). It should be noted that the latter process is zeroth order in O_2. The transition from low to high coverage will therefore occur (see Figs. 12 and 13) at about 1.5 V HRE on Ir in acid and alkaline solution, and at about 1.65 V HRE and 1.55 V HRE on

Rh in acid and alkali, respectively. From the difference in overpotential at constant current at high anodic potentials between Ir and Pt, it would seem that the transition point would be at about the reversible potential on the latter metal.

Although with such a scheme it can be shown that Tafel slopes under Temkin conditions should be $2.3RT/F$ for both anodic and cathodic processes,[158] it is clear that results on Pt in alkali and Rh in acid cannot be explained on this basis, since the potential where the slope changes is in each case higher than that on iridium, which must show the most negative ΔG value. The chemical step must be rate determining under these conditions. In acid solution the chemical step on Pt may be interpreted as being so slow that an alternative process occurs, with the "oxide" or "electrochemical oxide" paths becoming the preferred mechanisms. If the oxide path occurs, coverage will be potential independent, hence the electrochemical oxide path is most probable, with the discharge step rate determining in the cathodic direction under high coverage conditions, with very low reaction order. It must be admitted that this discussion is to a large extent speculative, but it seems to be at least reasonably consistent with the majority of the electrochemical evidence, although the concept of a chemical step on Pt in alkali and Rh in acid with a $2.3RT/F$ Tafel slope under high coverage conditions poses some difficulty. These reactions merit further careful investigation.

Few data are available on reaction rates in the cathodic direction on other metal oxides in alkaline solution. It is not clear whether the same mechanisms apply as in the anodic direction. In spite of this lack of theoretical background certain transition metal oxides—for example, nickel oxide doped with lithium to improve electronic properties and conductivity, have been widely used in alkaline fuel cells at elevated temperatures.[216] Oxides of cobalt have also been studied as electrocatalysts.[217] It has been shown that oxygen chemisorption on these semiconducting materials is related to hole conduction.[218,219] It has also been shown that correlations also appear to exist between oxygen chemisorption and the magnetic properties of nickel oxide.[221] Improved kinetic performances of NiO fuel cell electrodes at high temperatures have been linked to the antiferromagnetic–paramagnetic transition temperature.[220] Recently a group of oxides of the general formula[222] $(La, Sr)(Co, Ni)O_3$

have been studied,[223] whose room-temperature magnetic properties are similar to those of NiO above its Néel point. It was reported[223] that the perovskites doped with lanthanum alone showed cathodic reduction performances that were much the same as that of platinum when tested in dilute alkali at 25°C, whereas Sr–La-doped materials had sufficiently high rates at equilibrium to give reproducible reversible potential behavior, but showed poor performance at high cathodic overpotentials. This may be connected with a change in valence.

Limited emphasis in the past has been placed on the study of oxygen reduction rates and mechanisms on metal oxides. It seems imperative that this should be an area that must be carefully examined in the most fundamental way, in order to see which factors operate in determining reaction rates. Some progress has been made in the study of tungsten bronzes and related compounds, which will be discussed in a subsequent section. From the discussion in the next section it will become clear that large rate improvements using phase-oxide-free metals are improbable, so that oxides (along with organic semiconductors) appear to offer the only possible route to improved oxygen electrode performance in aqueous electrolytes.

VII. THE KINETICS AND MECHANISM OF OXYGEN REDUCTION ON PHASE-OXIDE-FREE METALS

1. General

In spite of the fact that intensive research has only been carried out in the last few years, the mechanism of oxygen reduction on the phase-oxide-free precious metals in acid solution is better understood than on oxide substrates. The mechanism in alkaline solution, though it is widely documented, is less easily interpreted.

2. Oxygen Reduction in Acid Solution

(i) Platinum

Early experiments established that, in general, only one oxygen reduction wave was noted in experiments at platinum electrodes, indicating that, unlike mercury, any hydrogen peroxide formed was rapidly consumed. Many workers noted the interfering effect of platinum oxides on the reduction process, since the latter have half-wave potentials that are of the same order as that for oxygen reduction.[224,226-229] Original results were somewhat conflicting with regard to the effect of surface oxides—in some cases, oxides were shown to inhibit reduction of oxygen[230-234]; in others an acceleration of the overall process in the presence of oxide was observed.[226]

The majority of workers have considered that a two-stage overall reduction process takes place in acid solution, with hydrogen peroxide as an intermediate.[232-235] It has been proposed that the hydrogen peroxide is either rapidly reduced, or spontaneously decomposes[236] on the electrode surface at the same potentials as its formation. Several workers, however, failed to detect hydrogen peroxide as a reaction product.[227,231]

Damjanovic and Bockris[166] first demonstrated that surfaces carrying phase oxide produced anodically and surfaces that were introduced into the electrolyte ($N\ HClO_4$) in an unambiguously phase-oxide-free condition showed quite different kinetic behavior. The latter electrodes show greater activity and a lower Tafel slope (about $2.3RT/F$) than the former. An extensive examination of hydrogen-reduced and cathodically reduced platinum electrodes in highly purified $N\ HClO_4$ has been carried out by Damjanovic and Brusic.[89] These authors noted the following characteristics.

(1) Even in solutions of highest purity the rest potentials of the electrodes were some 200 mV below the four-electron oxygen electrode equilibrium potential.

(2) Oxygen reduction Tafel slopes were approximately equal to $2.3RT/F$ in the 800–950 mV HRE range in solution saturated with oxygen at 1 atm pressure.

(3) Oxygen reduction at constant potential was first order, and the order with respect to H^+ ion was about $+3/2$ at constant potential.

(4) Exchange current values obtained by extrapolation to the reversible oxygen potential were about a factor of ten lower than on oxidized platinum electrodes. However, because of the difference in Tafel slope, the activity of the phase-oxide-free electrodes in the region where the $2.3RT/F$ Tafel slope is observed is one or two orders of magnitude greater than that on oxidized electrodes.

The mechanism that controls the overall rate of oxygen reduction on phase-oxide-free platinum clearly does not apply at anodic potentials, where the platinum surface has been shown to be oxidized,[159,160] nor is the overall equilibrium potential accessible. For these reasons the stoichiometric number cannot be determined by either the Parsons[137] or the Horiuti and Ikusima[138] method, and other criteria, principally reaction orders, must be used to determine the overall mechanism. Under Langmuir adsorption conditions a Tafel slope of $2.3RT/F$ can only be explained in terms of a chemical step following a one-electron transfer if a simple 1:1 coupled mechanism operates.[89,169] All charge transfer rate-determining steps are excluded by the unit oxygen reaction order. Certain more complex paths involving chemical rate-determining steps that do not follow charge transfer processes in quasiequilibrium are also formally possible under Langmuir adsorption conditions.[169] These may be excluded because of the experimental hydrogen ion reaction order. Any paths occurring under Langmuir conditions at high coverage are excluded by Damjanovic and Brusic's observation that coverage with reducible material on the electrode (expressed as the fractional coverage of adsorbed −O radicals) is less than about 0.3 under the conditions of the experiments.[89,164] The only other possible suggestion, that of a two-electron transfer step, may be excluded on activation energy grounds.[84] The above arguments are given in more detail in Refs. 89 and 169. It therefore appears that the reaction mechanism is inexplicable under Langmuir conditions.

Damjanovic and Brusic,[89] however, noted that under the experimental conditions medium coverages of adsorbed material were present on the electrode. The coverage–potential relationship for the adsorbate was approximately linear, and equilibrium coverage values were much the same at constant potential on electrodes in oxygen- and nitrogen-saturated solution (on potentiostatted electrodes). This may be adequately explained if it is suggested that the equilibrium oxygen coverage is derived from water discharge by the rapid equilibrium

$$M + H_2O \rightleftharpoons M-OH + H^+ + e^- \qquad (108)$$

which may be followed by

$$M-OH \rightleftharpoons M-O + H^+ + e^- \qquad (109)$$

where adsorption occurs under Temkin conditions. The same reactions were proposed by Böld and Breiter[237] to explain their potential sweep experiments on platinum electrodes in acid and alkaline solutions. Potential sweep experiments carried out in the potential range 0.8–1.0 V HRE show the chemisorbed oxygen film to be reversible under these conditions, and to become increasingly irreversible at higher potentials.[167] Coulometric data at higher potentials show increasing film thickness, accompanied by increasing irreversibility, so that the reduction potential is approximately constant and independent of formation potential or thickness.[159,165] Below 1 V HRE the O or OH radicals are merely chemisorbed, with the platinum atoms remaining in their lattice positions,[157,163] e.g., as shown by cyclic voltammetry and reflectance experiments of Conway et al.[169a,169b]

We can write the equilibrium for step (108) in the form (Frumkin isotherm)

$$\theta/(1 - \theta) = [H_2O][H^+]^{-1} \exp[(-\Delta G/RT)$$
$$- (r\theta/RT) + (FV/RT)] \qquad (110)$$

where ΔG is the free energy of adsorption of OH at $V = 0$, and r is the change of free energy of adsorption as θ increases from zero to one. The standard state for θ is taken as $\theta = 0$.

In the medium coverage range we can neglect, to a first approximation, the preexponential θ terms; thus at constant H_2O

$$r\theta/RT = -\ln H^+ + (FV/RT) - (\Delta G/RT) \qquad (111)$$

The Temkin isotherm [equation (111)] is supported by the pH dependence of Böld and Breiter's[237] potential scans (quoted in Ref. 89).

If we consider the rate-determining step in the oxygen reduction reaction to be

$$O_2 + H^+ + e^- \rightarrow O_2H_{ads} \qquad (112)$$

we can write a rate equation of the form [from equation (43)]

$$i = nFkpO_2[H^+](1 - \theta)\exp[-(1 - \beta)\Delta G_r/RT)$$
$$- (\beta \Delta G_p/RT)\exp(-\beta FV/RT)] \qquad (113)$$

where it is assumed that reaction product adsorption is on discrete sites on the bare metal, and where k is a metal-independent constant. ΔG_r and ΔG_p are the free energies of adsorption of the reactant (molecular oxygen) and the product (hydroperoxy radical). If we assume that the reaction product is present in low coverage (cf. coverage results in O_2 and N_2 saturated solution[89]), but that its free energy of adsorption is changed according to the $r\theta$ value that applies to the water discharge equilibrium in the Temkin range for OH adsorption, we have

$$i = nFkp_{O_2}[H^+]\exp[-(1-\beta)\Delta G_r/RT] - (\beta \Delta G_p/RT)$$
$$- (\beta r\theta/RT)\exp(-\beta FV/RT) \qquad (114)$$

where the standard state for the free energies of adsorption is taken to be $\theta \to 0$. If we now substitute equation (111), we obtain

$$i = nFkp_{O_2}[H^+]^{3/2}\exp[-(1-\beta)\Delta G_r/RT]$$
$$+ [\beta(\Delta G - \Delta G_p)/RT]\exp(-FV/RT) \qquad (115)$$

The reaction orders and Tafel slope of this expression are in agreement with experiment.[89] Two points should be noted:

(1) It has been assumed that no Temkin isotherm operates for the adsorption of molecular oxygen. This is in accordance with conclusions based on the variation of reaction order for O_2 with total coverage (or potential) under the conditions of experimental measurement.[169] Recently Bockris and Emi[238] have estimated the net free energy adsorption of O_2 molecules on a Pt surface at low coverage. The heat of physical adsorption was estimated using expressions for the attractive forces derived by Margenau and Pollard[239] and Bardeen,[240] with an allowance for repulsive interactions. An approximate value of -3.0 kcal mole^{-1} was obtained, for an O_2 molecule–metal surface distance of 2 Å. The heat of adsorption of the solvent molecules (i.e., H_2O) and that of any specifically adsorbed ions is likely to be considerably more negative than this figure. Bockris and Emi[238] also examine the possibility of O_2 molecules being chemisorbed on the electrode surface immediately before the rate-determining step. If the entropy changes for H_2O and O_2 adsorption are of the same order of magnitude, it is therefore reasonable to suppose a low net coverage of physically adsorbed O_2 molecules on the electrode surface. However, attempts to calculate the heat of chemisorption of O_2 on platinum by the method of Higuchi et al.[241] lead to positive values for all possible electronic structures, indicating considerably lower coverages if O_2 molecules

are presumed to be chemisorbed. On this basis, no $r\theta$ term will appear in the isotherm for oxygen molecules, since ΔG_r refers to the physically adsorbed state.

(2) Because of the approximations made in ignoring pre-exponential θ terms, the Tafel slope ($= 2.3RT/F$) is strictly valid only when $\theta/(1 - \theta) = 1$, i.e., at $\theta = \frac{1}{2}$.[285] However, for r/RT values of about 12, it can be shown that the expression for Tafel slope (with $\beta = \frac{1}{2}$) is within about 10% of $2.3RT/F$ at $\theta = 0.1$, and within 30% at $\theta = 0.05$. An r/RT value of 12 is of the same order of magnitude as that for oxygen radical adsorption by water discharge on platinum, provided that the value of fractional coverage at 1 V HRE (about 0.3 based on all available surface sites occupied by –O groups[89,164]) is taken to represent saturation ($\theta = 1$). This is reasonable in view of the observation that further adsorption of oxygen radicals beyond this coverage value results in the appearance of phase oxide.[157,160,163] As $\theta \to 0$ a Tafel slope of $2.3 \times 2RT/F$ should occur. This change has been detected.[242]

It can be seen from the above that Damjanovic and Brusic's analysis[89] gives a very satisfactory account of the kinetic characteristics of the oxygen electrode on oxide-free platinum in acid solution. Recently reduction kinetics on oxide-free platinum electrodes were examined in 85% orthophosphoric acid using techniques similar to those of Damjanovic and Brusic. The oxygen reaction orders, adsorption behavior, and Tafel slopes reported by Damjanovic and Brusic in N $HClO_4$ were confirmed in this medium.[169]

(ii) Reaction Products at Platinum Electrodes

Several authors have examined oxygen reduction in acid solution using the rotating ring-disk electrode,[232–234,243,244] which permits the continuous detection of readily oxidizable reaction intermediates. By making certain limiting assumptions, it is possible to establish criteria to determine whether this product (hydrogen peroxide, in the case of oxygen reduction) is produced as an intermediate in the main reaction, or as a product in a parallel process.[245] Hydrogen peroxide has been detected by several authors in the medium to high overpotential (0.7–0.0 V HRE) range in acid solution.[232–234,243,244] Muller and Nekrasov[243] have shown evidence that slow deactivation of the electrode in the medium overpotential range accompanies the production of hydrogen peroxide in dilute sulfuric acid solution. This observation was attributed to sulfate ion specific adsorption, despite the fact that the observed deactivation was a slow process. Damjanovic et al.[244] show that hydrogen peroxide production is minimal if the electrolyte is carefully purified, except at potentials below about 0.3 V, where atomic hydrogen adsorption occurs. In sulfuric acid without extra preparative treatment hydrogen peroxide is produced as a final product in a path parallel to the main reaction. The same situation occurs in purified acid in the hydrogen adsorption region. In concentrated phosphoric acid a similar deactivation effect to that noted by Muller and Nekrasov occurs. It has been suggested that this is due to impurity absorption, as in the case of dilute sulfuric acid, and polyphosphate was suggested as perhaps being responsible.[169] Other authors, however, note the fact that phosphate ion is adsorbed in this potential range.[246,247] However, at acid fuel-cell operating temperatures the amount of hydrogen peroxide detected is small, and it may be assumed that the direct reduction of oxygen molecules to water predominates.[169]

3. Oxygen Reduction on Phase-Oxide-Free Palladium and Rhodium in Acid Solution

Oxygen reduction on palladium in acid solution was reported by Sawyer et al.[226,228] to resemble that on platinum. Hoare[206,248] reported rest potentials of palladium electrodes in $2N$ H_2SO_4 to be somewhat less than those on platinum, and he obtained potentiostatic cathodic Tafel plots of slope equal to 102 mV/decade at room temperature on previously anodized electrodes. Damjanovic et al.[176] show Tafel behavior on phase-oxide-free palladium and platinum electrodes that is almost identical, although palladium was about 50 mV less active than platinum. Both metals show similar oxygen adsorption characteristics. Rhodium electrodes have been less studied. Hoare reported cathodic Tafel slopes varying between 85 and 303 mV/decade in $2N$ sulfuric acid, but his electrodes were probably not oxide free.[199] Gnanamuthu and Petrocelli[215] report 120 mV/decade, but again the surface condition is not well defined. Damjanovic et al. obtained cathodic Tafel slopes of about $2.3RT/F$ on indisputably phase-oxide electrodes.[198] Recently oxygen reduction was examined as a function of temperature in concentrated orthophosphoric acid on both phase-oxide-free palladium and rhodium electrodes.[249,250] Oxygen reduction was first order, and in both cases Tafel plots close to $2.3RT/F$ were obtained. On both metals, in the potential range of the Tafel plots, Temkin adsorption of oxygen radicals from water discharge occurs, as on platinum.[237,250–252] It is therefore reasonable to assume that the rate-determining step in oxygen reduction on Pt, Rh, and Pd in acid solution is the same.

4. Ruthenium, Iridium, and Osmium Electrodes

A Tafel slope of 80 mV/decade for oxygen reduction on ruthenium electrodes has been reported in NH_2SO_4 by Gnanamuthu and Petrocelli.[215] The same workers report slopes of 90 and 130 mV on iridium and osmium. The electrodes used were not in an unambiguously phase-oxide-free condition. Other oxygen reduction results have been reported on iridium, but the electrodes used had been previously anodized.[198,200] Recent work on phase-oxide-free electrodes[250,253,254] in concentrated orthophosphoric acid gave Tafel slopes close to $2.3 \times 2RT/F$ on all three metals for the oxygen reduction reaction. Reaction order with respect to oxygen molecules was unity at constant potential on Ru and Ir, and probably also on Os. These data are consistent with a primary charge transfer process involving molecular oxygen as the rate-determining step. Data on pH dependence are not at present available, but it is improbable that the process is

$$O_2 + e^- \rightarrow O_2^- \qquad (116)$$

in strongly acid solutions.* It is therefore very tempting to suggest that the same process applies on Ir, Ru, and Os under Langmuir conditions as on Pt, Pd, and Rh under Temkin conditions, i.e., the reaction

$$O_2 + H^+ + e^- \rightarrow O_2H_{ads} \qquad (117)$$

is rate determining. Heats of activation for the overall process were determined on all the above metals in 85% orthophosphoric acid under conditions of high surface purity.

---
*This redox-type process, involving a nonadsorbed superoxide ion, would be expected to have only a slight metal dependence.

5. Gold and Silver Electrodes

A number of studies on the oxygen reduction on gold electrodes have been conducted.[199,202,253,255-57] Tafel slopes close to $2.3 \times 2RT/F$ have been generally observed. Oxygen reduction has been shown to be first order at constant potential.[253] Potential scan experiments have shown that gold electrodes have negligible coverage of adsorbed oxygen radicals in the potential range where Tafel slopes are observed for oxygen reduction.[252,253] It has been shown that the pH dependence of the oxygen reduction reaction is approximately zero in the neutral pH range, becoming negative in strongly acid solutions.[215,256] These results indicate that the most probable rate-determining steps for oxygen reduction on gold electrodes are (pH > 2)

$$O_2 + e^- \rightarrow O_2^- \tag{118}$$

or

$$O_2 + H_2O + e^- \rightarrow (O_2H) + OH^- \tag{119}$$

and in strongly acid solutions

$$O_2 + H^+ + e^- \rightarrow (O_2H) \tag{120}$$

Krasil'shchikov[258] reported a pH-independent oxygen reduction mechanism on silver electrodes at pH values greater than 2, and suggested that reaction (120) was rate determining. Other studies on the effect of temperature, and the specific effect of different crystal faces on electrodes of this metal in acid solution, have been reported.[259,260] Bianchi et al.[261,262] have shown that silver electrodes give Tafel slopes of about $2.3 \times 2RT/F$ in 0.5 M H_2SO_4 solution at overpotentials greater than those on gold.[255] The results of Palous and Buvet[263] show no pH dependence on silver electrodes at pH values greater than about 2.0, in agreement with Krasil'shchikov,[258] but indicate that a negative pH dependence appears in strongly acid solutions. Recent work on the silver oxygen electrode in 85% orthophosphoric acid confirms the above observations.[249] Tafel slopes of $2.3 \times 2RT/F$ were observed, at overpotentials greater than those on gold,[253] in agreement with previous work.[255,261,262] Heats of activation on both gold and silver electrodes were determined.[249,250]

Results indicate that reaction (120) is also rate determining on silver electrodes in acid solution.[249]

6. Reaction Products—Effect of Impurities

Hydrogen peroxide has been detected as the reaction product on silver[258,259,261] and gold[255,256] oxygen electrodes in acid solution. It has been noted that gold electrodes are less affected by impurities than is platinum.[163,253] Ring-disk electrode studies have shown that hydrogen peroxide is a product of oxygen reduction in certain potential ranges on platinum[232–234,243,244] and rhodium[163,264] electrodes but, according to Damjanovic[163] the production of hydrogen peroxide is dependent on the presence of adsorbed impurities on the electrode surface, or alternatively (at low potentials) on the presence of adsorbed hydrogen. This is explicable if the stage following the preliminary charge-transfer reaction is of the form

$$O_2H_{ads} + H^+ + e^- \rightarrow 2OH_{ads} \tag{121}$$

(or perhaps $\quad O_2H_{ads} \rightarrow O_{ads} + OH_{ads}$) (122)

with an alternative process resulting in H_2O_2 as a product:

$$O_2H_{ads} + H^+ + e^- \mapsto H_2O_2 \tag{123}$$

The rates of the reactions involving breaking of the O–O bond will depend exponentially on $-2\beta \Delta G/RT$ [or on $-\beta(\Delta G + \Delta G')/RT$] assuming low to medium total coverage conditions, where ΔG and $\Delta G'$ are the free energies of adsorption of OH and O, respectively, on the electrode surface. The rate of the desorption process resulting in H_2O_2 does not contain such a term, and will thus not be favored on metals showing high ΔG (or $\Delta G'$) values. Nucleophilic impurities present in the system, which will be simultaneously adsorbed on the electrode with reaction products, will be expected to not only reduce the total number of active sites present for the splitting process to take place, but will also (by the induced dipole effect) reduce the standard free energy of adsorption of O, OH, and O_2H on the electrode surface. This will facilitate the formation of H_2O_2 as a reaction product, and by reducing the free energy of adsorption of the product of the primary charge transfer process (i.e., O_2H) they will strongly affect the overall rate of oxygen reduction. This model is in substantial agreement with experimental observation. After long periods of standing at the rest potential, deactivation due to phase oxide formation may be observed, especially on those metals that adsorb oxygen strongly.[250,253] Tafel plots observed in such cases depend on the isotherms for impurity desorption and irreversible oxide or adsorbed oxygen film reduction, and as such will have no mechanistic significance.[169] In comparative electrocatalytic studies of oxygen reduction on different electrode materials it is of utmost importance that the electrode surfaces under study are in a reproducible condition throughout the whole potential range studied, otherwise Tafel plots of spurious slope will be obtained.[169]

7. Electrocatalysis of Oxygen Reduction on Phase-Oxide-Free Metals In Acid Solution

It has been shown that there is a good correlation between equilibrium coverage of adsorbed oxygen at the rest potential for the oxygen electrode and the d-band vacancies for the platinum group metals.[265] The significance of d-band structure as a property indicative of heats or free energies of chemisorption has been pointed out in the earlier discussion on electrocatalysis at the hydrogen electrode. It has been pointed out that oxygen electrode activity in acid solutions does depend on those characteristics that determine heat of adsorption of reaction intermediates and that a volcano relationship appears to exist, with weak adsorbers (e.g., gold) showing poor activity with the same characteristic appearing in strong adsorbers, for example iridium.[157] Maximum rates are observed of metals with medium adsorption characteristics, for example platinum and palladium.

During recent work conducted in 85% (14.6 M) orthophosphoric acid, this volcano-type relationship was investigated in some detail. Volcano plots are obtained when i (at 25°C) at a potential of 800 mV HRE (extrapolated where necessary from the Tafel region) is plotted against metal parameters that are known to be related to heat of adsorption of reaction intermediates. Gold, Ag, Pd, Pt, Ir, Rh, Ru, Os, and four Pt–Ru solid solutions were studied.[169,249,250,253,254,266] Volcano-type plots showing log current density plotted against latent heat of sublimation of the pure metals, and against d-band vacancies, are shown in Figs. 14 and 15. The d-band

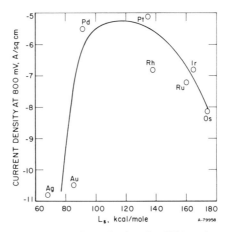

Figure 14. O_2 reduction in 85% orthophosphoric acid: Plot of i at $\eta = -460$ mV at 25°C against latent heat of sublimation (L_s values from Bond[98]). (From Ref. 3.)

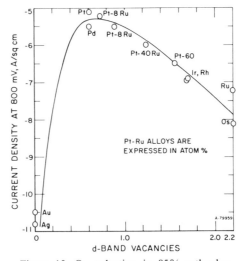

Figure 15. O_2 reduction in 85% orthophosphoric acid: Plot of i at $\eta = -460$ mV at 25°C against d-orbital vacancy values. d-Orbital vacancy values from Ref. 265: the value for osmium is considered to be the same as that for ruthenium and iron. (From Ref. 3.)

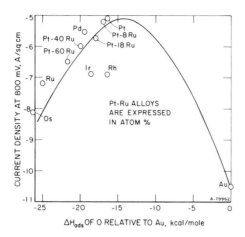

Figure 16. O_2 reduction in 85% orthophosphoric acid: Plot of i at $\eta = -460$ mV at 25°C against calculated M—O adsorbate bond strength using Pauling method. (M—O Bond strength values relative to that of Au). (From Ref. 3.)

vacancies in the Pt–Ru solid solutions are taken to depend linearly on atomic composition. The plots obtained are similar to those given by Kuhn et al.[148] for ethylene oxidation (Fig. 11). Plots of current density against other metal parameters (% d character and electronic work function) have been given elsewhere.[3] Since it appears that the rate-determining step on all the metals is the same, a plot of log current density against a calculated value of the heat of adsorption of the oxygenated reaction intermediate (O_2H) is also shown (Fig. 16).

The latter quantity has been calculated relative to the value for gold, assuming that Eley's[123] modification of Pauling's equation[121] holds for the adsorption of O_2H. The electronegativity of oxygen in the adsorbate has been taken to be that for atomic oxygen (3.5^{267}). Electronegativity values are from Gordy and Thomas[267] and latent heat of sublimation values from Bond.[98]

In all cases current densities in 85% orthophosphoric acid were about $2\frac{1}{2}$ orders of magnitude less than on the same metals (where published data exist[89,176,198]) in N perchloric acid at the same potential and temperature. This is partly a result of the effect of lower oxygen solubility values, but the major difference is to be accounted for by phosphate ion adsorption.[246,247]* Oxygen radicals are adsorbed, and are reduced, at higher HRE potentials on platinum group metals in concentrated phosphoric acid than in dilute perchloric acid.[169,237] This is consistent with the above view.

If the necessary rate correction is made to published data obtained in dilute acids for oxygen reduction on Pd–Au and Pt–Au alloy electrodes,[176] it appears that they can be accommodated on the volcano plot between Pt and Au. Similar data obtained on Pt–Rh alloys[269] will fall on the opposite side of the volcano in the same way as do Pt–Ru alloys.

8. Heats of Activation and Frequency Factors in the Oxygen Reduction Reaction

In the work conducted on the kinetics of oxygen reduction in 85% orthophosphoric acid it was possible to make a further examination of the volcano relationship by measuring heats of activation and Arrhenius preexponential factors on different electrodes studied. In order to avoid long extrapolations to the reversible potential, results on different metals were compared at an overpotential of 460 mV, which represents a reasonable compromise, giving a minimum of extrapolation in most cases. In addition, for the majority of metals studied it allowed results to be obtained within the Tafel region, to avoid any extrapolation inaccuracies due to small temperature variations of the experimental symmetry factor. Arrhenius plots under these conditions have been published.[270] For Au and Ag electrodes, which show similar oxygen adsorption characteristics in aqueous solution,[173] coverages with adsorbed oxygen are low in the Tafel region, and Langmuir adsorption of oxygenated reaction intermediates occurs. On platinum, rhodium, and palladium electrodes the activation energies were measured under conditions where Temkin adsorption of reaction intermediates takes place.[89,168,237,250–252] Similar effects occur on certain Pt–Ru alloy electrodes. For these substrates small corrections to the experimental

---

*Further evidence of the role of phosphate ions is shown by the fact that activation energies for oxygen reduction are lower in dilute acids (sulfuric) than in 85% orthophosphoric acid after corrections are made for the heat of solution of oxygen.

activation energies must be made for the change of value of the experimental transfer coefficient on going from Langmuir to Temkin conditions (i.e., from $\alpha = \frac{1}{2}$ to $\alpha = 1$). The transition potential, estimated from the onset of adsorption of oxygen radicals (in nitrogen-saturated solution 85% orthophosphoric acid) in potential scan experiments is about 850 mV HRE for Pt,[168] 700 mV on Rh,[250] and about 50 mV less than that for Pt on Pd electrodes.[252] For these metals the correction to the heat of activation may be written in the form

$$E_{act} = E'_{act} + \beta F(\eta - \eta') \tag{124}$$

where E'_{act} is the activation energy under the experimental conditions, and E_{act} is that under Langmuir adsorption conditions at low coverage. β is the symmetry factor (experimentally close to 0.5), η is the overpotential at which the E'_{act} value has been measured, and η' is the approximate value of the overpotential at which Temkin adsorption of oxygen radicals starts in the same electrolyte. A small correction was made to the E'_{act} values obtained on Pt–8 at% Ru and Pt–18 at.% Ru alloys, which have Tafel plots of intermediate slope.[266] No corrections were made to the E'_{act} values on the other metals and alloys studied. Table 1 shows the experimental activation energy values and Arrhenius preexponential values obtained. Values corrected for the effect of the adsorption isotherm are given in

Table 1
Experimental E_{act} and Preexponential Factors, O_2 Reduction in 85% Orthophosphoric Acid (at $= -460$ mV)

Electrode	E_{act}, kcal mole^{-1}	Preexponential factor
Au	20.1	5.46
Pt (oxide free)	13.7 (14.3)	5.13 (5.07)
Pt (active)	11.4	3.04
Ir	12.9	2.89
Rh	13.4 (12.4)	3.10
Ru	11.7	1.48
Pd	14.1	4.92
Ag	20.8	~4.9
Pt–8 at.% Ru	13.5 (14.0)	4.80
Pt–18 at.% Ru	13.4 (13.5)	4.47
Pt–40 at.% Ru	13.1	3.67
Pt–60 at.% Ru	12.7	2.94

parentheses. All data refer to the true overpotential at 1 atm oxygen partial pressure, but no corrections have been made for heat of solution of oxygen in the electrolyte.

9. Correlation between Heats of Activation and Estimated Heats of Adsorption

If changes in the free energy of adsorption of physically adsorbed oxygen molecules are ignored over the range of electrode materials considered, then so long as the $(1 - \theta)$ term in equation (113) is unimportant, the heat of activation will be an expression of the approximate form

$$E_{\text{act}} = K + \beta \Delta H \tag{125}$$

where E_{act} is experimental activation energy (under Langmuir conditions), K is a constant, β is the symmetry factor, and ΔH is the heat of adsorption of the reaction product of the rate-determining step (i.e., O_2H). A plot of experimental activation energy (corrected for Temkin adsorption as described above) against $\beta \Delta H$, relative to the value for gold, is shown in Fig. 17. β is given its experimental

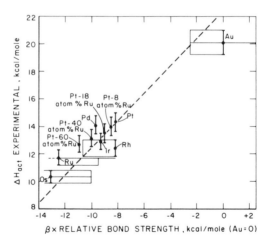

Figure 17. O_2 reduction in 85% orthophosphoric acid: Plot of experimental heat of activation (at $\eta = -460$ mV) against calculated value.[286]

value (close to 0.5), and the ΔH values are the same as those used in Fig. 16, calculated using the Pauling–Eley equation. Horizontal bars represent the comparative uncertainty in calculated $\beta\,\Delta H$, assuming electronegativity values used in the calculation have an uncertainty of ± 0.1 units. Vertical bars in each case represent the estimated uncertainty in experimental activation energy values. A line of 1:1 slope has been inserted on the plot, and it is clear that a good correlation based on equation (125) is observed. This plot shows that experimental volcano expressions in the oxygen reduction reaction cannot be due to the effect of the coverage $(1 - \theta)$ term in equation (113), on the lines suggested by Parsons[26] and Gerischer[82] for the hydrogen electrode reaction. If the maximum at platinum is a result of the general relationship

$$i = \frac{\text{const} \exp(-\beta\,\Delta G/RT)}{1 + \exp(-\Delta G/RT)} \quad (126)$$

where ΔG is the free energy of adsorption of a reaction product [cf. equations (14) and (32)], then for metals with ΔG more positive than for platinum ($\theta \to 0$), E_{act} will be given by a relationship of the same form as equation (125). However, for metals with ΔG more negative than platinum (i.e., $\theta \to 1$) the corresponding relationship will be

$$E_{\text{act}} = K - \beta\,\Delta H \quad (127)$$

A volcano plot in the activation energy values would therefore be expected. This is not observed.

An additional verification of the above concept is given by the potential scan or charging curve experiments on those metals that are strong oxygen adsorbers. In nitrogen-saturated solution none of the metals studied shows complete coverage with adsorbed oxygen radicals in the overpotential range where Tafel plots for oxygen reduction are observed.[250,253,254] Although coverages in oxygen-saturated solution under similar conditions have not in general been determined, it is extremely improbable that free energies of adsorption of an unstable reaction intermediate (O_2H) will be sufficiently close to the corresponding quantities for O and OH radicals to significantly change the total adsorbed oxygen coverage compared with that under an inert gas atmosphere.[89] In this connection it is interesting that strongly adsorbing metals (Os, Ir, Ru) do not (despite

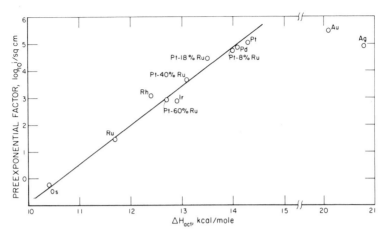

Figure 18. O_2 reduction in 85% orthophosphoric acid: Plot of experimental heat of activation (at $\eta = -460$ mP) against Arrhenius preexponential factor. (From Ref. 3.)

their medium coverages of adsorbed oxygen species) show a Temkin-type free energy of adsorption–coverage dependence for the oxygen electrode rate-determining-step product (i.e., O_2H), in contrast to Pt, Pd, and Rh. This seems to be best accounted for by the assumption that phase-oxide formation occurs in the potential range where Tafel slopes are observed on these metals. If patches of oxide develop around nucleation sites, adsorption of O_2H on the free metal surface can still be Langmuirian under low coverage conditions.

It should be noted that bond-strength calculations based on the Pauling equation and published electronegativity values cannot be fitted for silver electrodes on the same basis as those for other metals in Fig. 17. Experimental potential scan results on silver in alkaline solution suggest that the free energy of adsorption of oxygen radicals on this metal is much the same as that on gold,[173] despite the much lower apparent electronegativity value. This observation is in agreement with the observed activation energy values for oxygen reduction on silver and gold.[249,253]

In Fig. 18 the observed heats of activation and Arrhenius pre-exponential terms for oxygen reduction on each electrode material studied are plotted. It is clear that a "compensation effect" occurs between preexponential and heat of activation terms for the metals lying on the side of the volcano relationship between Pt and Os.

On the opposite side of the volcano (between platinum and gold) the preexponential factor remains virtually constant, but the overall rates fall due to the effect of the rising activation energy term.

10. The Compensation Effect

A number of explanations may be suggested for the compensation effect that is observed for oxygen reduction on phase-oxide-free noble metal electrodes. One obvious possibility is the effect of the potential drop across the diffuse double layer, which controls the concentration of H_3O^+ ions at the junction of the Helmholtz and diffuse double layers. This quantity depends on the electronic work function of the substrate.[92]

Increase in work function from Os to Pt would be expected to cause a rise in H^+ concentration at the initial site of reaction. However, in highly concentrated solution it is improbable that this effect can explain more than a fraction of a decade change in preexponential factor (see Section IV).

A second effect that undoubtedly is of much more importance is the effect of the number of available metal sites (i.e., fractional coverage with adsorbed material on the electrode surface). On strong oxygen-adsorbing metals (e.g., Ir, Ru, Os) no Temkin-type isotherm for adsorption of reaction intermediates is observed despite the medium to high coverages of adsorbed oxygen radicals derived from water discharge in the Tafel range for O_2 reduction. These coverages are apparently not strongly temperature dependent at a given potential, as can be determined by comparing potential scan experiments as a function of temperature. As discussed in the previous section, this probably results in the blocking of a large fraction of the available electrode surface by phase-oxide formation. It is probable that this effect accounts for part (perhaps 1–2 orders of magnitude) of the preexponential factor change in going from Pt to Os. It is, however, quite clear that it cannot apply to the case of Pt, Rh, and Pd, where Temkin adsorption of both the reaction intermediate (O_2H) and the adsorbed oxygen radicals derived from water discharge is observed. Under Temkin conditions extrapolated to constant V from equation (115), the reaction rate for oxygen reduction over a series of metals should depend primarily* on

---
*As discussed above (Ref. 238), free energy of adsorption of O_2 molecules will be metal independent.

$\exp[\beta(\Delta G - \Delta G_p)/RT]$, where ΔG is the free energy of adsorption of oxygen radicals (OH and O) on the metal surface, and ΔG_p is the free energy of adsorption of the product of the rate-determining step in oxygen reduction, O_2H. If, as seems reasonable from the Pauling equation, the difference in heat of adsorption between the O and OH radicals and O_2H should not vary to any marked degree on going from metal to metal, then the difference between the free energy terms can only vary as a result of entropy of activation changes. The fact that extrapolated rates at constant V on Pt, Pd, and Rh under Temkin conditions differ within a range of about two orders of magnitude seems to refute suggestions that O–O bond fission occurs within the rate-determining step,[238] i.e.,

$$2M + O_2 + H^+ + e^- \rightarrow M-OH + M-O \qquad (128)$$

where M is a metal adsorption site. If this were so, then the rate-determining-step products would be essentially identical with the adsorbate; hence extrapolated rates on all three metals would be expected to be much the same, except for small effects from the diffuse double layer and density of states terms.

Compensation effects have been noted in many catalytic systems. One of the first attempts to account for this effect is due to Constable (see Bond[98]). It assumes a surface that is heterogeneous. If we consider a surface with n_i sites of the ith type, on which the activation energy is E_i, then the total reaction rate on such a surface will be given by

$$v = K \sum_i n_i \exp(-E_i/RT) \qquad (129)$$

where K is constant. If the distribution of sites is exponential, such that

$$n_i = a' \exp(-E/c) \qquad (130)$$

where a' and c are constants and E is the energy of a particular site of the ith type. Using the relationship between adsorption energy for a reaction product and E_i [cf. equation (125)], we can write

$$E_i = K' - \beta E \qquad (131)$$

where β is a symmetry factor and K' is constant. Equation (130) can be rewritten

$$n_i = a \exp(E_i/b) \qquad (132)$$

where a and b are constants. Thus, from equation (129),

$$v = Ka \sum_i \exp gE_i \qquad (133)$$

where

$$g = (1/b) - (1/RT) \qquad (134)$$

If a continuous function between sites of limiting energies E_1 and E_2 is assumed, with E_2 small compared with E_1, we obtain

$$v = (Ka/g) \exp gE_1 \qquad (135)$$

or

$$\log v = \log(Ka/g) + (E_1/b) - (E_1/RT) \qquad (136)$$

The preexponential factor is thus proportional to the activation energy. The difficulty with an explanation along these lines is the fact that it would predict the wrong type of adsorption isotherm. It can easily be seen that if heat of adsorption is directly proportional to site energy, then a Freundlich isotherm results from the integration of equation (130) over the entire range of heat of adsorption values available on the surface of the material. This is contrary to the experimental observation that either Langmuir or Temkin-type isotherms occur.

An interesting alternative explanation (again for a gas-phase case) has been given by Zweitering and Roukens.[271] The rate of a gas adsorption process (according to absolute rate theory[25]) involving an immobile adsorbate is given by the expression

$$v_i = c_s c_g [h^4/8\pi^2 I(2\pi mkT)^{3/2}] \exp(-E/kT) \qquad (137)$$

For a mobile adsorbate the corresponding expression is

$$v_m = c_s(kT/h)[h/(2\pi mkT)^{1/2}] \exp(-E/kT) \qquad (138)$$

where c_g is the surface site concentration, c_s is the reagent concentration, and m and I are the mass and moment of inertia of the adsorbate, with E equal to activation energy. The ratio of v_m to v_i is considerable. If I is taken to be 10^{-39} cgs units, $T = 400°K$, and m (for O_2H) equal to $33/6 \times 10^{23}$ g, then

$$v_m/v_i \simeq 10^6/c_g \qquad (139)$$

The rate difference can therefore be formally ascribed to translational entropy of the activated complex. It seems probable, however, that vibrational degrees of freedom in the adsorbate will be excited before mobility occurs. A discussion of the compensation effect in gas-phase adsorption has been given by Everett,[272] who suggested that the entropy of adsorption may be directly dependent on the heat of adsorption. If adsorbate vibration occurs in classical energy wells, then the vibrational frequency (hence the vibrational entropy) depends on the curvature of the well at the potential minimum. Calculations assuming a classical simple Lennard-Jones and Devonshire bireciprocal potential field do not approach the order of magnitude required in gas-phase adsorption cases. In electrode reactions it does seem possible that vibrational (and hindered translational) entropies can explain the orders of magnitude change in preexponential factor (as in the oxygen reduction reaction), since they can be spread over many degrees of freedom if hydrogen bonding (and electrostatic effects) between the activated ion [H^+, in the case of reaction (112)] and its solvation sheath are considered. For a polyatomic activated complex we can write a rate equation in the form (cf. Ref. 25)

$$v = K \prod_n (1 - e^{-x})^{-1} \exp(-E_0/RT) \qquad (140)$$

where v is the rate constant, K is a constant, n is the number of vibrational degrees of freedom, $x = \beta\hbar\omega/kT$, where ω is the vibration frequency, and E_0 is the activation energy at absolute zero. Under the conditions of experimental measurements the Arrhenius preexponential factor will contain $\exp(+S^{\ddagger}/R)$, where $S^{\ddagger}$ is the entropy of the activated complex. The vibrational part of this entropy (and perhaps also the internal rotational part, if this is activated) will vary with substrate if ω varies. The vibrational contribution will equal[273]

$$S_v/R = \sum_n \{[xe^{-x}/(1 - e^{-x})] - \ln(1 - e^{-x})\} \qquad (141)$$

summed over n degrees of freedom. For small x (i.e., $\beta\hbar\omega \ll kT$), this may be written

$$S_v/R = \sum_n (1 - \ln x), \qquad x \ll 1 \qquad (142)$$

From the Morse equation for the potential energy distance curve for a chemical bond, neglecting the zero-point energy,[273] we have

$$D = h\omega/4\chi \qquad (143)$$

where χ is the anharmonicity constant for the adsorbed product of the rate-determining step, whose vibration frequency is ω, and D is the dissociation energy of adsorbed species. In order to give a variation of several orders of magnitude (summed over several vibrational degrees of freedom) in the preexponential term, in which the change in each S_v term is proportional to $-D$ (i.e., to E_{act}), it therefore appears to be necessary to assume that the anharmonicity term, χ is related to D by an exponential expression of the form

$$\chi = A \exp BD \qquad (144)$$

where A and B are constants. This is not impossible physically. Strictly, the above argument is not conclusive, since the expression for the vibrational partition function was derived on the basis of a linear harmonic oscillator model for the adsorbate, whereas the Morse equation assumes that the molecule is anharmonic, but to a first approximation it neglects harmonics above the first.

An alternative approach might be to describe the very loose vibrations of components of the transition state (in particular, those involved in the solvation sheath) as hindered translations. In such cases for each vibrational degree of freedom the corresponding partition function would be an expression of the form

$$f^{\ddagger} = [(2\pi mkT)^{1/2}/h]Q_u \qquad (145)$$

where m is the mass of the translating particle and Q_u is a one-dimensional analog of the configuration integral[273] in the partition function of a condensed phase. Q_u will be given by the expression

$$Q_u = \int e^{-u(r)/kT}\,dr \qquad (146)$$

where $u(r)$ is the potential energy–distance function for the particles undergoing the hindered translation. It is reasonable to suppose that the magnitude of Q_u is determined by the potential energy of the transition state as a whole, i.e., the metal–O_2H bond strength. In tightly bound cases Q_u will be small, and will become progressively larger as the motion of the particle approaches a free translation.

This method of approach may prove fruitful, though it requires a detailed knowledge of water structure in the double-layer region.

11. Consequences of the Compensation Effect

It seems likely that the observed compensation effect results from a combination of the vibrational entropy, mobility (hindered translational entropy), internal rotational entropy, and physical surface coverage factors discussed above. The effect of the diffuse double-layer potential drop (dependent on the electronic work function of the substrate) and on the density of states terms are probably much less significant. The entropy–heat of adsorption compensation effect discussed here has been noted before in electrochemical systems,[99,100,140a] and may be of general occurrence.

As a consequence of the compensation effect, it would appear that the oxygen electrode process on oxide-free noble metals is limited by thermodynamic considerations, and it therefore seems likely that no metal or alloy will show rates that are significantly greater than those on platinum, which lies at the volcano maximum for the primary charge transfer rate-determining step [reaction (112)]. Metals with activation energies greater than Pt show lower rates because of the effect of the activation energy term (at approximately constant preexponential factor), whereas metals with lower activation energies show lower rates because of the predominant effect of the preexponential term. There is evidence that higher rates are possible on metals with "activated" surfaces with a high density of defects.[274] These effects have been studied on active platinum surfaces, and they imply that under these conditions higher preexponential values occur for the same activation energy value compared with annealed surfaces. Such surfaces show lower activation energy values compared with annealed surfaces of the same metal.[274] This has been ascribed to increased heat of adsorption of intermediates at defects.[274] It has been suggested that the electronic work function at defects may not cancel in the rate equation, i.e., on a microscopic scale, a range of work functions may operate, according to the nature of the site, that vary on each side of the thermodynamic mean value.[275] Increasing surface defect concentration increases the number of sites with low work function, thereby increasing the preexponential factor. Detailed thermodynamic and quantum mechanical calculations based on this model would be of considerable value.

Although there is some hope in the possibility of increasing reaction rates by using phase-oxide-free metal catalysts with "active" surfaces in oxygen electrode electrocatalysis in acid solution, there seems to be little hope that such materials would remain stable. Aside from purely economic considerations inherent in the use of noble metal catalysts, it is clear that other types of catalyst are required if oxygen electrode rates in acid electrolyte are to be increased beyond present levels. Oxygen reduction on such materials would presumably have to follow a different mechanism and rate-determining step from that on the phase-oxide-free noble metals. Recently studies have been conducted on tungsten bronzes as possible acid-solution oxygen electrode catalysts.[157,276,277] These materials may perhaps best be regarded as a semiconducting WO_3 skin supported on a conducting tungsten bronze substrate. Though their activity in the pure state is poor, it approaches that of phase-oxide-free platinum when they are doped with traces of the latter metal.[157,278] Recent studies imply that the rate-determining step is first order in oxygen,[278] shows some pH dependence, and is a primary charge transfer. This appears to indicate that reaction (112) is rate determining on these substrates. They do not appear to be markedly active in concentrated phosphoric acid (the favored high-temperature acid fuel cell electrolyte), perhaps because of phosphotungstate formation. The role of the platinum dopant is at present obscure. There is an upper limit to the amount required to induce activity, and on an atom percent basis, the catalytic activity of the dispersed platinum is much higher than that in a pure platinum electrode. It is possible that the presence of the platinum reduces the gap width in the WO_3 skin, so that the surface becomes degenerate. The observed activity would then depend on the intrinsic adsorption properties of WO_3. More work is clearly required on bronzes and other acid-stable oxides, since they appear to be the most hopeful possibilities for future oxygen electrode catalysts.

Organic semiconductors, in particular transition metal phthalocyanines, have also been the subject of recent study. These materials have oxygen adsorption properties resembling those of hemoglobin, but have the disadvantage, from the electronic point of view, of high gap widths (about several eV). This problem may be avoided if they are deposited in the form of very thin layers on conducting substrates. Although oxygen electrode activity has been reported on such materials,[279] no mechanistic data are at present available.

VIII. THE OXYGEN ELECTRODE IN OTHER ELECTROLYTES

1. Alkaline Solutions

Early work by Berl[178] showed that the oxygen electrode on carbon in alkaline solution resulted in hydrogen peroxide or peroxide ions as the sole product of reaction, and that the O_2/H_2O_2 couple behaved reversibly under these conditions. On mercury electrodes, a $2.3 \times 2RT/F$ Tafel slope occurs in neutral solution in a pH-independent process, which changes to a slope of about 30–40 mV/decade in strongly alkaline solution, in a process that shows an apparent first-order $[H^+]$ ion dependence.[171] Similar results have been noted on silver electrodes. Gold electrodes apparently show a similar behavior,[163,256] although their Tafel behavior in the pH-independent (neutral) region is more complex.[257] Krasil'shchikov[192,258] suggested that this type of behavior could best be explained by supposing that a primary charge transfer of the type

$$O_2 + e^- \rightarrow O_2^- \qquad (147)$$

determined the rate in neutral solution, with a second step of the form

$$HO_2 + e^- \rightarrow HO_2^- \qquad (148)$$

determining the rate in alkaline solution. The latter is preceded by step (147) and the chemical step

$$O_2^- + H_2O \rightarrow HO_2 + OH^- \qquad (149)$$

in quasiequilibrium. This gives an adequate account of the pH dependence of step (148). It was suggested that this explanation applied to silver and mercury electrodes. An alternative explanation (for mercury) was offered by Bagotskii and Yablokova,[171] who suggested that the facts could best be explained if the overall process

$$O_2 + H_2O + 2e^- \rightleftharpoons HO_2^- + OH^- \qquad (150)$$

were reversible in alkali. This is reasonable, since the rate of the pH-independent forward reaction is favored by the change in potential on going from neutral to alkaline solution. The potential of an electrode at a given current density is then controlled by the molecular oxygen, hydrogen peroxide, and OH^- ion concentration in its vicinity. A recent study of oxygen reduction kinetics on gold electrodes in alkaline solution favors the same type of explanation.[163,256] However, on silver electrodes, which show generally similar adsorption properties to those of gold in the phase-oxide-free state,[173] the mechanism has been determined as that originally proposed by Krasil'shchikov.[174]

On platinum similar Tafel behavior to that on silver and gold is observed.[242] Hydrogen peroxide is detected as a reaction intermediate.[281] Other phase-oxide free metals that have been studied under conditions in alkaline solution show very similar activities in the Tafel range, despite differences in Tafel slope. It is tempting to suggest that, at least in many cases, the O_2/HO_2^- equilibrium [process (150)] occurs, with the potential of the electrode controlled by peroxide ion or hydrogen peroxide concentration at the electrode surface. If HO_2^- (or H_2O_2) concentration at the electrode surface is diffusion controlled, then a Tafel slope of $2.3RT/2F$ would be expected. If, however, simultaneous catalytic decomposition of H_2O_2 (but not of HO_2^-, which is much more stable) can take place with regeneration of part of the molecular oxygen consumed,[255] i.e.,

$$2H_2O_2 \rightleftharpoons H_2O + O_2 \qquad (151)$$

much more complex Tafel behavior would be expected. This reaction is probable on the more active metals. In addition, further (electrochemical) reaction of H_2O_2 via a two-electron process to water may occur. In both cases, the net result is an overall four-electron transfer process, as is normally observed in alkaline fuel cells. The actual behavior of an electrode material will therefore depend strongly on experimental conditions, for example, on the ratio of the electrode active area to the free volume of electrolyte. If this is large (as in the case of porous electrodes), reaction control by diffusion of peroxide ion or hydrogen peroxide is improbable under steady-state conditions, and the overall rate is probably determined by the rate of destruction of hydrogen peroxide, whether by a catalytic or electrocatalytic mechanism. Insufficient data exist to determine the precise contributions of the different processes in practical cases. However, the closeness of the activities of a wide range of phase-oxide-free metals in the experimental overpotential range in alkaline solution,[280] compared with corresponding data (where obtainable) in acid, seems to preclude simple electrocatalytic explanations based on free energies of adsorption.

2. Oxygen Electrodes in Nonaqueous Media

In aprotic solvents it has been shown that oxygen reduction is associated with the formation of superoxide ion as a reaction intermediate of comparatively high stability, following a one-electron charge transfer reaction of the same type as process (147).[282] In molten salts containing O^{2-} ion, or in which O^{2-} ion may exist as a result of chemical equilibria (for example, carbonates or nitrates) peroxide and superoxide ion can be produced by direct chemical reaction with oxygen molecules. The high stability of the superoxide ion in nitrate melts has been the subject of recent interest,[283] and it has been shown that oxygen electrode rest potentials in molten nitrates seems best explained by postulating a superoxide and peroxide equilibrium via chemical steps involving oxide ion and molecular oxygen.[284] Recent work in alkali metal carbonate melts confirms this concept.[270] Results are best explained by assuming that oxygen is present in the melt in the form of peroxide and superoxide ions, resulting from the equilibria

$$O_2 + 2CO_3^{2-} \rightleftharpoons 2O_2^{2-} + 2CO_2 \qquad (152)$$

$$O_2 + O_2^{2-} \rightleftharpoons 2O_2^{-} \qquad (153)$$

The relative amounts of O_2^{2-} and O_2^{-} ion depend on their relative stabilities, which in turn depend on temperature and the nature of the alkali metal ion present. Peroxide ion is reduced via a rapid electrochemical step to O^{2-}. Reduction of superoxide is slower, the rate depending on the stability of the ion.[270] Insufficient data are at present available to ascertain whether the electrochemical processes are of the redox or electrocatalytic type.

ACKNOWLEDGMENT

The author wishes to thank Marcel Dekker, Inc., for permission to reproduce Figures 14–16, 18; the Electrochemical Society for permission to reproduce Figures 12, 13; the American Institute of Physics for permission to reproduce Figure 9; and the North-Holland Publishing Co. for permission to reproduce Figure 17.

REFERENCES

[1] G. H. Young and R. B. Rozelle, in *Fuel Cells*, Ed. by G. H. Young, Reinhold, New York, 1960, Vol. 1, p. 23.
[2] J. O'M. Bockris and H. Wroblowa, *J. Electroanal. Chem.* **7** (1964) 428.
[3] A. J. Appleby, *Catalysis Reviews* **4**(2) (1970) 221.
[4] S. Srinivasan, H. Wroblowa, and J. O'M. Bockris, *Adv. Catalysis* **17** (1967) 351.
[5] V. G. Levich, in *Advanced Treatise of Physical Chemistry*—"*Electrochemistry*," Ed. by H. Eyring, D. Henderson, and Y. Jost, Academic Press, New York, 1970.
[6] R. W. Gurney, *Proc. Roy. Soc.* **A134** (1931) 137.
[7] W. Libby, *J. Phys. Chem.* **56** (1952) 852.
[8] J. Weiss, *Proc. Roy. Soc.* **A222** (1954) 128.
[9] A. J. Appleby, B. E. Conway, J. O'M. Bockris, and R. Sen, *MTP Internat. Rev. Sci., Phys. Chem. Series*, Vol. 6, Chapter 1, Butterworths, London, 1973.
[10] A. N. Frumkin, *Elektrokhimiya* **1** (1965) 394.
[11] R. Parsons, *Surface Sci.* **2** (1964) 418
[12] J. A. V. Butler, *Proc. Roy. Soc.* **A157** (1936) 423.
[13] H. Gerischer, *Z. Phys. Chem.* **26** (1960) 223.
[14] H. Gerischer, *Z. Phys. Chem.* **26** (1960) 325.
[15] R. Parsons and J. O'M. Bookris, *Trans. Farad. Soc.* **47** (1951) 914.
[16] B. E. Conway and J. O'M. Bockris, *J. Chem. Phys.* **26** (1956) 532.
[17] B. E. Conway and J. O'M. Bockris, *Proc. Roy. Soc.* **A248** (1958) 1394; *Electrochim Acta* **3** (1961) 340.
[18] B. E. Conway and J. O'M. Bockris, *Can. J. Chem.* **35** (1957) 1124.
[19] B. E. Conway, *Can. J. Chem.* **37** (1959) 178.
[20] J. O'M. Bockris and D. B. Matthews, *J. Chem. Phys.* **44** (1966) 298; *Proc. Roy. Soc.* **A292** (1966) 479; M. Salomon and B. E. Conway, *J. Chem. Phys.* **41** (1964) 3169.
[21] J. O'M. Bockris, S. Srinivasan, and D. B. Matthews, *Disc. Faraday Soc.* **39** (1965) 239.
[22] J. Horiuti and M. Polanyi, *Acta Physicochim. URSS* **2** (1935) 505.
[23] H. Eyring, *J. Chem. Phys.* **3** (1935) 107.
[24] M. G. Evans and M. Polanyi, *Trans. Faraday Soc.* **31** (1935) 875.
[25] S. Glasstone, K. J. Laidler, and H. Eyring, *Theory of Rate Processes*, McGraw-Hill, New York, 1941.
[26] R. Parsons, *Trans. Farad. Soc.* **54** (1958) 1053.
[27] N. F. Mott and R. W. Gurney, *Electronic Processes in Ionic Crystals*, Oxford University Press, New York, 1940.
[28] L. Landau, *Phys. Z. Sowjetunion* **3** (1933) 664.
[29] S. I. Pekar, *Investigations of the Electronic Theory of Crystals*, Fizmatgiz, Moscow, 1951.
[30] R. Platzman and J. Franck, *Z. Physik.* **138** (1954) 411.
[31] L. Landau, *Phys. Z. Sowjetunion* **1** (1932) 88; **2** (1932) 46.
[32] C. Zener, *Proc. Roy. Soc.* **A137** (1932) 696; **A140** (1933) 660.
[33] R. J. Marcus, B. J. Zwolinski, and H. Eyring, *J. Phys. Chem.* **58** (1954) 432.
[34] E. Sacher and K. J. Laidler, in *Modern Aspects of Electrochemistry*, Vol. 3, Ed. by J. O'M. Bockris and B. E. Conway, Butterworth, London, 1964.

[35] K. J. Laidler, *Can. J. Chem.* **37** (1959) 138.
[36] E. Sacher and K. J. Laidler, *Trans. Faraday Soc.* **59** (1963) 396.
[37] R. A. Marcus, *J. Chem. Phys.* **24** (1956) 966.
[38] R. A. Marcus, *J. Chem. Phys.* **26** (1957) 867; **38** (1963) 1858; **39** (1963) 1734.
[39] R. A. Marcus, *Can. J. Chem.* **37** (1959) 138; *Trans. Symp. Electrode Processes*, Ed. by E. Yeager, Wiley, New York, 1961, p. 239.
[40] N. S. Hush, *J. Chem. Phys.* **28** (1958) 962; *Z. Elektrochem.* **61** (1957) 734; *Trans. Faraday Soc.* **57** (1961) 557; *Progress in Inorganic Chemistry*, Ed. by F. A. Cotton, Interscience, New York, 1967, Vol. 8, p. 391.
[41] P. P. Schmidt and H. B. Mark, *J. Chem. Phys.* **43** (1965) 3291.
[42] P. P. Schmidt, *Aust. J. Chem.* **22** (1969) 673.
[43] J. Halpern and L. Orgel, *Disc. Faraday Soc.* **29** (1961) 32.
[44] R. A. Marcus, *Disc. Faraday Soc.* **29** (1961) 21; *J. Phys. Chem.* **67** (1963) 853, 2889; *J. Chem. Phys.* **43** (1965) 679.
[45] J. Jortner, *Rad. Res. Suppl.* **4** (1964) 24.
[46] V. G. Levich and R. R. Dogonadze, *Dokl. Akad. Nauk SSSR* **124** (1959) 123; **133** (1960) 158.
[47] R. R. Dogonadze, *Dokl. Akad. Nauk SSSR* **133** (1960) 1368; **142** (1962) 1108.
[48] V. G. Levich and R. R. Dogonadze, *Coll. Czech. Chem. Comm.* **26** (1962) 193.
[49] R. R. Dogonadze and Y. A. Chizmadshov, *Dokl. Akad. Nauk SSSR* **144** (1962) 1077; **145** (1962) 849; **150** (1963) 333.
[50] R. R. Dogonadze, A. M. Kuznetzov, and Y. A. Chizmadshov, *Zh. Fiz. Khim.* **38** (1964) 1195.
[51] A. M. Kuznetzov and R. R. Dogonadze, *Izv. Akad. Nauk SSSR, Ser.* 15, **10** (1964) 1885; **12** (1964) 2140.
[52] R. R. Dogonadze and A. M. Kuznetzov, *Elektrokhimiya* **1** (1965) 742, 1008; **3** (1967) 380.
[53] R. R. Dogonadze, A. M. Kuznetzov, and A. A. Chernenko, *Usp. Khim.* **34** (1965) 1779.
[54] V. G. Levich, in *Advances in Electrochemistry and Electrochemical Engineering*, Ed. by P. Delahay and C. W. Tobias, Interscience, New York, 1966, Vol. 4, p. 249.
[55] F. Fröhlich, *Adv. Phys.* **3** (1954) 325.
[56] G. R. Alcock, *Ann. Phys.* **5** (1956) 412.
[57] T. Holstein, *Ann. Phys. (N.Y.)* **8**, (1959) 343.
[58] H. Gerischer, in *Advances in Electrochemistry and Electrochemical Engineering*, Ed. by P. Delahay and C. W. Tobias, Interscience, New York, 1961, Vol. 1, p. 139.
[59] R. R. Dogonadze, in *Reactions of Molecules at Electrodes*, Ed. by N. S. Hush, Wiley, New York, 1971.
[60] R. R. Dogonadze, A. M. Kuznetzov, and V. G. Levich, *Electrochim. Acta* **13** (1968) 1025.
[61] V. G. Levich, R. R. Dogonadze, E. D. German, A. M. Kuznetzov, and Y. I. Kharkats, *Electrochim. Acta* **15** (1970) 353.
[62] B. E. Conway, J. O'M. Bockris, and H. Linton, *J. Chem. Phys.* (1956) 834.
[63] F. P. Bowden and K. E. W. Grew, *Disc. Faraday Soc.* **1** (1947) 86.
[64] J. O'M. Bockris, M. A. V. Devanathan, and K. Müller, *Proc. Roy. Soc.* **A274** (1963) 55.
[64a] B. E. Conway and H. P. Dhar, *Croatica Chem. Acta* **45** (1973) 109.
[65] D. H. Everett and W. F. K. Wynne Jones, *Trans. Faraday Soc.* **35** (1939) 1380.
[66] R. A. More O'Ferrall, G. W. Koeppl, and A. J. Kresge, *J. Am. Chem. Soc.* **93** (1971) 1.

References

[67] R. A. Marcus, *J. Phys. Chem.* **72** (1968) 891; R. A. Marcus and A. Cohen, *J. Phys. Chem.* **72** (1968) 4249.
[68] M. Salomon, C. G. Enke, and B. E. Conway, *J. Chem. Phys.* **43** (1965) 3989.
[69] H. S. Johnston, *Adv. Chem. Phys.* **3** (1960) 131.
[70] L. Van Hove, *Physica* **22** (1955) 517.
[71] R. Kubo, *J. Phys. Soc. Japan* **12** (1957) 570.
[72] X. de Hemptinne, *Soc. Chim. Belge (Memoires)*, 5th Ser., (1964) 2328.
[73] M. Temkin, *Zh. Fiz. Khim.* **15** (1941) 296.
[74] E. Gileadi and B. E. Conway, in *Modern Aspects of Electrochemistry*, Ed. by J. O'M. Bockris and B. E. Conway, Butterworth, London, 1964, Vol. 3, p. 347.
[75] J. Horiuti and G. Okamoto, *Sci. Pap. Inst. Phys. Chem. Res. Tokyo* **28** (1936) 231.
[76] G. Okamoto, J. Horiuti, and K. Hirota, *Sci. Pap. Inst. Phys. Chem. Res. Tokyo* **29** (1936) 213.
[77] J. Horiuti, *J. Res. Inst. Catal. Hokkaido* **4** (1956) 55; in *Transactions of the Symposium on Electrode Processes*, Ed. by E. Yeager, Wiley, New York, 1961, p. 17.
[78] M. Boudart, *J. Am. Chem. Soc.* **72** (1952) 1531, 3556.
[79] J. O'M. Bockris and D. A. J. Swinkels, *J. Electrochem. Soc.* **111** (1964) 736, 743.
[80] J. O'M. Bockris, E. Gileadi, and K. Müller, *Electrochim. Acta* **12** (1967) 1301.
[81] A. A. Balandin, *The Problems of Chemical Kinetics, Catalysis and Reactivity*, Academy of Sciences, Moscow, 1955, p. 462.
[82] H. Gerischer, *Bull. Chim. Soc. Belge* **67** (1958) 506.
[83] V. S. Bagotskii, L. N. Nekrasov, and N. A. Shumilova, *Usp. Khim.* **34** (1965) 1697.
[84] M. A. Vorotyntsev and A. M. Kuznetsov, *Electrokhimiya* **6** (1970) 208.
[85] T. P. Hoare, in *Proc. 8th Meeting CITCE, Madrid, 1956*, Butterworth, London, 1958, p. 439.
[86] H. Mauser, *Z. Elektrochem.* **62** (1958) 419.
[87] A. Damjanovic, A. Dey, and J. O'M. Bockris, *Electrochem. Acta* **11** (1966) 791.
[88] J. O'M. Bockris, *J. Chem. Phys.* **24** (1956) 817.
[89] A. Damjanovic and V. Brusic, *Electrochim. Acta* **12** (1967) 615.
[90] B. E. Conway, N. Marincic, D. Gilroy and E. Rudd, *J. Electrochem. Soc.* **113** (1966) 1144.
[91] A. N. Frumkin, *Z. Phys. Chem.* **A164** (1933) 121.
[92] J. O'M. Bockris, R. J. Mannan, and A. Damjanovic, *J. Chem. Phys.* **5** (1968) 1898.
[93] H. Gerischer, *Electrochim. Acta* **13** (1968) 1467.
[94] C. Kittell, *Introduction to Solid State Physics*, Wiley, New York, 1956.
[95] W. Mehl and F. Lohmann, *Electrochim. Acta* **13** (1968) 1459, 1467.
[96] M. Green, in *Modern Aspects of Electrochemistry*, Ed. by J. O'M. Bockris, Butterworth, London, 1959, Vol. 2, p. 343.
[97] G. Feuillade, J. Bouet, and B. Chenaux, *Electrochim. Acta* **15** (1970) 1527.
[98] G. C. Bond, *Catalysis by Metals*, Academic Press, New York, 1962.
[99] M. Breiter, *Ann. New York Acad. Sci.* **101** (Art 3) (1963) 709.
[100] L. I. Krishtalik and V. M. Tsionsky, *J. Electroanal. Chem.* **31** (1971) 363;
[100a] B. E. Conway and L. G. M. Gordon, *J. Phys. Chem.* **73** (1969) 3609.
[101] K. F. Bonhoeffer, *Z. Phys. Chem.* **A113** (1924) 199.
[102] H. Leidheiser, *J. Am. Chem. Soc.* **71** (1949) 3634.
[103] E. N. Khomatov, *Zh. Fiz. Khim.* **24** (1950) 1201.
[104] N. Ontani, *Sci. Reps. Inst. Tokuko Univ.* **A8** (1956) 399.
[105] S. G. Christov and N. A. Pargarov, *Z. Electrochem.* **61** (1957) 113.
[106] H. Fischer, *Z. Electrochem.* **52** (1948) 111.

[107] G. I. Volkov, *Zh. Fiz. Khim.* **29** (1955) 390.
[108] N. I. Kobozev, *Zh. Fiz. Khim.* **26** (1952) 112.
[109] A. K. Lorents, *Zh. Fiz. Khim.* **27** (1953) 317.
[110] J. O'M. Bockris, *Trans. Farad. Soc.* **43** (1947) 417.
[111] J. O'M. Bockris, *Chem. Rev.* **43** (1948) 525.
[112] J. O'M. Bockris and E. C. Potter, *J. Electrochem. Soc.* **99** (1952) 169.
[113] P. Ruetschi and P. Delahay, *J. Chem. Phys.* **23** (1955) 195.
[114] P. Ruetschi and P. Delahay, *J. Chem. Phys.* **23** (1955) 1167.
[115] J. O'M. Bockris and R. Parsons, *Trans. Faraday Soc.* **44** (1948) 860.
[116] J. O'M. Bockris and S. Ignatowicz, *Trans. Faraday Soc.* **44** (1948) 520.
[117] A. Hickling and F. W. Salt, *Trans. Faraday Soc.* **36** (1940) 1226.
[118] J. O'M. Bockris and A. M. Azzam, *Trans. Faraday Soc.* **48** (1952) 145.
[119] F. B. Bowden, *Proc. Roy. Soc.* **A128** (1930) 317.
[120] B. Kabanov and S. Iofa, *Acta Physicochim. URSS* **10** (1939) 17.
[121] L. Pauling, *The Nature of the Chemical Bond*, Cornell Univ. Press, Ithaca, New York, 1948.
[122] R. C. L. Bosworth, *Proc. Roy. Soc. N.S. Wales* **74** (1941) 538.
[123] D. D. Eley, *J. Phys. Coll. Chem.* **55** (1961) 1017.
[124] G. C. Bond, *Surf. Sci.* **18** (1969) 11.
[125] L. Pauling, *Proc. Roy. Soc.* **A196** (1949) 343.
[126] B. E. Conway and J. O'M. Bockris, *Naturwiss.* **43** (1956) 446.
[127] B. E. Conway and J. O'M. Bockris, *Nature* **178** (1956) 488.
[128] L. Pauling, *J. Am. Chem. Soc.* **69** (1947) 542.
[129] P. Delahay, *Double Layer and Electrode Kinetics*, Interscience, New York, 1965.
[130] F. Ludwig and E. Yeager, *J. Electrochem. Soc.* **113** (1966) 1109.
[131] J. O'M. Bockris and D. F. A. Koch, *J. Phys. Chem.* **65** (1961) 1941.
[132] B. E. Conway, *Trans. Roy. Soc. Can.* **54** (1960) 19.
[133] J. O'M. Bockris, I. A. Ammar, and A. K. M. S. Huq, *J. Phys. Chem.* **61** (1967) 879.
[134] S. Schuldiner, *J. Electrochem. Soc.* **106** (1959) 891; **107** (1960) 452.
[135] R. Parsons, *Trans. Faraday Soc.* **59** (1960) 1340.
[136] B. E. Conway, *Proc. Roy. Soc.* **A247** (1958) 400; **A256** (1960) 128; in *Transactions of the Symposium on Electrode Processes*, Ed. by E. Yeager, Wiley, New York, 1961, p. 267.
[137] R. Parsons, *Trans. Faraday Soc.* **47** (1951) 1332.
[138] J. Horiuti and M. Ikusima, *Proc. Imp. Acad. Tokyo* **15** (1939) 39.
[139] S. G. Christov, *Z. Elektrochem.* **62** (1958) 567; **64** (1960) 840; *Electrochim. Acta* **4**, (1961) 194, 306; **9** (1964) 575; *J. Res. Inst. Catalysis (Tokkaido)* **16** (1968) 169.
[140] B. E. Conway, E. M. Beatty, and P. A. D. DeMaine, *Electrochim. Acta* **7**, (1962) 39.
[140a] K. Gossner and U. Freyer. *Z. Phys. Chem. N. F.* **63** (1969) 132; K. Gossner and F. Mansfeld, *Z. Phys. Chem. N.F.* **63** (1969) 143.
[141] J. O'M. Bockris and E. C. Potter, *J. Chem. Phys.* **20** (1952) 614.
[142] J. O'M. Bockris and N. Pentland, *Trans. Faraday Soc.* **48** (1952) 833.
[143] J. O'M. Bockris, A. Damjanovic, and R. J. Mannan, *J. Electroanal. Chem.* **18** (1968) 349.
[144] P. Ruetschi and P. Delahay, *J. Chem. Phys.* **23** (1955) 556.
[145] A. Hickling and S. Hill, *Disc. Faraday Soc.* **1** (1947) 236.
[146] J. O'M. Bockris and A. K. M. S. Huq, *Proc. Roy. Soc.* **A237** (1956) 277.
[147] H. Dahms and J. O'M. Bockris, *J. Electrochem. Soc.* **113** (1964) 727.
[148] A. T. Kuhn, H. Wroblowa, and J. O'M. Bockris, *Trans. Faraday Soc.* **63** (1967) 1458.
[148a] J. Wojtowitz, D. Gilroy, and B. E. Conway, *Electrochim. Acta* **14** (1969) 1119.

References

[149] H. Kita, *J. Electrochem. Soc.* **113** (1966) 1095.
[150] C. F. Heath, *Proc. Am. Power Sources. Conf.* **18** (1964) 33; in *Proc. SERAI.*, Brussels, 1963, Vol. 1, p. 92; H. Binder, A. Kohling, and G. Sandstede, in *Hydrocarbon Fuel Cell Technology*, Ed. by B. Baker, Academic Press, New York, 1965, p. 91; L. W. Niedrach and D. W. McKee, *Proc. Am. Power Sources Conf.* **21** (1967) 6.
[151] G. A. Frysinger, in *Hydrocarbon Fuel Cell Technology*, Ed. by B. Baker, Academic Press, New York, 1965, p. 9; L. W. Niedrach and L. B. Wienstock, *Electrochem. Tech.* **3**, (1965) 270.
[152] B. S. Hobbs and A. C. C. Tseung, *Nature* **222** (1969) 556.
[153] M. Fleischmann, J. Koryta, and H. R. Thirsk, *Trans. Faraday Soc.* **63** (1967) 1261.
[154] J. W. E. Coenen, *Sur. Sci.* **18** (1969) 158.
[155] J. N. Butler, J. Giner, and J. M. Parry, *Surf. Sci.* **18** (1969) 140.
[156] W. M. Vogel and J. T. Lundquist, *J. Electrochem. Soc.* **117** (1970) 1512.
[157] J. O'M. Bockris, A. Damjanovic, and J. McHardy, in *Proc. 3rd Int. Symp. on Fuel Cells*, 1969, p. 15.
[158] A. J. Appleby, *J. Electroanal. Chem.* **24** (1970) 97.
[159] H. A. Laitinen and C. G. Enke, *J. Electrochem. Soc.* **107** (1960) 733.
[160] A. K. N. Reddy, M. A. Genshaw, and J. O'M. Bockris, *J. Chem. Phys.* **48** (1968) 671.
[161] T. Biegler and R. Woods, *J. Electroanal. Chem.* **20** (1969) 73.
[162] J. Horiuti, *J. Res. Inst. Catalysis Hokkaido* **4** (1956) 55; in *Transactions of the Symposium on Electrode Processes*, Ed. by E. Yeager, Wiley, New York, 1961, p. 17.
[163] A. Damjanovic, in *Modern Aspects of Electrochemistry*. No. 5, Ed. by J. O'M. Bockris and B. E. Conway, Plenum Press, New York, 1969, p. 369
[164] H. Wroblowa, M. L. B. Rao, A. Damjanovic, and J. O'M. Bockris, *J. Electroanal. Chem.* **15** (1967) 139.
[165] D. Gilroy and B. E. Conway, *Can. J. Chem.* **46** (1968) 875.
[166] A. Damjanovic and J. O'M. Bockris, *Electrochim. Acta* **11** (1966) 376.
[167] V. I. Tikhomirova, A. I. Oshe, V. S. Bagotskii, and V. I. Luk'yanycheva, *Doklady Akad. Nauk SSSR* **159** (1964) 644.
[168] A. J. Appleby and A. Borucka, *J. Electrochem. Soc.* **116** (1969) 1212.
[169] A. J. Appleby, *J. Electrochem. Soc.* **117** (1970) 328.
[169a] H. A. Kozlowska, B. E. Conway, and W. B. A. Sharp, *J. Electroanal. Chem.* **43** (1973) 9.
[169b] B. E. Conway and S. Gottesfeld, *J. Chem. Soc., Faraday Trans.*, I, **69** (1973) 1090.
[170] W. M. Latimer, *Oxidation Potentials*, Prentice-Hall, Englewood Cliffs, New Jersey, 1961.
[171] V. S. Bagotskii and I. E. Yablokova, *Zh. Fiz. Khim.* **27** (1953) 1663.
[172] A. J. Appleby, *J. Electroanal. Chem.* **27** (1970) 325.
[173] J. Giner, J. M. Parry, and L. Swette, in *Fuel Cell Systems II*, Ed. by B. Baker (Advances in Chemistry Series, 90), American Chemical Society, Washington, D.C., 1969, p. 102.
[174] D. Sepa, M. Vojnovic, and A. Damjanovic, *Electrochim. Acta* **15** (1970) 1335.
[175] A. Damjanovic, M. A. Genshaw, and J. O'M. Bockris, *J. Electroanal. Chem.*, **15** (1967) 173.
[176] A. Damjanovic and V. Brusic, *Electrochim. Acta* **12** (1967) 1171.
[177] J. P. Hoare, *J. Electrochem. Soc.* **109** (1962) 858.
[178] W. G. Berl, *Trans. Electrochem. Soc.* **83** (1943) 253.
[179] J. O'M. Bockris and L. F. Oldfield, *Trans. Faraday Soc.* **51** (1955) 249.

[180] S. Schuldiner and R. M. Roe, *J. Electrochem. Soc.* **110** (1963) 1142.
[181] A. Damjanovic and V. Brusic, *J. Electroanal. Chem.* **22** (1969) App. 1.
[182] A. J. Appleby, *J. Electroanal. Chem.* **35** (1972) 193.
[183] F. P. Bowden, *Proc. Roy. Soc.* **A126** (1929) 107.
[184] T. P. Hoar, *Proc. Roy. Soc.* **A142**, (1933) 628.
[185] J. P. Hoare, *J. Electrochem. Soc.* **110** (1963) 1019.
[186] N. Watanabe and M. A. V. Devanathan, *J. Electrochem. Soc.* **111** (1964) 615.
[187] J. O'M. Bockris and B. E. Conway, *Trans. Faraday Soc.* **45** (1949) 989; W. Visscher and M.A.V. Devanathan, *J. Electroanal. Chem.* **8** (1964) 127; N. P. Berezina and N. V. Nikolaeva-Fedorovich, *Elektrokhimiya* **3** (1967) 3. See also Refs. 89, 135, 146, 158, 175, 244, 256, and 281.
[187a] B. E. Conway, H. A. Kozlowska, and E. Criddle, *Anal. Chem.* **45** (1973) 1331.
[188] J. O'M. Bockris, in *Modern Aspects of Electrochemistry*, Vol. 1, Ed. by J. O'M. Bockris, Butterworths, London, 1954, p. 139.
[189] J. P. Hoare, *The Electrochemistry of Oxygen*, Wiley—Interscience, New York, 1968.
[190] J. O'M. Bockris, E. Gileadi, and G. F. Stoner, *J. Phys. Chem.* **73** (1969) 427.
[191] A. C. Riddiford, *Electrochim. Acta* **4** (1961) 170.
[192] A. I. Krasil'shchikov, in *Proceedings of the Electrochemical Conference*, Izd. Akad. Nauk SSSR, 1953, p. 71.
[193] V. I. Luk'yanycheva and V. S. Bagotskii, *Dokl. Acad. Nauk SSSR* **155** (1964) 160.
[194] J. P. Hoare, *J. Electrochem. Soc.* **112** (1965) 849; *Nature* **211** (1966) 703.
[195] Y. Yoneda, *Bull. Chem. Soc. Japan* **22** (1949) 266.
[196] H. P. Stout, *Disc. Faraday Soc.* **1** (1947) 246.
[197] R. Audubert, *Disc. Faraday Soc.* **1** (1947) 72.
[198] A. Damjanovic, A. Dey, and J. O'M. Bockris, *J. Electrochem. Soc.* **113** (1966) 739.
[199] J. P. Hoare, *Electrochim. Acta* **11** (1966) 203, 311.
[200] J. P. Hoare, *J. Electroanal. Chem.* **18** (1968) 251.
[201] S. Barnartt, *J. Electroanal. Chem.* **106** (1959) 991.
[202] A. K. M. S. Huq, Final Report, AD 414800, Contract DA-49-186-ORD-982, U.S. Department of Commerce, Washington, D.C., 1963.
[203] J. J. MacDonald and B. E. Conway, *Proc. Roy. Soc.* **A269** (1962) 419.
[204] P. Ruetschi and B. D. Cahan, *J. Electrochem. Soc.* **104** (1957) 406; **105** (1958) 369.
[205] W. H. Wade and N. Hackerman, *Trans. Faraday Soc.* **53** (1957) 1636.
[206] J. P. Hoare, *J. Electrochem. Soc.* **112** (1965) 1129.
[207] R. E. Meyer, *J. Electrochem. Soc.* **107** (1960) 847.
[208] B. E. Conway and P. Bourgault, *Can. J. Chem.* **37** (1959) 292; **40** (1962) 1690; *Trans. Faraday Soc.* **58** (1962) 593.
[209] A. I. Krasil'shchikov, *Zh. Fiz. Khim.* **37** (1963) 531.
[210] Y. A. Tur'yan and I. A. Gershkovich, *Zh. Fiz. Khim.* **34** (1960) 2654.
[211] A. L. L'vov and A. V. Fortunatov, in *Electrochemical Conf.*, Izv. Akad. Nauk SSSR, 1953, p. 71.
[212] M. Salomon, *J. Electrochem. Soc.* **114** (1967) 922.
[213] R. Kollrack, *J. Catalysis* **12** (1968) 321.
[214] K. I. Rozenthal and V. I. Veselovskii, *Dokl. Akad. Nauk SSSR* **111** (1956) 637.
[215] D. S. Gnanamuthu and J. V. Petrocelli, *J. Electrochem. Soc.* **114** (1967) 1036.
[216] F. T. Bacon, in *Fuel Cells*, Ed. by G. H. Young, Reinhold, New York, 1960, Vol. 1, p. 51; E. Yeager and A. Kozawa, in *Proc. 6th AGARD Combustion and Propulsion Colloquium*, Cannes, France, Pergamon, New York, 1964.
[217] G. Magner and M. Savy, *Compt. Rend.* **267** (1968) 944.
[218] O. K. Davtyan, *Zh. Fiz. Khim.* **38** (1964) 5.
[219] E. G. Misyuk, O. K. Davtyan, R. N. Stupichenko, and E. A. Kalyuzhnaya, *Elektrokhimiya* **2** (1966) 788.

[220] A. C. C. Tseung, B. S. Hobbs, and A. D. S. Tantram, *Electrochim. Acta* **15** (1970) 473.
[221] E. R. S. Winter, *J. Catalysis* **6** (1966) 35.
[222] D. B. Meadowcroft, *Magnetism and the Chemical Bond*, Interscience, New York, 1963, p. 236.
[223] A. C. C. Tseung and H. L. Beban, unpublished; presented at Fall 1970 Electrochemical Society Meeting, Atlantic City, New Jersey.
[224] J. Giner, *Z. Electrochem.* **63** (1956) 386.
[225] K. J. Vetter, *Electrochemische Kinetik*, Springer, Berlin, 1961.
[226] D. T. Sawyer and L. V. Interrante, *J. Electroanal. Chem.* **2** (1961) 310.
[227] J. J. Lingane, *J. Electroanal. Chem.* **2** (1961) 296.
[228] D. T. Sawyer and R. J. Day, *Electrochim. Acta* **8** (1963) 589.
[229] W. Vielstich, *Z. Instrumentenkunde* **71** (1963) 29.
[230] N. W. Breiter, *Electrochim. Acta* **9** (1964) 441.
[231] C. C. Liang and A. L. Juliard, *J. Electroanal. Chem.* **3** (1965) 930.
[232] E. I. Krushcheva, N. A. Shumilova, and M. R. Tarasevich, *Elektrokhimiya* **1** (1965) 730.
[233] L. Muller and L. N. Nekrasov, *Dokl. Akad. Nauk SSSR* **157** (1964) 416.
[234] L. Muller and L. N. Nekrasov, *J. Electroanal. Chem.* **9** (1965) 282.
[235] J. P. Hoare, *J. Electrochem. Soc.* **112** (1965) 602.
[236] G. Bianchi and T. Mussini. *Electrochim. Acta* **10** (1965) 455.
[237] W. Böld and M. Breiter, *Electrochim. Acta* **5** (1961) 145.
[238] J. O'M. Bockris and T. Emi, to be published.
[239] H. Margenau and W. G. Pollard, *Phys. Rev.* **60** (1941) 128.
[240] J. Bardeen, *Phys. Rev.* **58** (1940) 727.
[241] I. Higuchi, T. Ree, and H. Eyring, *J. Am. Chem. Soc.* **79** (1957) 1330.
[242] A. Damjanovic and M. Genshaw, *Electrochim. Acta* **15** (1970) 1281.
[243] L. Muller and L. M. Nekrasov, *Dokl. Akad. Nauk SSSR* **154** (1964) 437.
[244] A. Damjanovic, M. Genshaw, and J. O'M. Bockris, *J. Electrochem. Soc.* **114** (1967) 466.
[245] A. Damjanovic, M. Genshaw, and J. O'M. Bockris, *J. Chem. Phys.* **45** (1966) 4057.
[246] J. O'M. Bockris, B. D. Cahan, and G. E. Stoner, *Chem. Instrumentation* **1** (1969) 273.
[247] L. Formaro and S. Trasatti, *Electrochim. Acta* **15** (1970) 729.
[248] J. P. Hoare, *J. Electrochem. Soc.* **111** (1964) 610.
[249] A. J. Appleby, *J. Electrochem. Soc.* **117** (1970) 1373.
[250] A. J. Appleby, *J. Electroanal. Chem.* **27** (1970) 335.
[251] W. Bold and M. Breiter, *Electrochim. Acta* **5** (1961) 169.
[252] F. G. Will and C. A. Knorr, *Z. Elektrochem.* **64** (1960) 258, 270.
[253] A. J. Appleby, *J. Electroanal. Chem.* **27** (1970) 325.
[254] A. J. Appleby, *J. Electrochem. Soc.* **117** (1970) 1157.
[255] G. Bianchi, F. Mazza, and T. Mussini, *Electrochim. Acta* **11** (1966) 1509.
[256] M. Genshaw, A. Damjanovic, and J. O'M. Bockris, *J. Electroanal. Chem.* **15** (1967) 163.
[257] M. Bonnemay, C. Bernard, G. Magner, and M. Savy, *Compt. Rend.* **270** (1970) 1556.
[258] A. I. Krasil'shchikov, *Zh. Fiz. Khim.* **23** (1949) 332; **26** (1952) 216.
[259] A. U. Akopyan, *Izv. Akad. Nauk Arm. SSR, Khim. Nauki* **11** (1958) 441.
[260] V. N. Nikulin, *Zh. Fiz. Khim.* **35** (1961) 84.
[261] G. Bianchi, *Corrosion Anti-Corrosion* **5** (1957) 146.
[262] G. Bianchi, G. Caprioglio, F. Mazza, and T. Mussini, *Electrochim. Acta* **4** (1961) 232.
[263] S. Palous and R. Buvet, *Bull. Soc. Chim. France* (1962) 1606.

[264] E. I. Kruscheva, L. N. Nekrasov, N. A. Shumilova, and M. R. Tarasevich, *Elektrokhimiya* **3** (1967) 831.
[265] M. L. B. Rao, A. Damjanovic, and J. O'M. Bockris, *J. Phys. Chem.* **67** (1963) 2508.
[266] A. J. Appleby, *J. Electroanal. Chem.* **27** (1970) 347.
[267] W. Gordy and W. J. O. Thomas, *J. Chem. Phys.* **24** (1956) 439.
[268] R. Parsons, *Surf. Sci.* **18** (1969) 28.
[269] J. P. Hoare, *Electrochim. Acta* **14** (1969) 797.
[270] A. J. Appleby, in course of publication.
[271] P. Zweitering and J. J. Roukens, *Trans. Faraday Soc.* **50** (1954) 178.
[272] D. H. Everett, *Trans. Faraday Soc.* **46** (1950) 957.
[273] S. Glasstone, *Theoretical Chemistry*, Van Nostrand, New York, 1944.
[274] A. J. Appleby, *J. Electrochem. Soc.* **117** (1970) 641.
[275] A. Damjanovic, T. H. V. Setty, and J. O'M. Bockris, *J. Electrochem. Soc.* **113** (1966) 746.
[276] D. B. Sepa, A. Damjanovic, and J. O'M. Bockris, *Electrochim. Acta* **12** (1967) 746.
[277] A. Damjanovic, D. Sepa, and J. O'M. Bockris, *J. Res. Inst. Catalysis, Hokkaido Univ.* **16** (1968) 1.
[278] J. McHardy, B. Lovrecek, A. Damjanovic, and J. O'M. Bockris, Rep. No. 3, U.S. Army Mobility Equipment Research and Development Center, Ft. Belvoir, Virginia, Contract No. DA-44-009-AMC-469(T).
[279] R. Jasinski, *Nature* **201** (1964) 1212; H. Jahnke, Preprints 19th CITCE Meeting, Detroit, Michigan, 1968.
[280] M. Genshaw, V. Brusic, and A. Damjanovic, Report No. 11, Contract Da 36-039-SC 88921 (AD 459568), U.S. Army Electronics Research and Development, Ft. Monmouth, New Jersey, 1965.
[281] A. Damjanovic, M. Genshaw, and J. O'M. Bockris, *J. Electrochem. Soc.* **114** (1967) 1107.
[282] M. E. Peover and B. S. White, *Electrochim. Acta* **11** (1966) 1061.
[283] P. G. Zambonin and J. Jordan, *J. Am. Chem. Soc.* **91** (1969) 2225.
[284] P. G. Zambonin, *J. Electroanal. Chem.* **24** (1970) App. 25.
[285] T. Biegler and R. Woods, *J. Electroanal. Chem.* **20** (1969) 347.
[286] A. J. Appleby, *Surface Science* **27** (1971) 225.

Index

Acetonitrile, solvation of halides in, 42
Acid-base reactions in gas phase, 13
Activation entropy, 399
Activation parameters for oxygen reduction, tabulated, 457
Adsorbed H and proton transfer, 374
Adsorption
  of anions in double layer, 283
  congruency in, 260
  early theories of, 289
  effects on rates of reactions, 391
  Flory-Huggins isotherm for, 343
  of fluoride ion, 273
  general theory of, for organics, 299
  size factor in, 343
  theory, recent developments, 300, 301
Air-water interface, 246
Alkali
  chlorides, comparison of experimental and calculated properties, 225
  halides, entropies of molten, 205
  metal ions in gas phase, 5
Alkaline solution, oxygen electrode in, 468
Anion adsorption, 283
Anions, hydrogen bonding to, 37
Apparatus for high-pressure work, 58-62
Apparatus for shock experiments, 118
Aprotic solvents, solvation in, 41
Autoprotolysis in shock experiments, 123

Basicities of nitrogen bases in gas phase, 15
Born relation for compressibility, 98
Born relations for electron transfer, 379

Capacitance
  concentration dependence of, 277
  hump, 278-279
  of water monolayer, 306
Capacitance-potential relation, maxima in, 279
Capacities, inner layer, 316
Capacity hump, 336
Capacity of oriented dipoles, 319
Capacity for sodium fluoride, 310
Catalysis, electrochemical, 369
Cation and anion hydration, 21
  comparison of, 27
Compensation effect, 461
  theory of, 464
Compressibilities of alkali halides, 207
Compressibility, Born relation for, 98
Conductivities
  at high concentrations, 74
  of salts in shock-compressed water, 124
Conductivity
  at high pressures, 57, 75
  of electrolytes in shock waves, 117
  at infinite dilution, under pressure, 66
Configurational entropy, theory of, 177

Index

Congruency, double-layer theory and, 260
Coordination number for salts, 166
Corresponding-state data for salts, 200

Debye-Hückel Onsager parameters, 71, 72
Delocalization of charge, 100, 101
Deuterium effects in water, 53
Dielectric constant for water, 51
Dielectric constants of solvents, 56
Diffuse layer, 241
Diffusion constants, comparison between theoretical and experimental, 234
Diffusivity for water, 52
Dipole
 capacity, 319
 layer, energy of, 340
 orientations, 313
Dissociation of ion pairs, 107
Distribution of proton hydrates, 34
Double layer
 anion adsorption in, 283
 bibliography, 362-368
 in biological systems, 11
 diffuse, 241
 discussion of models for, 351
 fluoride ion in, 273
 model for, 240
 role of solvent in, 1
 solvent orientation at, 241
Double-layer
 capacitance, 277
 ionic models, 324
 ionic systems, 271
 model, review of, 288
 potential profile in, 292
 theory, general advances in, 360

Electrical variable, choice of, 260, 321
Electrocatalysis, 369
 bibliography of, 470-478
 on oxide-free metals, 453
 of oxygen reduction, 421
 various systems, 417

Electrochemical rates, adsorption effects on, 391
Electrode
 electron transfer at, 369
 processes, rate correlations for, 400
Electrolyte solutions at high pressures, 47
Electrolytes
 conductivity of, at high pressures, 57
 coordination numbers for, 166
 fused, 159
Electron
 delocalization, in aromatics, 100, 101
 pulsing, 8
 transfer at electrodes, 369
 transfer, outer sphere reaction of, 380
 transfer, simple harmonic oscillator model, 380
 transfer theory, electrostatic, 378
 transfer theory, quantum mechanical, 370
 transfer theory, thermal, 370
 transfer, transition state theories of, 385
 transmission coefficient, 398
Electronic
 structure and electrode material, 398
 work functions, 250
Electrostatic theory for electron transfer, 378
EMF measurements
 apparatus for, at high pressures, 89
 at high pressures, 87
Enthalpies
 of hydration, for single ions, 20
 of ions in gas phase, 18
Enthalpy of gas hydration reactions, 24
Entropies, of molten alkali halides, 205
Entropy
 of activation, 399
 configurational, for molten salts, 177
 excess experimental, 267
 and excess volume, 267
 for molten salts, 180
Equilibria
 ionization, at high pressures, 76

Index

Equilibria *(cont'd)*
  thermodynamics of, at high pressures, 78
Equilibrium constants for ion solvation in gas phase, 11-12
Excess entropy, 264
  experimental, in double layer, 267
Excess volume, 264
  experimental, 267
Excited states at high pressures, 106
Expansivity of alkali halides, 207
Experimental rate correlations in electrode processes, 400

Flory-Huggins isotherm problems, 343
Fluoride ion
  anomalous adsorption of, 273
  in double layer, 273
Fluoride salts, capacity of, 310
Free metal surfaces, reactions at, 443
Free-volume model, 173
Frequency factors
  for electron-transfer reactions, 415
  in oxygen reduction, 456
Frumkin isotherm, 358
Fusion, entropies of, 195

Gallium-water interface, 347
Gas-phase
  ion equilibrium, 6
  ions, lifetime of, 11
  solvation, 2
  solvation, relation to solution behavior, 3
Gold
  oxygen on, 424
  reduction of oxygen on, 451

Halide ions in acetonitrile and water, 42
Heat capacity of alkali halides, 206
Heat and entropy factors, compensation between, 466
Heats of activation
  and heats of adsorption, 458
  in oxygen reduction, 456

Heats of adsorption and heats of activation, 458
Heavy particle transfer, 386
High pressure
  apparatus for, 58-62
  electrolyte solutions at, 47
High-pressure
  apparatus for optical measurements, 91
  EMF apparatus, 89
  EMF measurements, 87
  errors in, 64
High pressures
  properties of solvents at, 55
  by shock, 111
  water at, 48
Hole models, 168
Hole theory, picture for, 167
Hump
  capacitance, 278
  capacity, 336
Hydrate distribution of proton, 34
Hydration
  energy, correlation between gas and liquid data, 28
  of hydrogen and hydroxyl, 32
  pressure effects on, 68
  single-ion enthalpies of, 20
  successive, 26
Hydrogen bonding
  to anions, Table of values, 38
  to ions, 37
Hydrogen evolution
  frequency factors for, 415
  kinetics, relation to metal properties, 404
  reaction, 402
  reaction, volcano plots for, 406
Hydrogen and hydroxyl ion hydrates, 32
Hydrogen ion solvation, 25
Hydrogen peroxide, role in oxygen electrode reaction, 468
Hydroxyl ion solvation, 25

Ice
  at high pressures, 53
  phase transitions of, 53

Impurity effects and reaction mechanism, 452
Interatomic distances for salts, 199
Interface, air-water, 246
Interionic distances in salts, 166
Intermediates in oxygen reduction and evolution, 428
Intermolecular
  forces in molten salt models, 182
  potentials, theory of, 182
Ion pairs, dissociation of, 107
Ion-molecule reactions, 4
Ion-solvent interaction in gas phase, 2
Ionic
  models, in double-layer theory, 324
  equilibrium, in gas phase, 6
  systems, in double layer, 271
Ionization
  constants, complete, 109
  constants, methods for determining, 87
  equilibria, tabulated, 131-150
  at high pressures, 76
  at high pressures of neutral molecules, 95
  processes, 4
  product of water under shock compression, 126
  successive, 102
  of water at very high pressure, 129
Iridium, reduction of oxygen on, 450
Isotherm, Frumkin, 358

Leonard-Jones-Devonshire theory, 195
Levich calculations, 382
Liquid structure and RDF, 165

Marshall-Quist theory of ionization, 109
Mass spectrometer
  diagram of, 9
  high-pressure, 5
  for ion equilibrium, 9
Mechanism of oxygen-electrode reactions, 439

Melting points of salts, 199
Membranes, double layer at, 11
Mercury
  adsorption of organics at, 287
  double layer at, 256
Metal properties and hydrogen evolution, 404
Metal surface, role of, in double layer, 248
Model
  free-volume, of fused salts, 173
  relation between free-volume and hole, 176
  significant structures, 179
Models
  hole, for molten salts, 168
  intermediate, for double layer, 325
Molecules, excited states of, at high pressures, 106
Molten-salt models, 159
Molten salts
  comparison of calculated and experimental data, 225
  diffusion constants of, 234
  Monte Carlo method for, 208
  specific conductance of, 181
  statistical mechanics of, 186
Monte Carlo method
  comparison of, with experimental data, 213
  for molten salts, 208

Nickel oxide surface, oxygen evolution at, 435
Non-aqueous media, oxygen electrodes in, 470

Optical measurements at high pressures, 91
Organic adsorption, general theory of, 299
Organic systems, adsorption on mercury in, 287
Orientation of dipoles, 313
Osmium, reduction of oxygen on, 450

Index

Outer sphere electron transfer, 380
Oxide-free surfaces, oxygen reduction on, 443
Oxygen electrode
  in alkaline solutions, 468
  and electrocatalysis, 421
  mechanism of, 439
  in non-aqueous media, 470
  on platinum oxide, 426
  theory of, 428
  in various electrolytes, 468
  on various surfaces, 430
Oxygen evolution
  and bond strengths, 418
  electrocatalysis of, 436
  and metal properties, 419
  on nickel, 435
Oxygen reduction
  frequency factors for, 456
  on gold and silver, 451
  heats of activation for, 456
  on oxide-free surfaces, 443
  on palladium, 449
  on rhodium, 449
  on ruthenium, iridium, and osmium, 450

Palladium, reduction of oxygen on, 449
Phase transitions for water and ice, 53
Platinum, reduction of oxygen on, 448
Platinum oxide, oxygen electrode on, 426
Potential profile in double layer, 292
Pressure effects
  bibliography, 151-158
  on hydration, 68
  on ionization equilibria, tabulated, 131-150
Pressure units, 47
Probability distribution for molten salts, 229
Protic and aprotic solvents, comparison of, 41
Proton, distribution of hydrates of, 34

Proton discharge, 371
  and adsorbed H, 374
  potential energy relations for, 374
Proton transfer
  in gas phase, 13
  and heat of adsorption, 377

Quantum mechanical electron transfer theory, 370

Radial distribution function (RDF), 160
  calculated, 216
  theoretical, 216-218
Rate correlations, experimental, 400
RDF, 160
  diagram of, 163
  and liquid structure, 165
Reaction rates, electrochemical, general factors, 397
Refractive index for water, 52
Ruthenium, reduction of oxygen on, 450

Salts
  melting and interatomic distances for, 199
  molten, models for, 159
Scaled particle theory, results for alkali halides, 203
Shock compression
  effects on water, 123
  effects on water ionization, 126
  measurement of, 115
Shock pressure technique, 111
Shock pressures, tabulated, 114
Shock wave apparatus, diagram of, 118
Shock waves
  and conductivity, 117
  generation of, 113
  theory of, 112
Significant structures model, 179
Silver, reduction of oxygen on, 451

Simple harmonic oscillator in electron transfer, 380
Single ions, solvation energy of, 18
Single-ion enthalpies of hydration, 20
Size factor in adsorption, 343
Sodium hydrates in gas phase, 21
Solvation energies in gas and liquid, 28
Solvation of ions, total energies of, 18
Solvation in protic and aprotic solvents, 41
Solvent
  displacement in adsorption, 297
  excess, 262
  near interfaces, 244
Solvents
  dielectric constants of, at high pressures, 56
  properties of, at high pressures, 55
Specific conductance of molten salts, 181
Spectrophotometry at high pressures, 91
Statistical mechanics
  of molten salts, 186
  nonequilibrium, for molten salts, 221
Successive hydration, 26
Superoxides in oxygen reduction, 470
Surface excess
  of entropy, 264
  of solvent, 262
  of volume, 264
Surface tension, scaled particle theory of, 204

Theoretical solvation energies, 30, 31
Time profiles for ion lifetimes, 11
Transition state theories for electron transfer, 385
Transmission coefficient, electron, 398
Tunnel transfer, 370

Units of pressure, 47

Van't Hoff plots for gas-phase ion equilibria, 23

Vaporization, entropies of, 195
Vibrational spectrum for water, 53
Viscosity
  at high pressures, 75
  shear, for molten salts, 234
  of water, 49
Volcano plots
  and bond energies, 412
  and heat of adsorption data, 411
  for hydrogen evolution reaction, 406
  theory of, 407
  and work function, 412
Volcano relations in oxygen reduction, 454
Volume change for ionization, tabulated, 131-150
Volume for molten salts, 180

Water
  activity in ionization equilibria, 109
  density of, 49
  dielectric constant of, 51
  diffusivity of, 52
  at high pressures, 48
  ionization of, in gas, 4
  orientation, capacitance for, 306
  orientation and work function, 254
  properties of, at high pressures, 49
  properties of, at high pressures, tabulated, 50
  properties of, under shock compression, 123
  refractive index of, 52
  vibrational spectrum of, 53
  viscosity of, 49
Water-air interface, 246
Water-gallium interface, 347
Work function
  and orientation of water, 254
  and surface properties, 250
  and zero charge potential, 252

Zero charge potential and work function, 252